The English Zechstein and Related Topics

GEOLOGICAL SOCIETY SPECIAL PUBLICATION NO. 22

The English Zechstein and Related Topics

EDITED BY

GILL M. HARWOOD

Lecturer in Sedimentology,
University of Newcastle upon Tyne

AND

DENYS B. SMITH

Consultant Geologist,
Newcastle upon Tyne

1986

Published for the Geological Society by
Blackwell Scientific Publications
OXFORD LONDON EDINBURGH
BOSTON PALO ALTO MELBOURNE

Published by
Blackwell Scientific Publications
Osney Mead, Oxford OX2 0EL
8 John Street, London WC1N 2ES
23 Ainslie Place, Edinburgh EH3 6AJ
52 Beacon Street, Boston, Massachusetts 02108, USA
667 Lytton Avenue, Palo Alto, California 94301, USA
107 Barry Street, Carlton, Victoria 3053, Australia

First published 1986

Printed in Great Britain by
The Alden Press Ltd, Oxford

DISTRIBUTORS

USA and Canada
Blackwell Scientific Publications Inc.
PO Box 50009, Palo Alto
California 94303

Australia
Blackwell Scientific Publications
(Australia) Pty Ltd
107 Barry Street, Carlton,
Victoria 3053

British Library Cataloguing
in Publication Data

The English Zechstein and related topics.
— (Geological Society special publication, ISSN 0305–8719; no. 22)
1. Geology, Stratigraphic—Permian 2. Geology—England
I. Harwood, Gill M. II. Smith, Denys B. III. Geological Society of London IV. Series
551.7′56′0942 QE674

ISBN 0–632–01067–3

Contents

Preface

This volume is the result of a discussion workshop (EZ 82) with field excursions and oral presentations which was based at Nottingham University in the spring of 1982. The editors would like to thank the numerous reviewers who helped them in their task of compiling this volume and those who helped arrange the workshop, particularly Graham Aplin, Jack Pattison, Tim Pettigrew and Jean Pearson. Tim Pettigrew helped with some preliminary editing. One of us (GMH) would like to apologize for any delay in reviewing and editing manuscripts caused by recent moves back and forth across the Atlantic. Both editors hope that this volume will help workers on the Zechstein and other evaporite basins to understand some of their problems better and that it will stimulate further Zechstein research.

Gill M. Harwood
Denys B. Smith

The English Zechstein and related topics—introduction

G.M. Harwood & D.B. Smith

This volume is the printed record of EZ 82, a workshop on the results of recent research on strata of the English Zechstein and related topics. The workshop comprised a six-day field study meeting on Permian rocks at outcrop in north-east England and a three-day indoor discussion meeting in the Department of Geology at the University of Nottingham. Twenty-one research and review papers were presented at this meeting. Most of these were written up for this volume which has been further supplemented by the inclusion of papers submitted by two Polish workers who were unable to attend the workshop.

The field meeting commenced with visits to the classic exposures of Upper Permian reef-margin and slope carbonate rocks in County Durham and adjoining areas and continued to exposures of nearshore and shelf carbonates and continental sediments in Yorkshire, Derbyshire and Nottinghamshire. A guide to the exposures visited, which also included background information on their stratigraphical, environmental and palaeogeographical significance, was prepared for the participants; copies of this guide may be obtained from The Secretary, Department of Geology, The University of Nottingham, Nottingham NG7 2RD (1986 price £3.50 including postage).

The pattern of the workshop was based on that of a highly successful meeting of Zechstein workers in Poland in 1978 (Depowski *et al.* 1981) at which the growing school of Polish workers, amongst others, presented their research results to an international audience with research interests in the Upper Permian. About the same time, the discovery, possible sourcing potential, plus exploitation of hydrocarbons in Zechstein strata in Poland, Denmark, Germany, The Netherlands, England and the southern North Sea had stimulated Fuchtbauer & Peryt (1980) to put together a collection of papers dealing primarily with Zechstein carbonate rocks (*Contr. Sedim.* **9**, 1980). A steady trickle of papers on many aspects of the nature and evolution of Permian strata in Europe, together with the processes that shaped them, was also appearing in other journals, providing further evidence of increased interest in these sometimes enigmatic rocks. The EZ 82 workshop brought together both Zechstein workers from many parts of Europe plus researchers who were working on similar carbonate-evaporite basins from the USA to the USSR, a combination which created a lively discussion meeting.

In a volume dealing with so many diverse aspects by many authors we claim no unifying theme than that of relevance to the Zechstein. We have divided the papers herein on a geographical basis, with papers 2–11 dealing primarily with topics within the United Kingdom, papers 12–15 within Germany and papers 16–21 within Poland with the final contribution, paper 22, describing Upper Permian sediments from the Urals. Within each geographical section the papers are presented, as far as is possible, in stratigraphical order.

The first section, United Kingdom papers, commences with revision of nomenclature for the Upper Permian strata in eastern England (Smith, Harwood, Pattison and Pettigrew). This revision has been thought necessary for some time, particularly as strata in different Zechstein cycles had similar, or confusing, nomenclature in different areas of eastern England. The EZ 82 workshop provided the impetus to produce a revised nomenclature, which is detailed here and includes definitions and type localities.

The remaining papers in the United Kingdom section are mostly concerned with carbonate strata at, or near, outcrop within eastern England. One exception is the paper by Turner and Magaritz, who have carried out detailed mineralogical, chemical and stable isotope analyses on one core from the Marl Slate, taken just offshore of County Durham. They detail a close correlation between lithological and geochemical parameters, including stable isotope values. Sapropelic units, at the base of sub-cycles within the Marl Slate, are characterized by high contents of iron, quartz silt and organic carbon with strongly negative oxygen isotope values. Turner and Magaritz conclude these are indicative of freshwater influxes which promoted productivity, water stratification and regional anoxic conditions, thus forming the base of each sub-cycle. They also conclude that the increase in base metals (notably copper and zinc) away from the sapropelic units indicates these metals to be more associated with the marine conditions re-established in the upper parts of each sub-cycle.

Schweitzer presents an excellent systematic review of all the species of higher plants that have been described from the Zechstein strata of England and Germany and includes one new species. This flora is characterized by conifers which had reached a surprisingly advanced stage of polygenetic development and resemble some Recent conifers much more closely than Lower Permian flora. Conifer woods display distinct growth rings indicating rhythmic climatic changes, perhaps on an annual cycle. Schweitzer demonstrates that the Zechstein transgression was a significant turning point in floral evolution in northern Europe and that floral development was uniform in both German and English areas of the Zechstein basin. The most abundant plant remains have been discovered in the Marl Slate and the equivalent Kupferschiefer.

Well preserved specimens of the conodont *Merrillina divergens*, separated and described by Swift, have been recovered from English Zechstein Cycle 1 carbonate rocks (EZ1 Ca) in north-east England. Swift reports a uniform collection from each locality which can be assigned to just six elements. This low diversity is typical of Zechstein Cycle 1 (Z1) strata throughout England, Germany and Poland and the recognition of a symmetry transition between the elements is one factor used to conclude that these elements are the apparatus of a single species. Swift surmises that the dominance of a single species reflects difficult living conditions in a wholly, or partially, enclosed basin with fluctuations in sea-level and salinities.

The upper member, the Sprotbrough Member, of the Zechstein Cycle 1 carbonate, the Cadeby Formation (EZ1b Ca—the former Upper Subdivision of the Lower Magnesian Limestone) contains evidence of a former ooid shoal complex which was present over most of its outcrop area. Kaldi describes large-scale cross-stratified bedforms within this complex, including hummocky cross-stratification, fan-shaped bedding, convex-upwards cross-stratification and contemporaneous erosion surfaces. He presents data on palaeocurrent styles and azimuths and concludes that the predominant current direction was from the north-east, but that periodic storms produced spillover lobes orientated towards the north-west. Kaldi adds that although sandwave topography must have controlled local currents by channelling or refracting flows, the exact manner in which the sandwaves originated could not be determined.

Diagenesis within the Cadeby Formation (EZ1 Ca) is the subject of two papers. In the first, Harwood distinguishes the timing of different diagenetic events from penecontemporaneous through burial diagenesis to those events associated with uplift of the formation during the Cenozoic. She includes evidence that shows early, penecontemporaneous events to be facies-restricted whereas later events were more pervasive. These later events, during early burial, were dominated by dolomitization and subsequent partial replacement by evaporites. Harwood concludes that three major dolomite types are present within the Cadeby Formation and that these formed in penecontemporaneous hypersaline lagoons, from reflux of hypersaline fluids after initial deposition and cementation and from mixed water circulation in some nearshore areas. She documents several near-surface events, including anhydrite to gypsum transformation with later gypsum dissolution, leaching of the carbonates and dedolomitization. Kaldi concentrates his paper on the upper, Sprotbrough Member of the Cadeby Formation. He defines five types of dolomite within this member, penecontemporaneous supratidal dolomite, primary dolomite precipitates within semi-restricted lagoons, early and late dolomite cements within the ooid shoal facies and local dolomite cements within open shelf carbonates and a later pervasive, replacive, coarsely crystalline dolomite, the result of mixing of fresh and marine waters. Harwood and Kaldi thus differ as to the cause of the replacive dolomitization phase, but unite in that this was the major diagenetic event within the Cadeby Formation. Kaldi also documents the formation of evaporites both within supratidal sediments and as a later replacive phase. He concludes that porosity evolution was dependent both on the nature of the primary sediments and on their diagenetic history. Together the papers present a good introduction to the complex diagenetic history of these sediments.

In a different approach to the Cadeby Formation, Harwood and Smith describe mineralization occurrences throughout the formation at outcrop. They demonstrate that the majority of mineralized localities cluster within four geographical areas. An underlying structural control is envisaged as a channelway for mineralizing fluids in three of these areas; in the fourth, mineralization appears to be the result of upwards leakage of formation fluids. Minerals present include lead, zinc and copper sulphides with barite and minor fluorite; occurrences are in vugs and breccias of sedimentary origin with extremely rare examples of vein mineralization.

An extensive thin bed, the Trow Point Bed, at the top of the Zechstein Cycle 1 carbonate (EZ1 Ca) in north-east England forms the subject of Smith's paper. This bed is present on the lower part of the basin margin slope and is continuous over an uneven surface, resulting from a submarine slide, with a local relief of up to 15 m. The bed contains oncoids, peloids, algal laminations and columnar stromatolites and is similar, and is in a comparable stratigraphic position, to deposits described from Germany, Poland and the southern North Sea. Smith concludes that the similarities of these occurrences suggest that, although both German and Polish equivalents are considerably thicker, they are possible correlatives. He further surmises that the Trow Point Bed accumulated in water depths of 25–100 m in oxic conditions and may have been much more extensive than is apparent at outcrop today.

Evidence for present-day dissolution of gypsum from both Zechstein Cycle 1 and Cycle 2 evaporites in North Yorkshire is given by Cooper. The timing of the subsidence resulting from this dissolution within the last few hundred years has been documented from newspaper articles, unpublished notebooks and previous published papers; subsidence during Pleistocene glaciations is evidenced by Late Devensian and Flandrian lacustrine deposits over some tens of square kilometres. Cooper demonstrates the linearity of many of the subsidence hollows, which he concludes is due to a subsurface joint-controlled cave and fissure system where active dissolution is taking, or has taken, place. He also demonstrates that the eastern limit of the subsidence-prone belt is marked by the transition of the secondary gypsum to anhydrite with depth; the western limit is where the gypsum, or its residue, crops out. Cooper's paper is the last in the United Kingdom section of this volume and makes a valuable contribution to this aspect of the Zechstein strata.

The remaining sections of the volume contain papers on the Zechstein elsewhere in northern Europe, plus one paper on the carbonate-evaporite deposits of the Urals. The German papers are contributed by two authors, Paul and Richter-Bernburg. In the first of his two papers, Paul considers the effects of a basin and schwellen topography, inherited from the Variscan regime, on sedimentation within the lower Zechstein cycles in Germany. He shows that although the basinal Kupferschiefer facies are similar throughout the Zechstein, different facies, including red shales and fossiliferous dolomitized carbonates, are present around the basin rim and on schwellen within the basin. In the Zechstein Cycles 1 and 2 carbonates (Z1 and Z2 Ca) reefs and other biogenic sediments dominate the basin margins and biogenic sediments were also abundant on the schwellen (or palaeohighs). Paul further describes how evaporite deposition in both cycles was much greater on the palaeohighs and the basin rims, increasing the relief between palaeohigh and basin floor with consequent sediment instability.

In a second paper Paul describes the stratigraphy of the Lower Werra Cycle Carbonate (Z1 Ca) in West Germany. This paper was submitted after the EZ 82 workshop and is a direct consequence of the field excursions before the oral presentations. In visits to exposures in Yorkshire, Paul was shown the Hampole Discontinuity, first described by Smith (1968). This seemed similar to some German exposures and, on return to his native country, Paul carried out preliminary fieldwork resulting in documentation, here, of the 'Solhops Discontinuity'. Paul concludes this to be equivalent to the Hampole Discontinuity in eastern England. Furthermore, he takes it to represent an extensive regression of the Zechstein Sea, leading to subaerial exposure of marginal areas of the Zechstein basin. This paper is included in this volume both because it resulted from the workshop itself and because of its implications on basin-wide changes in sea-level.

Richter-Bernburg's two papers concentrate on the evaporites within the Zechstein sequence. The author has worked with Zechstein material for many years and has probably seen more Zechstein sediments, both at outcrop and in core, than any other researcher in this field. In the first of two papers he discusses some of the problems of anhydrite sedimentation in Zechstein Cycles 1 and 2 (Z1A and Z2A). He documents five subcycles in the Werraanhydrit (Z1A) throughout Germany which are correlatable with similar evaporites within the southern North Sea. Laminites near the base of the Basalanhydrit (Z2A) can be traced throughout the German Zechstein basin, although higher within the sequence they are abruptly replaced by flaser, or more nodular, anhydrite. This change takes place at a lower position near the basin margins, a factor Richter-Bernburg suggests may indicate an alteration in the depositional environment in those areas. Richter-Bernburg also introduces three new descriptive terms for use in study of evaporite sequences in his paper.

One correlation of Zechstein cycles between Germany, Denmark and England is presented in Richter-Bernburg's second paper. Using criteria including similarities in distribution, thickness and lithology of the evaporites, along with a change in colour, he suggests that the strata of the Don and Aislaby Groups of eastern England are equivalent in age to those of the Werra Cycle in Germany and that those of the Teesside Group are correlatives of the German Stassfurt sequence. This correlation is at variance with that of both Smith *et al.* (1974) and Taylor (1980) and provoked active discussion when it was presented; short discussion contributions are included after the paper.

Although two of the Polish authors were able

to attend the EZ 82 workshop, others found it impossible to do so. We have included two additional papers in this section from these authors both because the authors had originally hoped to present these in Nottingham and because of their relevance to Zechstein research. The first of these papers is by Oszczepalski, and is one of a series of three papers on the Copper Shale in the area of the Foresudetic Monocline in south-western Poland. Oszczepalski, working with a series of cores, has used microfacies analysis combined with geochemical studies to define three major facies within the Copper Shale; a deep shelf facies, a shallow shelf facies and a nearshore facies. Shorewards the Copper Shale is replaced by carbonate and/or terrigenous sediments. A great deal of detailed research has been compiled in this paper as Oszczepalski combines both inorganic and organic geochemical results with personal core descriptions in his analysis.

In contrast, in his two papers, both presented at Nottingham, Tomaszewski concentrates on the base metal enrichment of the Copper Shale. In the first of his papers, he comments on the genesis and structure of the polymetallic ores in the area of the Foresudetic Monocline. Also working with shallow cores, he has produced detailed maps showing the distribution of the types of ore, the different mineral zones and the lead content within the area, together with cross-sections showing the vertical distribution of the mineralization. In the second paper, on the sedimentary environments of the lowest Zechstein sediments, Tomaszewski presents data from samples from underground galleries in the Lubin copper mine, which have been analysed for their metal content and mineralogical components. With additional data from the mine's geological service, he has constructed isopach maps for the Border Dolomite and Copper Shale and also shows the amount of mineralization in the underlying white sandstones. Taken together, Tomaszewski's papers present a wealth of hitherto unpublished data on the extent of mineralization in the Lubin area of the Foresudetic Monocline and discuss some of the current ideas on its genesis.

Peryt contributes a discussion on the chronostratigraphical and lithostratigraphical correlations of the Zechstein Limestone (PZ1 Ca) throughout Poland, which unfortunately he was unable to present personally in Nottingham. He recognizes three sub-cycles within this carbonate, each representing progressive shallowing upwards. In areas around the basin margin the upper parts of these subcycles contain evidence of subaerial exposure. Peryt uses these exposure surfaces as chronostratigraphic markers which he traces across the carbonate platform sediments and throughout many of the slope and basin deposits. He comments that throughout some areas of Poland, in Germany and in England there is only one exposure surface recognized (the 'Solhops Discontinuity' in Germany and the Hampole Discontinuity in eastern England) but is unable to determine whether this coincides with the boundaries between his first and second sub-cycles, or between the second and third, or whether one of these sub-cycles is not developed in the other areas of the Zechstein basin.

The macrofauna present in the first three Zechstein cycles in Poland is summarized by Karwowski and Klapcinski. They distinguish four biozones with characteristic faunal assemblages, which typify palaeogeographical areas within the Zechstein carbonates. Preservation of the macrofauna is variable, in part dependent on the host rock and its diagenetic history. Fauna discussed include bryozoans, brachiopods and bivalves with gastropods, cephalopods and echinoderms being poorly represented. The authors include a faunal list showing stratigraphic occurrence.

In a separate paper, Klapcinski describes the Leine Anhydrite (PZ3A) within Poland. He recognizes three lithostratigraphic units within the anhydrite, principally based on its textural characteristics. The spherulitic structure of units within the anhydrite is described, with examples shown in the text. This texture is recognizable throughout much of the Leine Anhydrite in Poland. Klapcinski concludes that the presence of this spherulitic texture in the anhydrite may provide a useful and extensive marker unit within the Zechstein basin.

The final paper in this volume deals with a carbonate-evaporite succession outside the Zechstein basin, but of a similar age. Chuvashov presents a brief description of sediments from the Kungurian basin of the Urals, concentrating on the carbonates present. He compares these to similar sediments within the Zechstein basin and also provides a stratigraphic summary which helps to depict the evolution of the Kungurian basin.

In compiling this volume we have rejected some manuscripts initially presented at the EZ82 workshop, in addition to including three papers not orally presented at the meeting. We have tried to include as many of the foreign papers as is practicable as little research in this field is published in English; perhaps the references included will encourage some readers to consult the German, Polish and Russian literature. We realize that this represents a very small contribution to research over the whole Zechstein basin and that there are many areas of relevant research that are not included. Perhaps the greatest omissions are the areas of East Germany, Denmark, the northern Netherlands and the southern North Sea. We also realize that many questions remain unanswered and that many of the papers included in this volume perhaps raise more questions than they answer. We are very conscious of the problems yet to solve and of whole fields that have been largely ignored in research. We hope that this book will go some way towards stimulating interest, and hence further research, on Zechstein strata throughout the basin.

References

DEPOWSKI, S., PERYT, T.M. & PIATKOWSKI T.S. 1981. *Proceedings of International Symposium on Central European Permian.* Wydawnicta Geologiczne, Warszawa.

FUCHTBAUER, H. & PERYT, T.M. 1980. The Zechstein Basin with emphasis on carbonate sequences. *Contr. Sedim.* **9,** Schweizerbart'sche Verbuchhandlung, Stuttgart.

SMITH, D.B. 1968. The Hampole Beds—a significant marker in the Lower Magnesian Limestone of Yorkshire, Derbyshire and Nottinghamshire. *Proc. Yorks. geol. Soc.* **36,** 463–74.

SMITH, D.B., BRUNSTROM, R.G.W., MANNING, P.I., SIMPSON, S. & SHOTTON, F.W. 1974. Correlation of the Permian rocks of the British Isles. *Spec. Rep. geol. Soc. Lond., No. 5.*

TAYLOR, J.C.M. 1980. Origin of the Werraanhydrit in the U.K. Southern North Sea—a reappraisal. In: FUCHTBAUER, H. & PERYT, T.M. (eds) *The Zechstein Basin with Emphasis on Carbonate Sequences. Contr. Sedim.* **9,** 91–114. Schweizerbart'sche Verbuchhandlung, Stuttgart.

GILL M. HARWOOD, Department of Geology, University of Newcastle upon Tyne, Newcastle upon Tyne NE1 7RU.

DENYS B. SMITH, 79 Kenton Road, Newcastle upon Tyne NE3 4NL.

United Kingdom

A revised nomenclature for Upper Permian strata in eastern England

D.B. Smith, G.M. Harwood, J. Pattison & T.H. Pettigrew

SUMMARY: Revised names of some Upper Permian strata in eastern England are proposed and defined, together with new names for some strata previously without formal names; the new and revised names are based on nominated type localities and are intended to replace existing names that were confusing or inappropriate. We propose that the names Lower and Upper Magnesian Limestone in the Yorkshire Province be replaced by the new names Cadeby (Magnesian Limestone) Formation and Brotherton (Magnesian Limestone) Formation respectively and that the Lower and Middle Magnesian Limestone in the Durham Province should become the Raisby (Magnesian Limestone) and Ford (Magnesian Limestone) formations respectively. The Middle Marls are renamed the Edlington Formation in both provinces and the new names Sherburn (Anhydrite) Formation, Sneaton (Halite) Formation and Littlebeck (Anhydrite) Formation are proposed for the Upper Anhydrite, Upper Halite and Top Anhydrite formations. The term Roxby Formation is proposed to replace the present name Upper (or Saliferous) Marls.

The nomenclature and classification of English Zechstein (Upper Permian) strata on land in eastern England evolved gradually with discoveries and new interpretations. The scheme in use most recently is by Smith *et al.* (1974) in which many of the traditional names were retained. Further research and refinements of earlier work, however, have shown that the retention of several of the traditional stratigraphical names is now more confusing than enlightening, especially to workers in other countries, and we therefore propose a revised nomenclature (Table 1) in which several of the main Upper Permian rock units are redefined and renamed on a geographical basis. These proposed new names are used throughout this volume wherever possible, but to assist the transition the previous names are shown in parentheses in Table 1 and also on first mention in each paper.

The proposed new or modified names are defined in the next section, in which the stratigraphical position of the marine strata is defined using the English Zechstein cycle abbreviations of Smith (1970 *et seq.*). The main change proposed is the discontinuation of the terms Lower, Middle and Upper Magnesian Limestones, which have been used in different senses and for rocks in different Zechstein cycles in the Durham and Yorkshire provinces; this has led to considerable confusion, especially near the mutual boundary at the Cleveland Axis in North Yorkshire (Fig. 1). The proposed new names are based on nominated type sections in accordance with current stratigraphic practice; they retain some links with the existing nomenclature to facilitate literature reference during a change-over period, but we anticipate future simplification as the new names become familiar. Other changes proposed mainly comprise the substitution of geographically-based names for some evaporite deposits in upper parts of the sequence and the shortening of some other names. The stratigraphical relationships of the Permian strata in the Yorkshire Province are shown diagrammatically in Fig. 2; the names of equivalent strata in the Durham Province are shown in Table 1.

Another major change is in the use of the term 'marls'. We propose that the formal name Lower Marls be discontinued, mainly because the rock thus designated is neither a marl nor a stratigraphical unit but is a facies variant of the Wetherby Member of the Cadeby Formation; we do not, however, wish to recommend discontinuance of the portrayal of this facies on maps, where an informal facies name would remain appropriate. The name Edlington Formation, recommended for the former Middle Marls, is somewhat unsatisfactory in that the formation is atypically thin and lacking in bedded evaporites at the type locality of New Edlington Clay Pits. The formation is so varied, however, that no single borehole section could be said to be representative and the section at New Edlington is to be permanently preserved as a Site of Special Scientific Interest and will thus be readily accessible for study. The almost unexposed red argillaceous strata currently known as the Upper (or Saliferous) Marls, though considered by some authors (e.g. Rhys 1974) as Triassic, are renamed the Roxby Formation in both provinces.

Definitions of proposed new names

Cadeby (Magnesian Limestone) Formation (EZ1 Ca)

Former name: Lower Magnesian Limestone of

From HARWOOD, G.M. & SMITH, D.B. (eds), 1986, *The English Zechstein and Related Topics*, Geological Society Special Publication No. 22, pp. 9–17.

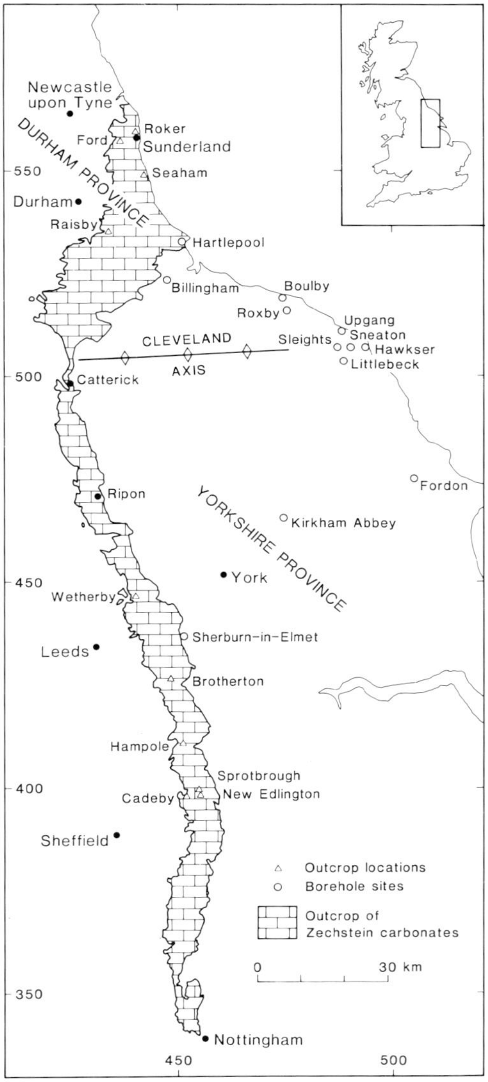

FIG. 1. Map of eastern England showing the location both of type localities at outcrop and of type bore hole cores for subsurface formations.

Yorkshire (except extreme north), Humberside, Derbyshire and Nottinghamshire.
Stratigraphical definition: Upper Permian marine carbonate strata from latitude of Catterick (North Yorkshire) southwards, lying between the Marl Slate *sensu-stricto* (previously defined), where present, at the base, and the Edlington Formation (defined here) at the top; in the subsurface of Humberside and east Yorkshire it is overlain by the Hayton Formation. The top of the Marl Slate is taken at the top of the highest finely laminated stratum; where the Marl Slate is absent the base of the Cadeby Formation is taken at the base of the lowest marine carbonate stratum overlying basal Permian continental deposits or Carboniferous strata. The formation is divisible into two members (the lower, Wetherby Member, and upper, Sprotbrough Member) throughout much of the outcrop but this division is less clear in the subsurface; the junction of the two members is at the top of the lower dolomite of the Hampole Beds (Smith 1968) or, where the lower dolomite is absent, at the Hampole Discontinuity. *Type locality*: Cadeby Quarry, U.K. Ordnance Survey Grid Reference SE 5200, Cadeby, South Yorkshire, and adjacent areas. Thickness exposed at Cadeby Quarry about 35 m: lowest beds (about 7 m) not exposed but proved in boreholes; highest strata (probably not exceeding 5 m) eroded. Wetherby Member mainly comprises generally parallel-bedded, buff-coloured moderately fossiliferous oolitic dolomite with scattered bryozoan-algal patch reefs: Sprotbrough Member mainly comprises very pale grey, sparsely fossiliferous, finely oolitic dolomite and is characterized by large-scale cross-stratification. The intervening Hampole Beds are atypically thick (up to 4 m) at Cadeby Quarry and comprise multicoloured dolomite and mudstone, the latter with scattered remains of land plants. See definitions of members and diagnostic faunas.
Lateral variation: Both members of the formation comprise several facies (see definition of members for details). Its thickness ranges to a maximum exceeding 100 m in the subsurface east of the outcrop. At outcrop it is 20–60 m thick over a wide area but thins gradually southwards and westwards and, more sharply, with onlap, against eminences on the underlying unconformity. Thins sharply to 5–10 m on basin floor E of shelf margin under east Yorkshire and adjoining areas (Fig. 2).
Biota: See lists in definitions of component members.

Cadeby (Magnesian Limestone) Formation, Wetherby Member (EZ1a Ca)
Former names: Compact Limestone; Lower Magnesian Limestone, Lower Subdivision.
Stratigraphical definition: Upper Permian marine carbonate strata, south of the latitude of Catterick, lying between the Marl Slate *sensu stricto* and the top of the lower dolomite of the Hampole Beds or, where the lower dolomite is absent, at the Hampole Discontinuity. Top not readily distinguishable where the Hampole Beds are absent or atypical, especially north of Ripon, south of Mansfield and in the subsurface.

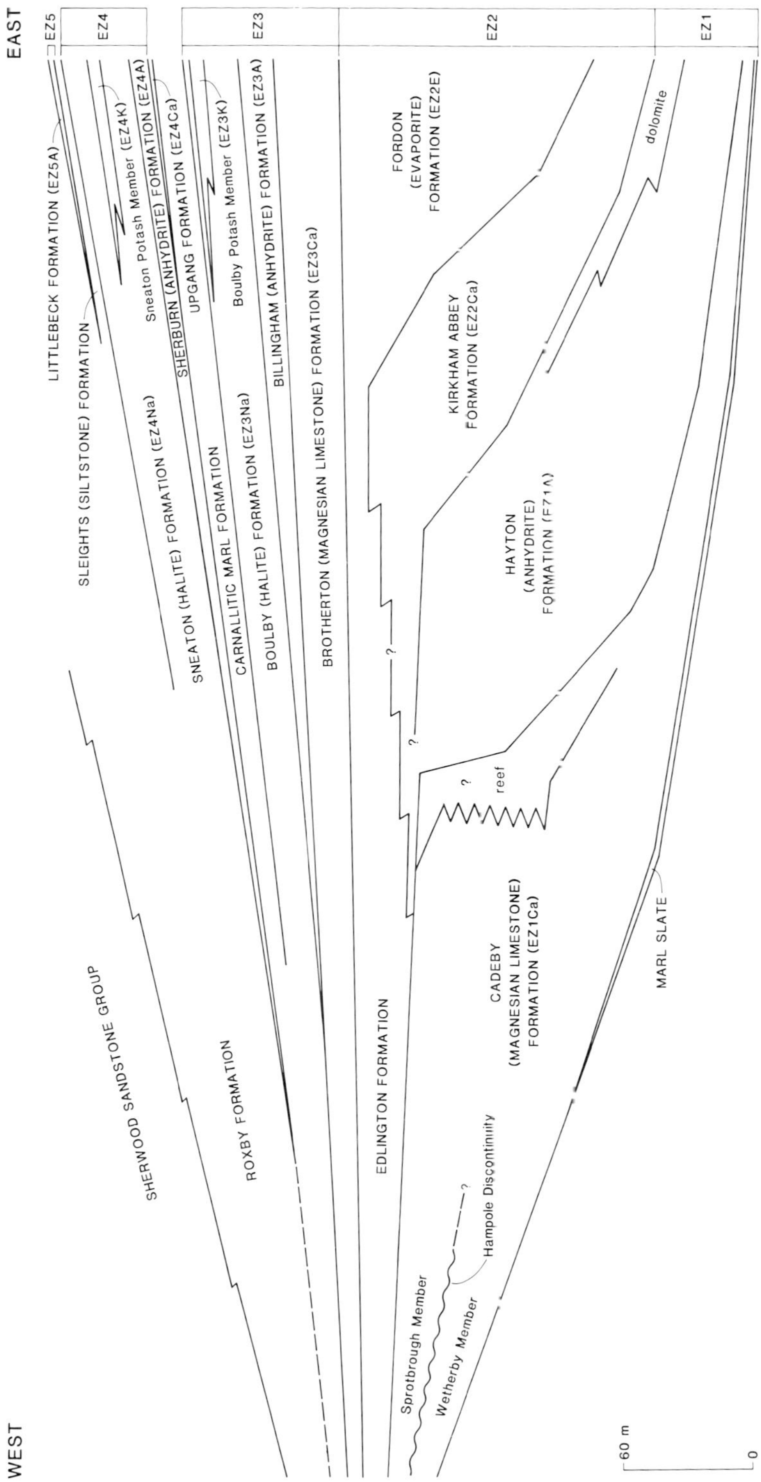

FIG. 2. Stratigraphical relationships of Upper Permian strata in the Yorkshire Province at outcrop and in the subsurface.

TABLE 1. A revised nomenclature for Upper Permian strata in eastern England

Group	Cycle	YORKSHIRE PROVINCE (outcrop)	DURHAM PROVINCE (outcrop and subsurface)	YORKSHIRE PROVINCE (subsurface)	NORTH SEA, GERMANY THE NETHERLANDS	Cycle
ESKDALE	EZ5	ROXBY FORMATION (Upper Marls)	ROXBY FORMATION (Upper or Saliferous Marls)	ROXBY FORMATION (Upper or Saliferous Marls) LITTLEBECK FORMATION (Top Anhydrite) SLEIGHTS (SILTSTONE) FM	ZECHSTEINLETTEN GRENZANHYDRIT	Z5
STAINTONDALE	EZ4		SHERBURN (ANHYDRITE) FM (Upper Anhydrite)	SNEATON (HALITE) FORMATION; including SNEATON POTASH MEMBER (Upper Halite and Potash) SHERBURN (ANHYDRITE) FM (Upper Anhydrite) UPGANG FORMATION CARNALLITIC MARL FM	ALLER HALIT and EQUIVALENTS PEGMATITANHYDRIT and EQUIVALENTS ROTER SALZTON	Z4
TEESSIDE	EZ3	BROTHERTON (MAGNESIAN LIMESTONE) FORMATION (Upper Magnesian Limestone)	BILLINGHAM (ANHYDRITE) FM (Billingham Main Anhydrite) SEAHAM FORMATION (part of Upper Magnesian Limestone)	BOULBY (HALITE) FORMATION; including BOULBY POTASH MEMBER (Middle Halite and Potash) BILLINGHAM (ANHYDRITE) FM (Billingham Main Anhydrite) BROTHERTON (MAGNESIAN LIMESTONE) FORMATION (Upper Magnesian Limestone)	LEINE HALIT and EQUIVALENTS HAUPTANHYDRIT PLATTENDOLOMIT GRAUER SALZTON	Z3
AISLABY	EZ2	EDLINGTON FORMATION (Middle Marls)	South of Province: EDLINGTON FORMATION (Middle Marls) East of Province: FORDON (EVAPORITE) FORMATION and SEAHAM RESIDUE ? ROKER (DOLOMITE) FORMATION (Hartlepool and Roker Dolomite) ? CONCRETIONARY LIMESTONE FM (parts of Upper Magnesian Limestone)	FORDON (EVAPORITE) FM KIRKHAM ABBEY FORMATION	STASSFURT SALZE HAUPTDOLOMIT	Z2
DON	EZ1	CADEBY (MAGNESIAN LIMESTONE) FM (Lower Magnesian Limestone): SPROTBROUGH Mbr (Upper Subdivision); WETHERBY Mbr (Lower Subdivision)	HARTLEPOOL (ANHYDRITE) FM FORD (MAGNESIAN LIMESTONE) FM (Middle Magnesian Limestone) RAISBY (MAGNESIAN LIMESTONE) FM (Lower Magnesian Limestone) MARL SLATE FORMATION	HAYTON (ANHYDRITE) FM CADEBY (MAGNESIAN LIMESTONE) FORMATION MARL SLATE FORMATION	WERRAANHYDRIT WERRADOLOMIT and ZECHSTEINKALK KUPFERSCHIEFER	Z1

Type locality: Disused railway cutting at site of former Wetherby Station (SE 397483), West Yorkshire, where about 16 m thick; other sections in disused railway cutting near Ingbarrow Farm (SE 387496) Wetherby. Comprises peritidal mainly parallel-bedded pale buff altered dolomite, with abundant small and medium-scale channels and small-scale cross-stratification; scattered bivalves; minor erosion surfaces at many levels; some algal lamination, especially in uppermost metre below Hampole discontinuity; scattered evidence of bioturbation. Base formerly exposed (Edwards *et al.* 1950, p. 38) but not now visible. Scattered quartz sand grains in lower about 4 m of dolomite.
Lateral variation: Ranges from 0 to 40 m at outcrop (commonly 12–20 m), increasing in subsurface (eastwards). Several major facies variants, including near-shore peritidal and extensive open shelf carbonate rocks (mainly altered bioclastic and oolitic dolomites with some pisoids with scattered to abundant bryozoan-algal patch reefs), restricted shelf/lagoonal carbonate rocks, shelf-margin slope and basin-floor carbonate rocks and widespread grey argillaceous limestone and dolomite (the former Lower Marl, sometimes erroneously termed Marl Slates). Quartz sand present in places, especially in Nottinghamshire. Thin algal-laminite (lower dolomite beds of Hampole Beds) at top in places.
Biota: Fossil content dependent on facies. Biota in littoral to sub-littoral facies largely comprises the bivalves *Bakevellia binneyi*, *Liebea squamosa*, *Permophorus costatus* and *Schizodus obscurus*; these forms occur widely but are most abundant in coquinas near the base of the member and, with ramose bryozoa (especially *Acanthocladia anceps* and *Thamniscus dubius*), in patch reefs. Offshore facies have more diverse faunas, including foraminifera, bryozoa, brachiopods, bivalves and gastropods. The 'Lower Marl' facies contains remains of terrigenous conifers and pteridosperms. Cyanophytic bacteria (blue-green algae) contributed to the formation of algal-laminites.

Cadeby (Magnesian Limestone) Formation: Sprotbrough Member (EZ1b Ca)
Former name: Small grained limestone; Lower Magnesian Limestone, Upper Subdivision.
Stratigraphical definition: Upper Permian marine carbonate strata from the latitude of Catterick southwards and lying between the top of the lower dolomite bed of the Hampole Beds (Smith 1968), or the Hampole Discontinuity where the lower dolomite bed is absent, and the base of the Edlington Formation (defined here) in the west and the Hayton (Anhydrite) Formation in the east. Locally rests directly on Carboniferous strata.
Type locality: Boat Lane Quarry (SE 535016), Sprotbrough, South Yorkshire. Exposed thickness about 17 m, uppermost beds (probably no more than 5 m) eroded; comprises pale cream, large-scale cross-stratified altered oolitic dolomite with a sparse bivalve fauna.
Lateral variation: Ranges generally from 8 to 40 m at outcrop, thickening eastwards towards the shelf-margin. Characteristic rock type at outcrop is coarsely cross-stratified off-white to cream fine-grained oolitic dolomite of high-energy shoal facies, passing west and south into parallel-bedded inner shelf oolitic dolomites and eastwards into lower energy, partly pelleted outer shelf, slope and basin floor fine-grained pale buff dolomite; intense bioturbation in places. Algal laminites in uppermost 1–4 m of shelf facies in places at outcrop and widely present in subsurface; locally interbedded with evaporites or their residues. Quartz sand throughout member in many southern outcrops.
Biota: Generally sparse but locally contains a diverse bryozoan-brachiopod-bivalve fauna comparable with that of the reef facies of the Ford Formation. Elsewhere bivalves predominate amongst macrofossils and brachiopods, crinoids and bryozoans are rare and generally absent. Microfossils include milioloid foraminifera. Many burrows in places.

Raisby (Magnesian Limestone) Formation (EZ1a Ca)
Former name: Lower Magnesian Limestone of County Durham, Tyne & Wear, Cleveland, and northernmost North Yorkshire.
Stratigraphical definition: Upper Permian marine carbonate strata from latitude of Catterick (North Yorkshire) northwards, lying between the Marl Slate (previously defined) and the Ford (Magnesian Limestone) Formation (defined here). The base is generally sharp and is taken at the top of the highest finely laminated stratum of typical Marl Slate lithology; we do not support Magraw's (1975) view that the base is commonly gradational by alteration and recommend that his term 'Passage Beds' be abandoned. At outcrop the top is taken at the incoming of oolitic and pisolitic dolomite or equivalent reef of the Ford Formation; farther east it is taken at the base of a thin group of transitional strata (Magraw *et al.* 1963) or at the top of the Trow Point Bed (see Smith, *this volume*).
Type locality: Raisby Quarries (NZ 3435), County Durham, where the full thickness of about 48 m is exposed. Strata mainly of cream, brown and grey fine-grained dolomite but with thick grey fine-grained limestone in lower part; thin-bedded and sparingly to richly fossiliferous in lower half, medium- to thick-bedded and only

slightly fossiliferous in upper half where overlain by oolitic dolomite of the Ford Formation.
Lateral variation: Ranges generally from 20 to 50 m but thins against underlying eminences, especially in Cleveland and adjoining areas; thickest (locally 73 m) in N–S belt in eastern County Durham, but thins sharply into offshore area. Characteristically of parallel-bedded buff and grey fine-grained altered dolomite with abundant replacive gypsum and anhydrite in the subsurface; thin argillaceous dolomite beds and small-scale cross-lamination common in lower part, less so above; evidence of basinward submarine sliding at several levels, especially near base and top.
Biota: Varied but only locally abundant fauna, largely consisting of foraminifers, bryozoans, brachiopods and bivalves; characteristic species include *Agathammina pusilla*, *Horridonia horrida* and *Bakevella binneyi*. Burrows abundant in lower beds.

Ford (Magnesian Limestone) Formation (EZ1b Ca)
Former name: Middle Magnesian Limestone of County Durham, Tyne & Wear, Cleveland and northernmost North Yorkshire.
Stratigraphical definition: Upper Permian marine carbonate strata from the latitude of Catterick northwards, lying between the Raisby Formation (defined here) and the Edlington Formation (defined here) in the west and the Hartlepool (Anhydrite) Formation (or its residue) in the east.
Type locality: Ford Quarry and Claxheugh Rock (NZ 362572), Ford, Sunderland where at least 35 m thick. Lowest beds probably absent through onlap against markedly uneven substrate, highest beds (thickness unknown) eroded. Two facies (reef and back-reef) present with interdigitate generally vertical contact (see Smith 1970, fig. 17); reef comprises mainly massive buff and brown altered dolomite with a bryozoan framework in most parts but with dominant algal encrustations, laminae and some stromatolites near the top; back-reef comprises bedded buff mainly fine-grained altered dolomite with scattered submarine slide masses and a sparse foraminiferal-chonetid brachiopod-molluscan fauna.
Note: Because of the great lateral variation in the Ford Formation, there is no one place or borehole where typical representatives of all the main facies can be seen or documented. Ford Quarry and Claxheugh Rock together expose sections of the two main facies and comprise the most representative group of exposures; the reef here yields an atypically sparse fauna, however, because of widespread texture-destroying dolomitization and the back-reef strata differ from the cross-stratified peloidal dolomite present in most areas immediately landward of the reef.
Lateral variation: Extreme; forms broad wedge, thickness known locally to exceed 120 m and possibly 150 m but thins very sharply to basin floor where less than 1 m. Two main facies, reef and back-reef, with minor basin-floor facies; reef facies comprises narrow (generally 400–800 m) shelf-margin bryozoan-algal dolomite up to 120 m thick; back-reef facies mainly comprises oolitic and sparingly pisolitic dolomite up to 100 m thick, with evidence of former replacive evaporites; the basin floor facies comprises a sharp marginal wedge of reef talus and an extensive thin bed of oncoids and algal stromatolites (the Trow Point Bed, Smith, *this volume*). Also probably includes thick (15–30 m) algal-stromatolitic strata (the Hesleden Dene Stromatolite Biostrome, Smith 1981) and perhaps overlying oolitic dolomite (< 10 m).
Biota: Each facies has its characteristic biota. The low reef yields the richest English Zechstein faunas including milioloid foraminifers, bryozoans (e.g. *Fenestella retiformis*, *Synocladia virgulacea*), brachiopods (e.g. *Horridonia horrida*, *Pterospirifer alatus*), gastropods (e.g. *Mourlonia antrina*), bivalves (e.g. *Bakevellia ceratophaga*) and nautiloids (*Peripetoceras freieslebeni*). Higher reef assemblages show a progressive upward decline in number of species (especially among the brachiopods), although not necessarily of individuals, with gastropods and bivalves forming a higher proportion of the shelly fauna. Cyanophytes (blue-green algae) abundant in upper part of reef. Back-reef faunas are dominated by gastropods and bivalves; *Astartella vallisneriana* is a characteristic species. Most of the macrofossils recorded from basinal facies are reef-derived species; probably autochthonous forms include nodosariid foraminifera and chonetoid brachiopods.

Roker (Dolomite) Formation (EZ2 Ca); *name amendment.*
Former name: Hartlepool and Roker Dolomite; proposed amendment permits nomination of a single type section and eliminates duplication of Hartlepool, already in use in Hartlepool (Anhydrite) Formation.
Type locality: Sea cliffs, Roker, Sunderland, extending from mouth of Roker Park (NZ 407592) to Cannon Ball Rocks (NZ 407596). Exposed thickness about 20 m, top not exposed. Comprises sparingly fossiliferous cream shallow-water peritidal oolitic dolomite with subordinate thin beds of fine-grained dolomite; is generally parallel-bedded but features abundant small-

scale cross-bedding and several erosion surfaces, contemporaneous breccias and evidence of soft-sediment deformation. Base stepped against and interdigitates with the underlying Concretionary Limestone.

Lateral variation: Is thickest (up to 80 m) in offshore area of Durham Province. Lithology generally similar to type locality but oolites less dominant. Some beds contain concentrations of ostracods.

Biota: Limited, dominated by the bivalves *Schizodus obscurus* and *Liebea squamosa* with locally abundant small gastropods; ostracods common in some beds. Generally indeterminate. Plant debris at some levels.

Edlington Formation

Former name: Middle Marls, Permian Middle Marls etc.

Stratigraphical definition: Upper Permian alluvial, lagoonal and marine strata in NE England lying between the Cadeby Formation (defined here) and the Brotherton Formation (defined here).

Type locality: New Edlington Clay Pits (SK 529987), New Edlington, South Yorkshire. Thickness about 8 m: comprises reddish-brown mudstone and subordinate pale greyish-green siltstones and sandstones with veins of fibrous gypsum, scattered nodules of alabastrine gypsum and some halite pseudomorphs. Several thin lenses and beds of dolomite at top may form part of overlying Brotherton Formation.

Lateral variation: Only locally absent, up to 55 m thick in belt from York to Retford, thins farther east where the formation may pass into the Hayton (Anhydrite) Formation, Kirkham Abbey and Fordon (Evaporite) Formations (Fig. 2). At outcrop the Edlington Formation is generally similar to type locality but is in places sandier and contains lenses of aeolian sandstone in south. In subcrop and at outcrop near Ripon it contains abundant disseminated (secondary) and bedded evaporites (mainly gypsum), locally up to 70% of sequence, and thin dolomite beds (some relatively extensive and dividing the formation into 3 or more cyclic units).

Biota: Remains of land plants preserved locally in greyish-green beds.

Brotherton (Magnesian Limestone) Formation (EZ3 Ca)

Former names: Brotherton Beds; Upper Magnesian Limestone.

Stratigraphical definition: Upper Permian marine carbonate strata south of the Cleveland Axis, lying between the Edlington Formation (defined here) and the Billingham (Anhydrite) Formation (name amended here) or its residue (or the Carnallitic or Rotten Marl where the Billingham Formation is absent).

Type locality: Brotherton (SE 4825), North Yorkshire. Thickness about 20 m; no single nominated type section but many separate disused quarries, mostly in uppermost 12 m of the formation. Lower beds (about 8 m) are said to be of dolomite (Edwards *et al.* 1940), higher beds are of grey uniformly fine-grained thin-bedded limestone with a patchy bivalve fauna and innumerable remains of *Calcinema* on some bedding planes. Small-scale sedimentary structures, especially ripple marks, abundant at many levels.

Lateral variation: Thickens progressively eastwards to about 65 m in the subsurface of Humberside and eastern North Yorkshire but commonly 10–20 m thick at outcrop. Lithology generally uniform, as at type locality, but dark grey to almost black in deep subsurface. Basal beds argillaceous (equivalent to Grauer Salzton) in eastern North Yorkshire and adjoining areas. Partly oolitic and oncolitic or with thin beds of grey, red and purple mudstone in some surface exposures. Uppermost 1–2 m widely algal-laminated, with interbedded nodular anhydrite in eastern North Yorkshire.

Biota: Almost restricted to the alga *Calcinema permiana* and the bivalves *Liebea squamosa* and *Schizodus obscurus*; all three are locally abundant. Some ostracods and small gastropods have also been recorded.

Billingham (Anhydrite) Formation (EZ3A); *name amendment*

Former name: Billingham Main Anhydrite; former name necessary to permit distinction from a younger anhydrite bed also present at Billingham; creation of new name for the younger anhydrite (see below) permits shortening of the existing name.

Sherburn (Anhydrite) Formation (EZ4A)

Former name: Upper Anhydrite

Stratigraphical definition: An Upper Permian anhydrite formation lying between the Upgang Formation (previously defined) and the Sneaton (Halite) Formation (defined here).

Type locality: Sherburn Mine (No. 2 shaft, SE 51663440), Sherburn in Elmet, North Yorkshire, where worked for gypsum. At type locality is pale grey and alabastrine and is weakly bedded except in roof of workings where locally ripple-laminated; passes laterally into finely crystalline anhydrite east of mine workings.

Lateral variation: A persistent formation, generally 2–4 m thick on Tees-side, 5–15 m thick in Yorkshire and Humberside, thins in extreme east.

Commonly a coarse, grey to buff or pink finely crystalline anhydrite rock, distinguished by the present of many courses of upright-fibrous pseudomorphs (in anhydrite, halite or sylvite) probably after early gypsum. Abundant uneven laminae of dolomite or magnesite. Cross-lamination fairly common in upper parts, where locally purplish-grey, and locally near base.
Biota: None known.

Sneaton (Halite) Formation, including Sneaton Potash Member EZ4H
Former name: Upper Halite, including Upper Potash Member
Stratigraphical definition: a lithologically varied Upper Permian halite formation lying between the Sherburn (Anhydrite) Formation (named here) and the Sleights (Siltstone) Formation (previously defined).
Type locality: Sneaton (E4) Borehole (NZ 904078), Sneaton, North Yorkshire, depth 1228.7–1263.4 m; comprises five lithological units (A–E, Smith 1971, p. 223) including the Sneaton Potash member (C), depth 1243.0–1248.2 m.
Lateral variation: Dies out westwards at a solution edge and thickens gently eastwards to 55 m in easternmost North Yorkshire and Humberside (Fig. 2). Four of the five lithological units are relatively persistent but the Sneaton Potash is present mainly in a belt a few kilometres wide near the present coastline.
Biota: None known.

Littlebeck (Anhydrite) Formation (EZ5A)
Former name: Top Anhydrite
Stratigraphical definition: A thin (0.8–1.5 m) anhydrite unit in coastal districts of North Yorkshire, lying between the Sleights (Siltstone) Formation (previously defined) and the Roxby Formation (defined here). Generally grey and finely crystalline, commonly unevenly laminated.
Type locality: Littlebeck (F3) Borehole (NZ 877050), Littlebeck, North Yorkshire, depth 1197.4–1198.3 m where pinkish-grey with discontinuous laminae of reddish-brown silty 'marl'.
Lateral variation: Wedges out gradually inland, thickens eastward to 1–2 m at the present coast. Cored in only a few boreholes, where generally fine-grained and with closely-spaced faintly-pink haematitic partings.
Biota: None known.

Roxby Formation
Former names: Upper Permian Marl, Saliferous Marl
Stratigraphical definition: A formation of reddish-brown mudstones, siltstones and subordinate sandstones, lying between the top of the Littlebeck Formation and the gradational base of the Sherwood Sandstone Group; generally anhydritic or gypsiferous in lower part, where commonly blocky and less clearly rhythmic than in higher part.
Type locality: Staithes S20 Borehole (NZ 760180), Boulby, Cleveland, depth 920.5–1049.8 m. Borehole S20 is the pilot bore for the Boulby Potash Mine (Woods 1973), near Staithes, but the names Boulby and Staithes are already in stratigraphic use. A complete core is retained by the British Geological Survey.
Lateral variation: Lithologically fairly uniform throughout NE England where it appears to pass diachronously into the lower part of the Sherwood Sandstone Group. It is more than 120 m thick in east Cleveland, North Yorkshire and Humberside, but thins gradually westwards and southwards and is absent in parts of South Yorkshire and north Nottinghamshire.
Biota: None known.

References

EDWARDS, W., WRAY, D.A. & MITCHELL, G.H. 1940. Geology of the country around Wakefield. *Mem. geol. Surv. G.B.* HMSO, London.

——, MITCHELL, G.H. & WHITEHEAD, T.H. 1950. Geology of the district north and east of Leeds. *Mem. geol. Surv. G.B.* HMSO, London.

MAGRAW, D. 1975. Permian beds of the offshore and adjacent coastal areas of Durham and south eastern Northumberland. *J. geol. Soc. London*, **131,** 397–414.

——, CLARKE, A.M. & SMITH, D.B. 1963. The stratigraphy and structure of part of the south-east Durham coalfield. *Proc. Yorks. geol. Soc.* **34,** 153–208.

SMITH, D.B. 1968. The Hampole Beds—a significant marker in the Lower Magnesian Limestone of Yorkshire, Derbyshire and Nottinghamshire. *Proc. Yorks. geol. Soc.* **36,** 463–77.

—— 1970. The palaeogeography of the English Zechstein. In: RAU, J.L. & DELLWIG, L.F. (eds) *Third Symposium on Salt*, Vol. 1, 20–3. Northern Ohio Geological Society.

—— 1971. Possible displacive halite in the Upper Permian Evaporite Group of northeast Yorkshire. *Sedimentology*, **17,** 221–32.

—— 1981. The Magnesian Limestone (Upper Permian) reef complex of northeastern England. In: TOOMEY, D.F. (ed.) *European Fossil Reef Models. Spec. Publ. Soc. econ. Paleont. Miner., Tulsa*, **30,** 161–86.

—— (*this volume*). The Trow Point Bed—a deposit of Upper Permian marine oncoids, peloids and columnar stromatolites in the Zechstein of N.E. England.

——, BRUNSTROM, R.G.W., MANNING, P.I., SIMPSON, S. & SHOTTON, F.W. 1974. A Correlation of Permian rocks in the British Isles. *Spec. Rep. geol. Soc. London*, No. 5, 45 pp.

D.B. SMITH, 79 Kenton Road, Gosforth, Newcastle upon Tyne NE3 4NL.
G.M. HARWOOD, Department of Geology, The University, Bedson Building, Newcastle upon Tyne NE1 7RU.
J. PATTISON, British Geological Survey, Keyworth, Nottingham NG12 5GG.
T.H. PETTIGREW, Sunderland Museum and Art Gallery, Borough Road, Sunderland.

Chemical and isotopic studies of a core of Marl Slate from NE England: influence of freshwater influx into the Zechstein Sea

P. Turner & M. Magaritz

SUMMARY: Detailed mineralogical, chemical and stable isotope analyses have been made of single core through the Marl Slate in order to investigate the Zechstein transgression and its associated mineralization. The core (VT8), drilled by the National Coal Board off the coast of NE England, contains three subfacies: (1) dark grey, organic-rich, sapropelic siltstones; (2) organic–dolomite laminites; (3) massive finely crystalline dolostones. Two cycles of deposition which include the sequence sapropel→laminite→dolostone are present although the lower one (112 cm) is much thicker than the upper (32 cm).

There are some close correlations between the lithological and geochemical variables, including the stable isotope values. The sapropelic units are characterized by high Fe content (as iron sulphide) and strongly negative $\delta^{18}O$ values in associated carbonate minerals. Quartz silt and organic carbon are also higher in these units but diminish upwards in each depositional cycle. Each cycle of deposition appears to have been initiated by an influx of freshwater into the Zechstein basin which promoted surface productivity, water stratification and regional anoxic conditions. The relative change in $\delta^{13}C$ to more positive values higher in the Marl Slate is independent of lithological variation and indicates progressive enrichment of ^{13}C in the Zechstein Sea in response to the precipitation of large amounts of organic matter.

The distribution of base metals in the Marl Slate in this core strongly suggests that they were introduced by freshwater influx. Although Pb shows no clear relationship to the depositional cycles, Cu and Zn contents are relatively low in the lower sapropelic parts of the cycles and become higher in the upper parts. We take this to indicate that Cu and Zn sulphides were precipitated interstitially when more 'normal' marine conditions were re-established in the later part of each cycle.

The Marl Slate is a silty sapropelic dolomite containing sulphide minerals which include chalcopyrite, sphalerite and galena. Over much of the Zechstein basin the Marl Slate is less than 80 cm thick (Smith 1980) but in NE England it is commonly 1–2 m thick. The Marl Slate shows considerable lithological variation; generally quartz silt is abundant towards the base and in the upper part well-developed carbonate laminites occur. There is no bioturbation and the biota, which includes foraminifera, nautiloids and well-preserved fish, indicates deposition in an anoxic environment (Demaison & Moore 1980).

The Marl Slate is the initial deposit of the Zechstein transgression, which has been discussed in detail by Smith (1970a, 1979, 1980). The evidence suggests that the Rotliegendes desert basin of the Southern North Sea was below the contemporary sea-level and was rapidly submerged as the Boreal Ocean flooded in from the north. This is consistent with the preservation of dune bed forms in unconsolidated sands beneath the Marl Slate. The height of these ridges in County Durham is up to 60 m and it follows that the initial Zechstein Sea must have been of substantial depth, probably up to 250 m (Smith 1980).

It is not clear what the causes of sea-level change were which initiated the Zechstein transgression, although most authors consider the initial and later Zechstein transgressions to be the result of eustatic changes in sea-level. These may have been brought about by major deglaciations in the southern polar region (Smith 1970b, p. 21; Crowell 1978; Crowell & Frakes 1970).

The trace metal (particularly base metal) content of the Marl Slate/Kupferschiefer is greater than the average black shale (Hirst & Dunham 1963; Wedepohl 1980) and there has been considerable discussion regarding the source of these metals and how their enrichment took place. Davidson (1962) regarded the mineralization as epigenetic whereas Hirst & Dunham (1963) suggested that the metals may have derived from Palaeozoic mineralization to the west and beneath the Marl Slate and supplied to the Zechstein Sea via submarine springs. However, most authors now acknowledge that metals were probably present in Zechstein seawater and deposited within the Marl Slate as part of the biochemical cycle (Brongersma-Sanders 1965). Diagenetic reactions subsequently played an important role in producing the characteristic features of the mineralization.

From HARWOOD, G.M. & SMITH, D.B. (eds), 1986, *The English Zechstein and Related Topics*, Geological Society Special Publication No. 22, pp. 19–29.

In this paper we report the investigation of a complete sequence of the Marl Slate from a borehole drilled off the NE coast of England by the National Coal Board. The core is from VT8 borehole (Grid Ref. NZ 4862 5095) and details of the full sequence are given by Magraw (1978). The part of the core analysed is 144 cm thick and stable isotope and chemical analyses have been made at 1 cm intervals throughout.

By comparing the stable isotope variations ($\delta^{13}C$ and $\delta^{18}O$) with chemical and lithological changes we show that the metal concentration is correlated with the isotopic composition of associated carbonate minerals and subtle lithological variation. A model is proposed in which we relate marine transgression, freshwater influx and the development of anoxic conditions in the early Zechstein Sea to the characteristic metal sulphide mineralization of the Marl Slate.

Analytical methods

The core was divided into 144 1-cm lengths which were then ground to powder in an agate mortar. Part of the powder was used for carbon and oxygen isotopic measurements using the conventional phosphoric acid method. The CO_2 gas was analysed using Atlas M86 and Varian 250 mass spectrometers and the results are given using the δ notation relative to PDB standard:

$$\delta^{18}O = \frac{((^{18}O/^{16}O)\ \text{sample} - (^{18}O/O^{16})\ \text{standard})}{(^{18}O/^{16}O)\ \text{standard}} \times 1000.$$

Reproducibility of duplicate samples is better than 0.2‰.

The organic carbon (C_{org}) and carbonate carbons were determined by barium hydroxide absorption and titration using thymolphthalien as indicator. The precision of the method was checked using synthetic samples of known concentration.

All the other major and trace elements were determined using atomic absorption and ICPS techniques. Both lithium metaborate and acid digestion methods were used.

The remaining portions of the sample powders were all subjected to standard X-ray diffraction analysis.

Lithological variation of the VT8 core

In the VT8 core the Marl Slate rests unconformably on the Lower Permian Yellow Sands and passes gradationally upwards into the Raisby Formation. It is not possible to place a precise boundary between the two units although the Marl Slate (s.s.) as reported in this paper is probably no more than 1 m thick. The part of the core analysed is 144 cm thick and is a complete sequence of the Marl Slate and part of the overlying Passage Beds of Magraw (1978).

Three lithotypes can be distinguished in the core although the boundaries between these are gradational. The fundamental difference between the lithotypes is in the content and distribution of organic matter.

Lithotype 1 (Fig. 1a)

Lithotype 1 comprises organic-rich, silty dolomicrite. Organic carbon content is generally between 8 and 10% and quartz 25 and 35%; the latter occurs as small silt lenses associated with other detrital minerals (mainly micas). In thin section this lithotype has a distinct lamination with discontinuous lenses of ferroan dolomicrite (60 μm) alternating with thinner organic laminae. The lamination is extremely fine, averaging 80 μm, and is not obvious in hand specimens.

An important feature of this lithotype is the occurrence of small micronodules of dolomite with length-slow chalcedony and metal sulphides. They are considered to be the replacements of anhydrite crystals (Turner *et al.* 1978).

Lithotype 2 (Fig. 1b)

Where the content of organic carbon is less than 6% the organic laminae are farther apart and the rock has an obviously laminated appearance. The carbonate laminae consist of hypidiotopic dolomite, commonly ferroan, with a crystal size of 5–100 μm, rarely micritic dolomite and organic layers (10–20 μm). The laminae are generally less than 100 μm in thickness but a few are up to 5 mm. Many of the thicker laminae are grouped close together in 'packets' with intervening units of predominantly thinner laminae.

The quartz silt content of lithotype 2 decreases sharply to about 10% in parallel with the decrease in organic carbon.

Lithotype 3 (Fig. 1c)

In this lithotype organic matter is not present as discrete laminae but is distributed unevenly in flakes. The overall appearance is of a homogeneous finely crystalline dolostone which reveals little or no lamination. The average grain size is about 10 μm and the dolomite is generally non-ferroan. In the upper part of Cycle 2 in this facies there are two discrete pale grey unlaminated

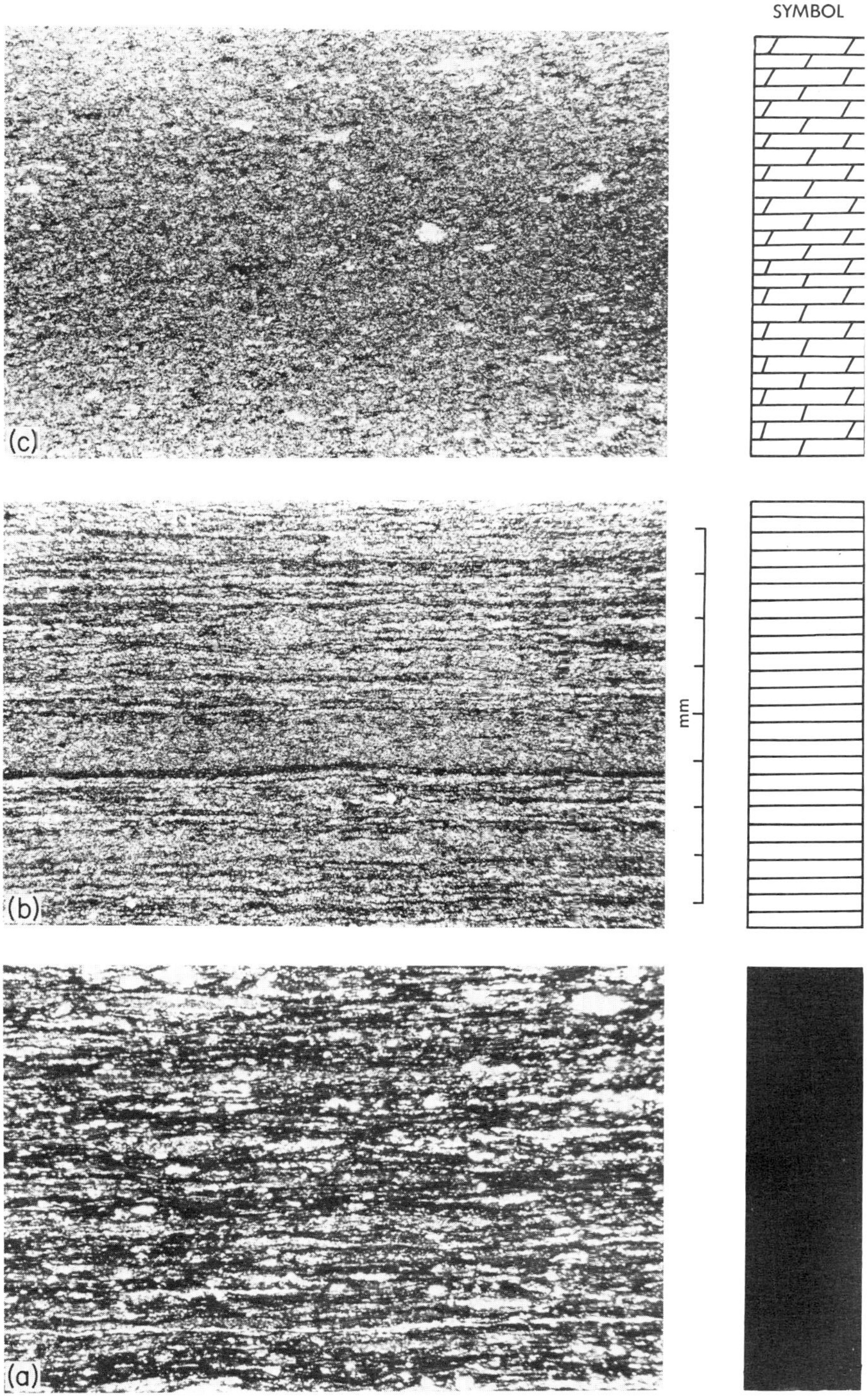

FIG. 1. Photomicrographs illustrating the characteristic features of Marl Slate lithotypes. They are shown in stratigraphical order, from the bottom upwards. (a) Sapropelic silty dolomicrite, (b) laminite, (c) massive finely crystalline dolostone.

dolostone beds up to 2.5 cm thick with a general porcellanous appearance.

The Marl Slate in Core VT8 has two cycles of sedimentation, each comprising the sequence sapropel→laminite→massive dolostone (Fig. 2); the lower cycle is 112 cm thick and makes up the main part of the Marl Slate. It is succeeded by a similar, but much thinner (32 cm) cycle which begins with a very thin, but distinct sapropel (1.5 cm) overlain by further laminite and massive finely crystalline dolostone.

Both of these cycles show the characteristic features of accumulation in anoxic water as described by Demaison & Moore (1980). These include the presence of sapropels with over 10% organic carbon, the absence of a benthic fauna and bioturbation and the preservation of very fine laminae. The units overlying the sapropel are characterized by thicker laminae, greater total carbonate content and lesser amounts of quartz and organic carbon.

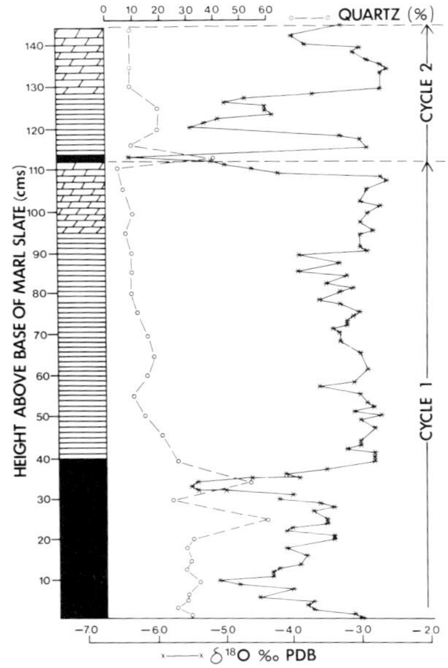

FIG.2. Variation in $\delta^{18}O$ and percentage quartz in the VT8 core.

Variation in $\delta^{18}O$ and quartz

The variation in $\delta^{18}O$ and quartz content in the VT8 core is shown in Fig. 2. The remarkable feature of the variation in $\delta^{18}O$ are the peaks of negative values at 11, 33, 112, 120, 127 cm above the base. The peaks at 11 and 33 cm lie within the sapropel of Cycle 1 and the large negative peak at 112 cm corresponds to the thin sapropel at the base of Cycle 2.

The average value of $\delta^{18}O$ in the lower sapropelic unit is about −4.0‰ and enrichment in $\delta^{18}O$ to about −3‰ occurs at the sapropel/laminite boundary. The variation in quartz content is antithetic to $\delta^{18}O$; in the lower sapropel unit the quartz content is about 35‰ and drops to 10‰ in the overlying laminite.

There is no evidence that the $\delta^{18}O$ values have been modified by later diagenesis or recrystallization. Indeed, the close correspondence between $\delta^{18}O$ variation and the detrital mineral variation suggests that the isotopic variation is primary.

Variation in $\delta^{13}C$ and organic carbon

Variations in $\delta^{13}C$ values and organic carbon content in the VT8 core are shown in Fig. 3. In the sapropel of Cycle 1 $\delta^{13}C$ values begin at +2.3‰ and become enriched in ^{13}C up to +4.0‰. From this point upwards there is a very sharp decrease in ^{13}C, returning to the values seen at the base of the sapropel; this minimum lies about 5 cm below the $\delta^{18}O$ minimum (Fig. 2). Overlying this minimum there is a gradual enrichment of the carbonate in ^{13}C up to a value of +4.5‰. The boundary between the sapropel and laminite of Cycle 1 corresponds approximately to the mid-point in this trend. Throughout the rest of the core $\delta^{13}C$ values become steadily more positive upwards, reaching values of about +5.0‰ near the top. Such highly positive $\delta^{13}C$ values are characteristic of all later Zechstein marine carbonates (Bolz 1979; Margaritz & Schulze 1980; Magaritz *et al.* 1981). In this part of the core the negative peaks in $\delta^{18}O$ are accompanied by relatively negative $\delta^{13}C$ values; the peak at 112 cm corresponds to sapropel 2.

The organic carbon content shows a similar trend although less data are available. From the high content (exceeding 10%) in the lower part of sapropel 1, a general upwards decrease accompanies the change from the sapropel to the laminite of Cycle 1. The sapropel at the base of Cycle 2 shows a return to a higher organic carbon content.

Variation in Ca and Mg

The variation of Ca and Mg content is shown in Fig. 4. The general features of Ca and Mg variation are broadly similar, reflecting the fluc-

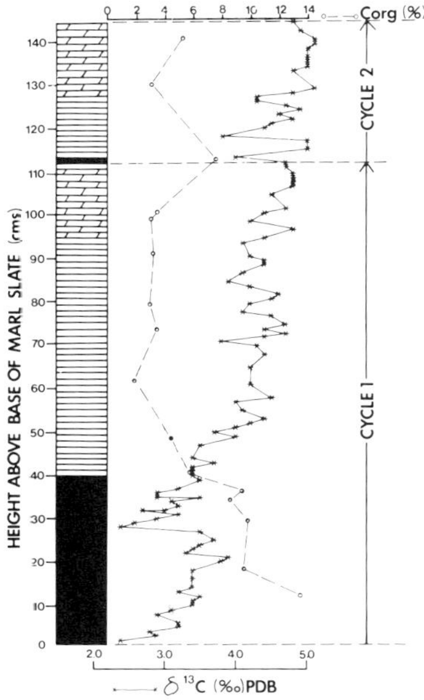

FIG. 3. Variation in $\delta^{13}C$ and organic carbon (C org) in the VT8 core.

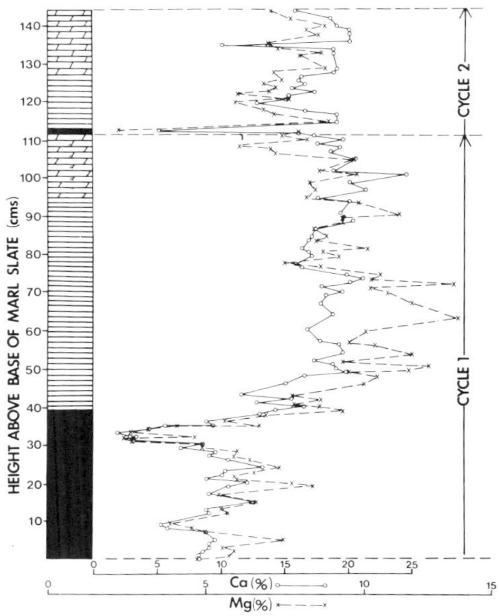

FIG. 4. Variation in Ca and Mg in the VT8 core.

tuation in dolomite content. XRD analysis of all the 1-cm samples shows that dolomite is the only carbonate in the VT8 core.

Sapropel 1 has the lowest content of both elements and there is a sharp increase between 34 and 50 cm above the base. This coincides with the change in the $\delta^{13}C$ values (Fig. 3) and the transition between the sapropel and laminite of Cycle 1. In the Cycle 1 laminite there is a close similarity between Ca/Mg variation and the variation in $\delta^{18}O$, indicating that where dolomite is least abundant it also tends to be depleted in ^{18}O. Above the Cycle 1 laminite the Ca and Mg content is more constant, although with minor perturbations.

Parts of the core are characterized by very low Ca and Mg content; they partly correspond to the maxima in quartz content and the minima in $\delta^{18}O$ values although the greatest excess of Mg over Ca occurs in the laminite of Cycle 1 (Fig. 2). Within these parts there are indications that there are contributions to the Mg content other than that from dolomite, suggesting a detrital source for the additional Mg. This source is most likely to be clay minerals, including chlorite and montmorillonite (Hirst & Dunham 1963).

There is considerable textural variation in the dolomite. Ferroan dolomicrite dominates the sapropelic units and non-ferroan to ferroan dolospar is important in the laminite and massive dolostones. However, all the dolomites are calcian in composition, in keeping with evidence from other Zechstein dolomites (Turner *et al.* 1978).

Variation in iron, arsenic and selenium

Iron is present in the Marl Slate principally as pyrite (FeS_2) and as $FeCO_3$ in ferroan dolomites. Minor contributions come from chalcopyrite ($CuFeS_2$) and iron in clay minerals. The distribution of Fe in the VT8 core (Fig. 5) is broadly related to the two cycles of deposition. The highest iron contents (7.1%) are found 10 cm above the base of the Marl Slate in sapropel 1. Taken as a whole, Cycle 1 shows a gradual upwards decrease in Fe to less than 1% in the dolostone at the top of the cycle. There are perturbations in the iron content, with minor peaks at 10, 33, 73 and 121 cm which correspond to minima in $\delta^{18}O$ and maxima in quartz content.

Arsenic and selenium occur in trace concentration in the Marl Slate with maximum values of 71 and 36 ppm respectively (Fig. 6). Their behaviour is very similar, with the highest concentrations occurring in the sapropels of Cycles 1 and 2. There is a sharp drop in both the Se and As concentration across the sapropel/laminite transition of Cycle 1. This drop coincides with the increase in Ca and Mg (Figs 4 and 6).

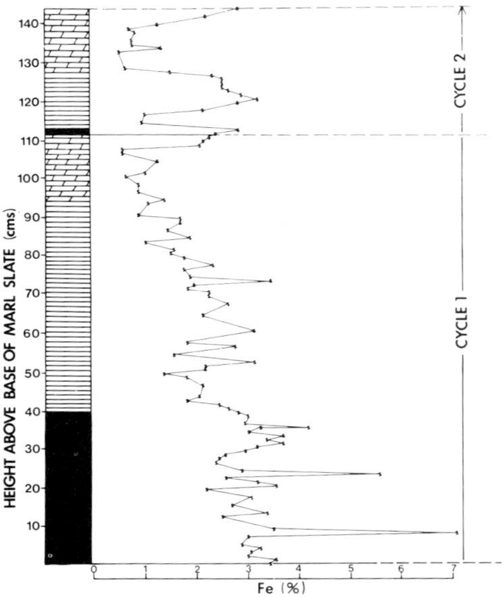

FIG. 5. Variation in Fe in the VT8 core.

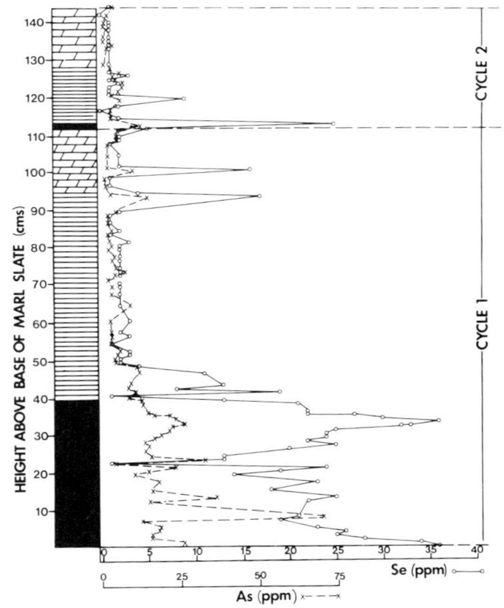

FIG. 6. Variation in As and Se in the VT8 core.

Peaks in Se occur at the base and at 33, 93, 100, 112 and 120 cm above the base of the Marl Slate. Some of these correspond to the low $\delta^{18}O$ values.

These siderophile elements are clearly related to the sulphide distribution and their peak concentration is parallel to that of iron. In view of their low concentration there is no need to postulate their presence in other phases.

Variation in copper, zinc and lead

Cu and Zn are present in the Marl Slate as the sulphides chalcopyrite and sphalerite. These sulphides are present as thin lenses and laminae. The distribution of Cu and Zn in the VT8 core is shown in Fig. 7. The data show some scatter within short vertical intervals. The peaks in these perturbations correspond to specimens that contain thin sulphide laminae. However there is a clear general trend shown by Cu and Zn in both Cycles 1 and 2.

At the base of Cycle 1 Cu (200 ppm) and Zn (100 ppm) are low in the sapropelic unit and increase progressively upwards, reaching a maximum in the upper part of the Cycle 1 dolostone with Cu about 900 ppm and Zn about 500 ppm (Fig. 7). The maximum values of both Cu and Zn are found 108 cm above the base, corresponding precisely to the start of the change in $\delta^{18}O$ values immediately prior to the base of the second cycle.

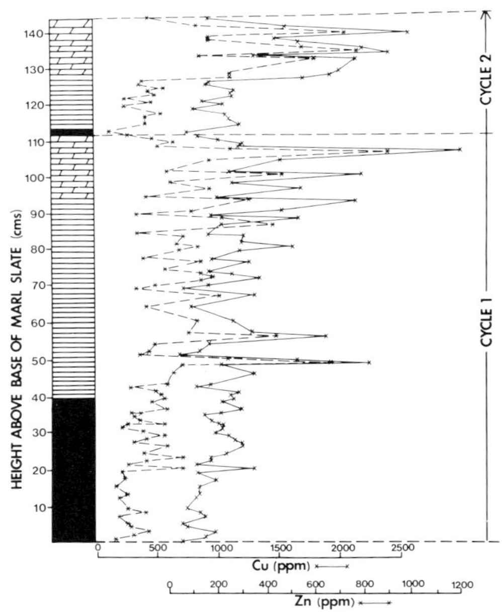

FIG. 7. Variation in Cu and Zn in the VT8 core.

Cycle 2 starts with low Cu (350 ppm) and Zn (200 ppm) values and at 129 cm there is a sharp rise in the content of both elements. Again this sharp increase corresponds to the increase in $\delta^{18}O$ shown in Fig. 2.

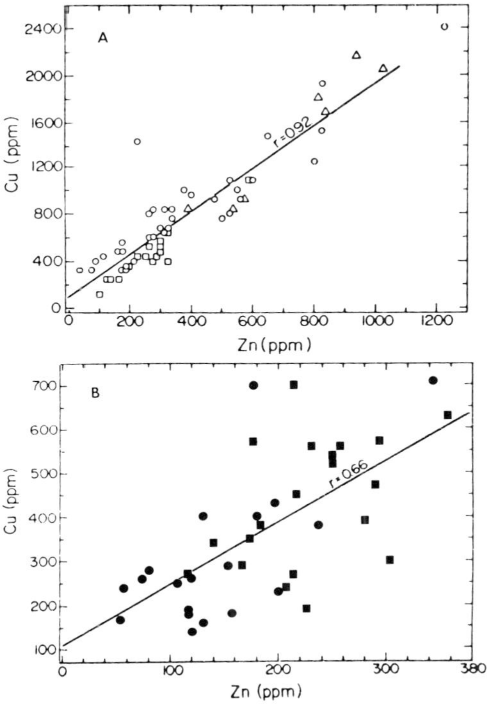

FIG. 8. Comparison of Cu versus Zn values in the upper part of the VT8 core (A) with those in the lower part of the core (B). Note that the correlation is better in the upper part of the core. Various parts of the core are indicated as follows: ● Cycle 1 sapropel; ■ Cycle 1 sapropel→laminite transition zone; ○ Cycle 1 laminite and dolostone; □ Cycle 2 sapropel and laminite; △ Cycle 2 dolostone.

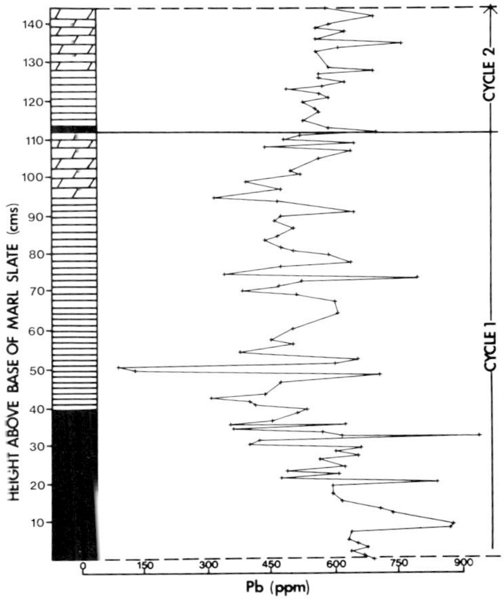

FIG. 9. Variation in Pb in the VT8 core.

Generally there is good positive correlation between Cu and Zn. In detail, however, there is a difference between the lower 50 cm and the rest of the core. In this lowermost part the correlation is less pronounced ($r=0.66$) (Fig. 8) whilst the upper part correlation is significantly better ($r=0.92$).

The distribution of Pb, which is present mainly as galena, is shown in Fig. 9. There is considerable variation, particularly in Cycle 1, and some of the perturbations are caused by galena in sulphide-rich laminae; Pb ranges from 91 ppm to a maximum of 941 ppm. In Cycle 1 the Pb concentration decreases in the sapropel from about 650 ppm to about 400 ppm at the top.

From this level upwards there is a general gradual increase and no obvious break between Cycles 1 and 2.

Discussion

The variation of $\delta^{18}O$ reflects the isotopic composition of the water in which the carbonate was formed. In general, seawater and evaporated water are enriched in ^{18}O whereas continental fresh water is depleted in ^{18}O. We argue that the pattern observed in $\delta^{18}O$ values (Fig. 2) is related to admixture of those two types of water during the deposition of the Marl Slate. The minima in $\delta^{18}O$ values indicate periods when the influx of fresh water into the basin was at its maximum. The fact that the peaks embrace several 1-cm samples and show well-defined trends within individual lithological units indicates that these features are probably primary. Furthermore the beginning of Cycle 2 exactly coincides with a period of inferred freshwater influx and the rest of the minima show a close relationship with high contents of detrital components (quartz, iron, clay minerals); this suggests that the changes in $\delta^{18}O$ probably reflect the isotopic composition of surface waters in which the initial carbonates were precipitated.

In general the sapropelic units have more negative $\delta^{18}O$ values than the laminites or dolostones. A similar relationship has been described in cores from the same area (Magaritz *et al.* 1981). This association of sapropel and carbonates with low $\delta^{18}O$ values (fresh water) is well-known, particularly in the Pleistocene of the Eastern

Mediterranean area (Vergnaud-Grazzini *et al.* 1977; Thunnell *et al.* 1977). In this model, freshwater influx leads to a stratified water column and promotes anoxic conditions; in this regime sediments with high organic and detrital contents accumulate. The fact that the $\delta^{18}O$ content of the Marl Slate has recorded those freshwater events indicates that the original carbonate minerals must have formed in the upper freshwater mass in a manner similar to that seen in the Black Sea (Degens & Stoffers 1980).

The changes in carbon isotopic composition within the Marl Slate towards highly enriched ^{13}C values have been discussed by Magaritz *et al.* (1981) and Magaritz & Turner (1982). This enrichment in ^{13}C reflects a global change in the carbon isotope budget of the ocean which is related to the removal of large quantities of organic material. The spikes of slightly less enriched $\delta^{13}C$ values which correspond to minima in $\delta^{18}O$ values are related to the influx of bicarbonate depleted in ^{13}C from the continent; this association of ^{18}O and ^{13}C depletion is a characteristic feature of continental water. The fact that a number of constituents show marked enrichment at times of freshwater influx indicates their continental source. Quartz, which is of obvious detrital origin in the Marl Slate, shows a good correlation with the minima in $\delta^{18}O$ values (Fig. 2) and large quantities were introduced into the basin during the flooding events. Iron shows a similar enrichment in the sapropelic sediments (corresponding with minima in $\delta^{18}O$ values). This inter-relationship is very clearly seen in Cycle 2 (Figs 2 and 5) where iron maxima coincide with $\delta^{18}O$ minima. We suggest that this indicates that iron, like quartz, was introduced into the sea during freshwater flooding.

Arsenic and selenium show a similar pattern to the iron, a high content during the freshwater sapropelic events. This is related to the greater rate of formation of sulphide minerals during the freshwater anoxic events.

We conclude from these data that during maximum freshwater flooding detrital quartz, iron and clay minerals were transported into the basin in larger quantities than when fully marine conditions prevailed. Although carbonate sedimentation was continuous during these periods, the proportion of carbonate is relatively low. The major increase in carbonate percentage takes place in the transition between the sapropel and laminate of Cycle 1 following the $\delta^{18}O$ minima at 34 cm.

The deposition of the Marl Slate was seasonal (Oelsner 1959; Brongersma-Sanders 1965, 1971). This seasonality is reflected in the rock by the presence of alternating carbonate-rich and organic-rich layers. The organic-rich layers are considered to be the product of seasonal phytoplankton blooms (Brongersma-Sanders, 1971). In sapropel 1 the distance between the organic layers is about 80 μm and they are continuous and bifurcate in such a way that the carbonate layers are lenticular.

As the $\delta^{13}C$ values become enriched in ^{13}C there is an accompanying decrease in the organic carbon content (Fig. 3). This is consistent with the model explaining the change in carbon isotope composition (Magaritz & Turner 1982). This model relates change in $\delta^{13}C$ values of carbonate minerals to the precipitation of large amounts of organic material enriched in ^{12}C. The carbonate reservoir and subsequently precipitated carbonate minerals are thus enriched in ^{13}C and show higher $\delta^{13}C$ values. The removal of organic material depletes the water mass in essential nutrients such as P and N which causes a reduction in productivity. This reduction in productivity causes a corresponding decrease in the number of phytoplankton blooms and results in a greater separation of organic layers as seen in the laminite units. In some years no organic layer may have been formed. With transition to the massive dolostones the organic matter becomes more disseminated and there are no discrete organic laminae.

The initially deposited carbonate phases in the Marl Slate were calcium carbonates which were subsequently converted to dolomite. The grain size distribution of the dolomite is largely controlled by the nature of the organic layering. In the sapropels the fine organic laminae impose very fine grain size and in units with thicker carbonate layers the dolomite attains a larger crystal size.

The fact that very detailed fluctuations in both carbon and oxygen are preserved today imposes the constraint that during dolomitization no large-scale mixing of components took place. The preservation of the original calcium carbonate isotopic record in dolomite has been observed in many other cases since the pioneering work of Degens & Epstein (1964). Their model, in which the bicarbonate ion is transferred as a unit to dolomite, seems best to explain our data. If there were, for example, oxygen exchange between this bicarbonate ion and the interstitial water, then variation in $\delta^{18}O$ values of this water on a scale of a few centimetres must exist. Such variations could exist only for a very short time because of diffusion and certainly not long enough for the time-scale of dolomite formation. The preservation of other sedimentological and chemical characteristics in these zones of low $\delta^{18}O$ would be very difficult to explain by any other mechanism.

The Marl Slate represents a relatively short time interval, probably less than 20,000 yr (Oelsner 1959; Hirst & Dunham 1963). The anoxic events are of much shorter duration; sapropel 1 is considered to have lasted 5,000 yr (Magaritz & Turner 1982) and sapropel 2 only a few hundred years. Such time intervals are closely comparable with the Pleistocene anoxic events in the Eastern Mediterranean (Thunnell *et al.* 1977). The freshwater flooding peaks also represent time intervals of less than 1,000 yr. It is important to note that the sharp change of the carbon isotopic composition which characterizes Zechstein carbonates as a whole also took only a few thousand years to occur which indicates a drastic change in ocean-wide chemical cycles.

The base metal data presented here contrast with the data reported by Turner *et al.* (1978) from the D4 core (NZ 5100 5201) which is only 3 km to the NE. In VT8 a general increase in Cu from the base to the top is observed whereas in D4, although the absolute values at the base are comparable, there is a general decrease. The Pb in VT8 shows an upward decrease in sapropel 1 whereas in D4 the pattern is reversed with Pb increasing upwards. The Zn in VT8 increases gradually upwards with a maximum spike to the base of laminite 1. A similar spike is seen in D4 but there is no subsequent increase there. In other chemical and isotopic characteristics, including quartz content, Ca/Mg variation (Turner *et al.* 1978) and carbon and oxygen isotope composition (Magaritz *et al.* 1981), the cores are very similar.

The differences in the base metal data between the two cores indicate that the sulphides could not have been formed in the water column (cf. Spencer & Brewer 1973); the differences in the distribution of base metal sulphides therefore probably reflect chemical conditions in interstitial water. Slight variation in Eh can explain such data (Garrels & Christ 1965); since chalcopyrite forms at a more negative Eh than galena we suggest that in the VT8 core the interstitial Eh conditions became progressively more negative whereas in the D4 core they become progressively less negative.

Conclusions

We make the following conclusions:

(1) Isotopic and chemical variations in the Marl Slate are related to two cycles of deposition each including the sequence: sapropel→laminite-→finely crystalline dolostone.

(2) We agree with most earlier authors that anoxic conditions dominated during the deposition of the Marl Slate. We believe that these conditions resulted from substantial freshwater influx into the basin which promoted surface productivity and the burial of large amounts of organic material.

(3) The isotopic record was probably established in carbonates precipitated in the surface water mass. This record was not altered during subsequent dolomitization.

(4) There is a marked transition in the isotopic composition of dolomites to values highly enriched in ^{13}C. This took place in less than 5,000 yr and persisted throughout much of the later Zechstein

(5) The distribution of base metal sulphides, particularly those of copper and lead, shows some differences from that seen in a nearby core (D4). We believe that metals were supplied by freshwater influx, were incorporated in the biogeochemical cycle and precipitated interstitially as sulphides. There is no evidence that metal sulphides were precipitated at the anoxic boundary layer.

ACKNOWLEDGMENTS: We thank S. Kazas, M. Feld and I. Morovinski for isotope sample preparation and measurements, M. Whitehouse-Yeo and K. Meneer for AA determinations and D. Wraith for the As and Se determinations. Cathy Kennedy typed the manuscript and Beverley Parker drafted the diagrams. Our paper has benefited from discussions with many colleagues at the EZ 82 meeting.

References

BOLZ, R. 1979. *Mineralogie, Geochemie und Isotopengeochemie des Stinkschiefers im Germanischen Becken.* Unpublished PhD Thesis, University of Heidelberg.

BRONGERSMA-SANDERS, M. 1965. Metals of Kupferschiefer supplied by normal sea water. *Geol. Rdsch.* **55,** 365–75.

——1971. Origin of major cyclicity of evaporites and bituminous rocks: an actualistic model. *Mar. Geol.* **11,** 123–44.

CROWELL, J.C. 1978. Gondwanan glaciation, cyclothems, continental positioning and climatic change. *Am. J. Sci.* **278,** 1345–72.

CROWELL, J.C. & FRAKES, L.A. 1970. Phanerozoic glaciation and the causes of ice ages. *Am. J. Sci.* **268,** 193–224.

DAVIDSON, C.F. 1962. The origin of some strata-bound sulphide ore deposits. *Econ. Geol.* **57,** 265–74.

DEGENS, E.T. & EPSTEIN, S. 1964. Oxygen and carbon isotope ratios in coexisting calcites and dolomites

The land flora of the English and German Zechstein sequences

H.-J. Schweitzer

SUMMARY: All species of higher plants described from the Upper Permian Zechstein and related rocks of Germany and England are reviewed, except for those thought to be dubious.

The plants described are: *Neocalamites mansfeldicus* Weigelt, *Sphenopteris dichotoma* Althaus, *S. patens* (Althaus) Geinitz, *S. geinitzi* (Geinitz) Göppert, *S. kukukiana* Gothan et Nagalhard, *Callipteris martinsi* (Kurtze) Zeiller, *Taeniopteris eckardti* Kurtze, *Pseudoctenis middridgensis* Stoneley, *Plagiozamites bellii* nov. spec., *Sphenobaiera digitata* (Brongniart) Florin, *Pseudovoltzia liebeana* (Geinitz) Florin, *Ullmannia bronni* Göppert, *U. frumentaria* (Schlotheim) Göppert, *Quadrocladus solmsi* (Gothan et Nagalhard) Schweitzer, *Q. orobiformis* (Schlotheim) Schweitzer, *Culmitzschia florini* Ullrich and *Rhenania reichelti* Schweitzer. Of these *Pseudoctenis middridgensis* and *Plagiozamites bellii* are known only from England and *Sphenopteris geinitzi*, *Quadrocladus solmsi*, *Q. orobiformis*, *Culmitzschia florini* and *Rhenania reichelti* only from Germany.

In the Upper Permian there were two plant associations; a hygrophilic *Neocalamites*-Sphenopterid association and a xerophytic *Callipteris*-conifer association. Distinct growth rings in the conifer woods indicate rhythmic changes in the climatic conditions.

When the Zechstein Sea transgressed, the Palaeozoic flora had already changed considerably. The tree-sized arborescent lycopods, horsetails, ferns and pteridosperms that had predominated during the Carboniferous and the Lower Permian had disappeared almost completely and were followed by the conifers which characterize the Zechstein flora; almost all of the Zechstein fossil flora is coniferous. The conifers already had reached a surprisingly advanced stage of phylogenetic development; in contrast to the Lower Permian Walchiales, they closely resemble some Recent conifers.

Ginkgophytes, which first appeared in the Lower Permian, are more numerous in the Zechstein sequences but are represented by only one species. Compared with these two groups of higher plants, the pteridophytes, pteridosperms and the cycads (or bennettitids) comprise a minor part of the Zechstein flora.

The rapid decline of eusporangiate ferns during the Lower Permian and the gradual evolution of the leptosporangiate ferns, coupled with the extinction of the cordaites during the Upper Permian and the appearance of the cycads and probably the bennettitids, show that the Zechstein transgression was a significant turning point in floral evolution. From the palaeobotanical point of view it is therefore justifiable to consider this floral change as the beginning of the Mesophytic.

In the whole of the coastal belt around the German and English Zechstein basins (including that of the Bakevellia Sea) the flora was very uniform; it comprises only the fossilized remains of plant fragments which by chance drifted into the seas, and the inland vegetation is unknown.

The most varied and abundant accumulations of plant remains occur in the Lower Zechstein, especially in the Kupferschiefer of Middle Germany (Mansfeld, Gera, Richelsdorf) and the Lower Rhenish border (Kamp-Lintfort), and in the Marl Slate of County Durham, NE England (Fig. 1) where the most famous locality is Middridge Quarry (NZ 249 252) (see Bell *et al.* 1979, for references).

The Zechstein flora is not rich in species. Only about 40 different spore-species and about 25 species represented by megafossils have been described. Most of the remains are rather poorly preserved impressions, especially those from Middle Germany. During the last 25 years, however, better preserved specimens have been found in the Rhineland and in England (Stoneley 1958); from these it has been possible to study the cuticular structure of leaves and in some specimens even the anatomical structure of stems, leaves and fructifications. This has led to a more refined systematic classification of the Zechstein conifers.

Systematic review

Pteridophytes and pterophylls

Amongst pteridophytes, the lycopods and ferns are represented only by sparingly scattered spores. A spore recently collected from the Hilton Plant Beds of Cumbria belongs to the genus *Osmundacidites* Couper and is the only undoubted member of the Osmundaceae (Royal Ferns)

The horsetails, by contrast, are represented by

From HARWOOD, G.M. & SMITH, D.B. (eds), 1986, *The English Zechstein and Related Topics*, Geological Society Special Publication No. 22, pp. 31–54.

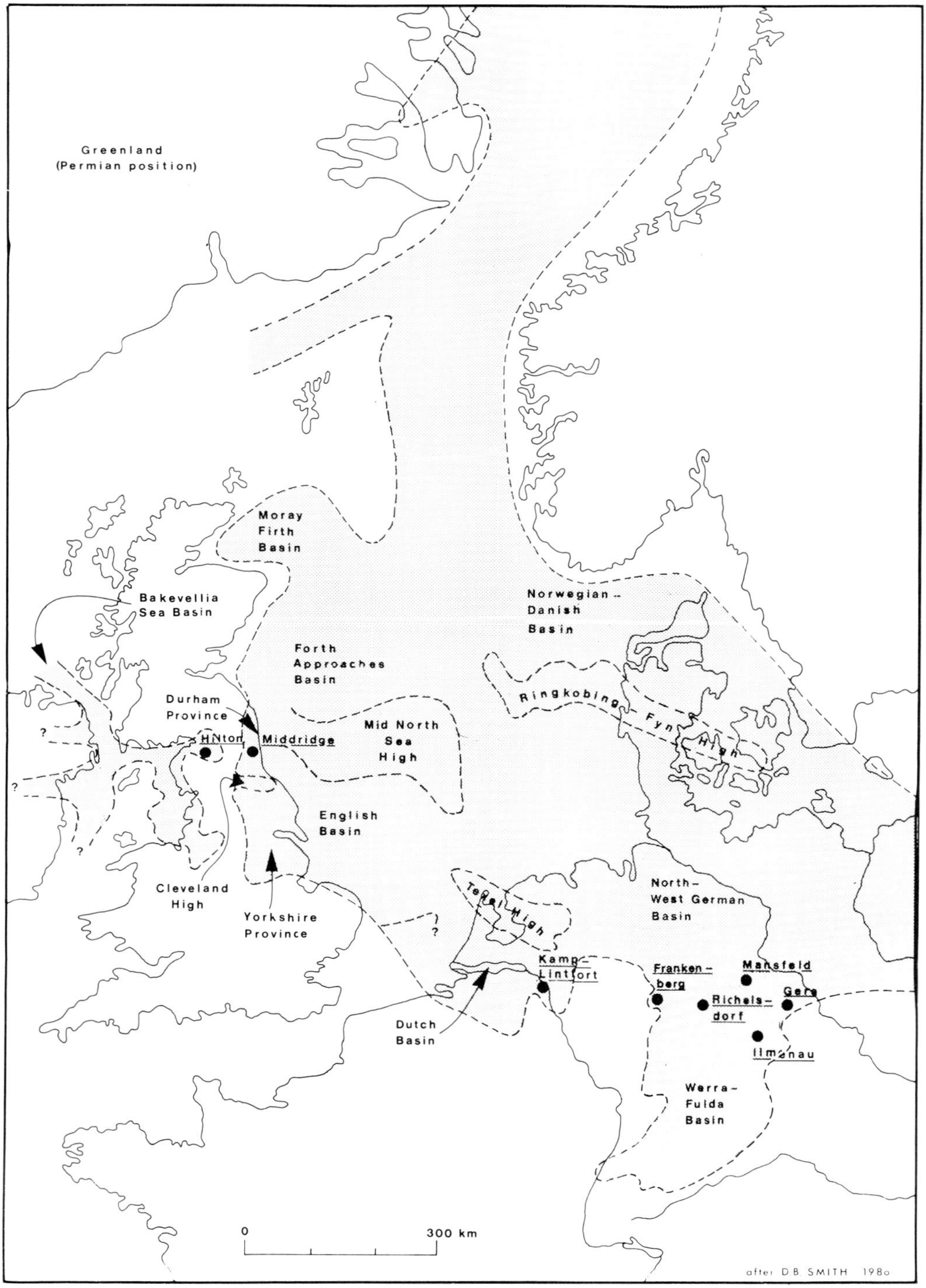

FIG. 1. Generalized sketch map of the German and English Zechstein basins (after Smith 1980), showing the main fossil plant localities.

megafossils which all appear to belong to the species *Neocalamites mansfeldicus* (Fig. 2). The axes of this small plant are hollow and their surface is characterized by ribs that pass the nodes in a straight line, as in the Lower Carboniferous genus *Archaeocalamites*. The leaves are undivided, relatively broad and much shorter than in the Upper Triassic and Jurassic species of *Neocalamites*. Though rare, *N. mansfeldicus* was distributed all around the margins of the German and English Zechstein basins and has been collected in Middle Germany (Mansfeld, Richelsdorf) as well as in the Rhineland and at Middridge Quarry, England. It is possible that the plant is related to other contemporary equisetes of the Angara province, which have been described by Zalessky (1927) as *Paracalamites*. The stems of *Paracalamites* show the same features as those of *N. mansfeldicus*, but the foliage is unknown and no definite determination is possible. A definite determination is also impossible for some equisetalean stem fragments from the Copper Sandstone of the SW Urals (Orenburg). Unbranched axes which belong most probably to *N. mansfeldicus* were formerly described by Geinitz (1880) as ? *Calamites kutorgai* and were later transferred to the form-genus *Paracalamites* by Stoneley (1958).

Even though there is no evidence of true fern megafossils, there occur fern-like fronds referable to five different genera. The species of *Sphenopteris* and *Callipteris* are probably pteridosperms, but the species of *Taeniopteris* and *Pseudoctenis* are more likely to be cycads. The systematic position of *Plagiozamites* is completely uncertain.

Of the Upper Permian sphenopterids, almost every single frond had formerly been described as an individual species. A revision of all the originals, however, has shown that there are only four species (Schweitzer 1960); these are shown in Figs 3–6. The number of known specimens of each species and their distribution are shown in Table 1.

Each of the other three genera are represented by one species. *Callipteris martinsi* (Fig. 7), although rare, is the most common and occurs at nearly all the Zechstein plant localities. In contrast to the sphenopterids, it has rather thick cuticles; every cell of the epidermis bears a large papilla. The stomata are deeply immersed and their vestibules are covered by the papillae of the auxiliary cells. All these features are characteristic of xerophytic plants, indicating that *C. martinsi* must have grown in dryer biotopes or must have been influenced by saline conditions. The stomata are unevenly distributed over the whole surface of the pinnules, but on the upper side they are less numerous than on the lower, where the veins commonly form a network. New findings of the female fructification *Autunia* in connection with fronds of *Callipteris conferta*, the index fossil of the Lower Permian (Kerp 1982), reveal that *C. martinsi* most probably belongs to the Peltaspermales, as suggested by Townrow (1960).

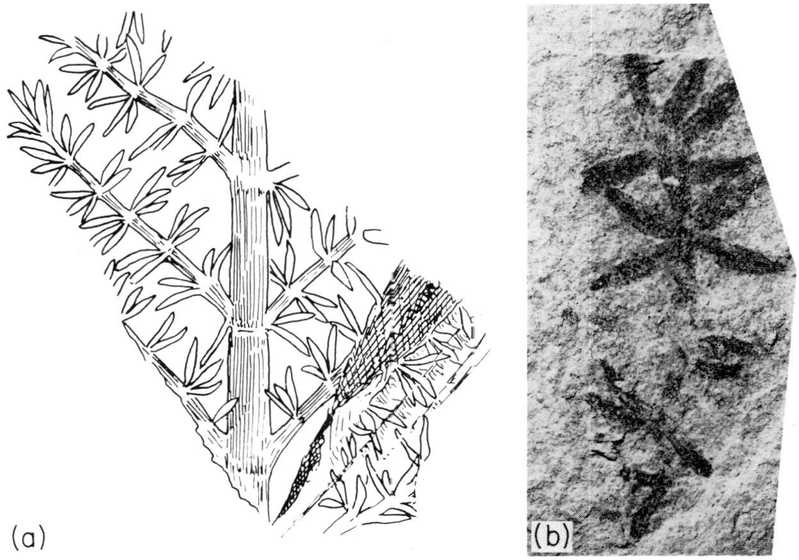

FIG. 2. *Neocalamites mansfeldicus* Weigelt. (a) Type-specimen. Kupferschiefer, Mansfeld (after Weigelt 1928), ×0.5. (b) Marl Slate, Middridge, Co. Durham; Sunderland Mus. collection no. TWCMS: B2809; ×1.

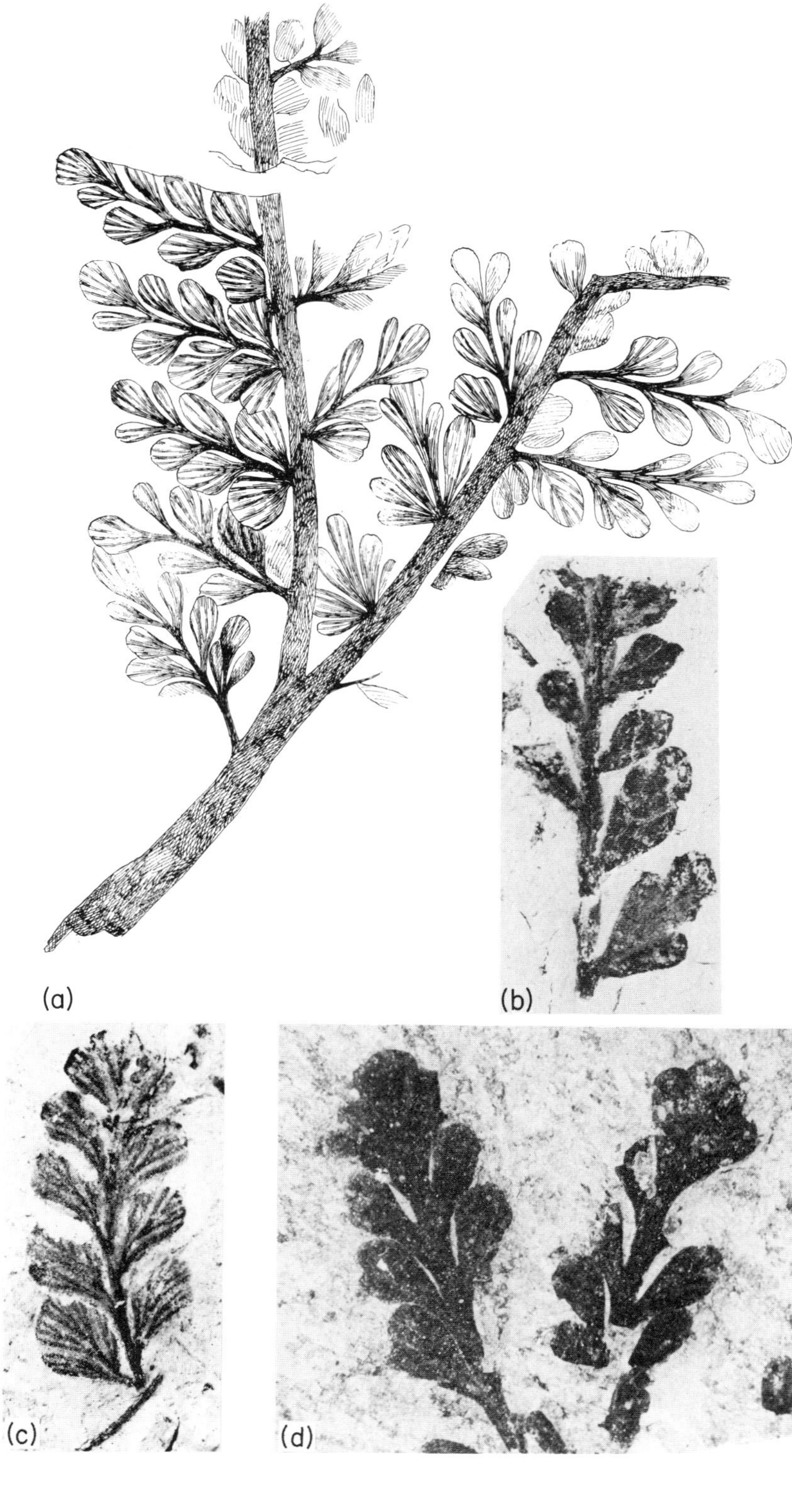
(a)
(b)
(c)
(d)

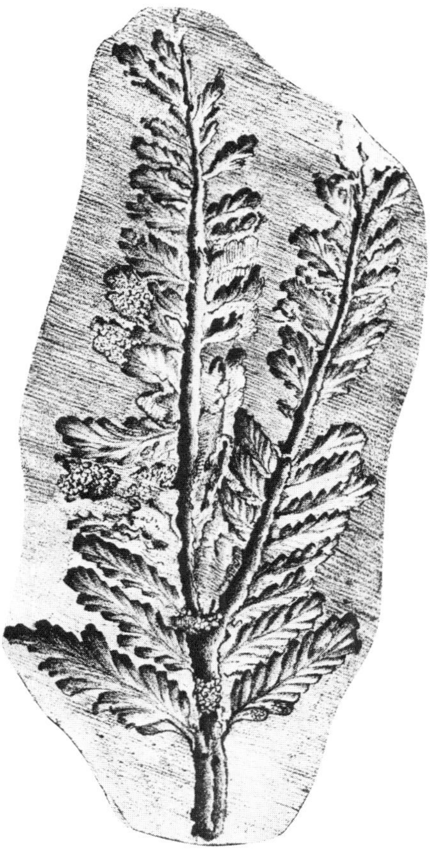

FIG. 4. *Sphenopteris geinitzi* (Geinitz) Göppert. Kupferschiefer, Ilmenau (after Geinitz 1848); ×0.6.

Taeniopteris eckardti (Fig. 8) is known only from tongue-shaped leaves up to 20 cm long with a pronounced midrib and distinct simple lateral veins. It is relatively common only in the region surrounding Mansfeld. Only two other specimens are recorded in the literature (Schuster 1933; Stoneley 1958), one from Saxonia (Schuster 1933) and one from Middridge (Stoneley 1958; Bell *et al.* 1979); a further specimen was discovered at Middridge in 1981 (NZ 249 252) by Dr C. Cleal and is now lodged at Sunderland Museum (TWCMS: G46 98).

From Middridge Quarry (NZ 249 252) came two of the most remarkable Zechstein plants *Pseudoctenis middridgensis* (Fig. 9) and *Plagiozamites bellii* nov. spec. (Fig. 11), both solitary discoveries.

Pseudoctenis middridgensis is the oldest incidence of a genus that is rather common in the Jurassic. The cuticular structure is not preserved but the pinnae are traversed from the base to the apex by four to six parallel veins and not by two as described by Stoneley (1958, p. 321).

The genus *Plagiozamites*, 1894, founded by Zeiller on Lower Permian material from Alsace appears already in the Upper Carboniferous and is mainly distributed in the Cathaysian flora of China. Some younger species have also been recorded from the Triassic of south-eastern Asia. The systematic position of the genus is uncertain and it is also not sure if all its species belong to the same group of plants. Formerly the genus was generally regarded as belonging to the cycads; however, according to Halle (1927, p. 226), it seems more probable that at least some members belong to the Noeggerathiales, a peculiar plant group of questionable affinities. It cannot even be excluded that certain species, for instance the new Zechstein plant, may belong to the conifers. In the opinion of Halle, which is supported here, *Plagiozamites* may represent shoots with simple leaves and not pinnate fronds. The phyllotaxis resembles that of the modern conifer genus *Metasequoia*. It can be assumed that the leaves originally had been attached helically, but are, due to the plagiotropic direction of the ultimate shoots, spread out in one plane thus forming two irregular rows one on each side of the axis as it is shown in Fig. 11(b). The leaf bases are broad and obliquely inserted and the venation of the large leaves is fan-shaped.

A diagnosis of the new Zechstein species, *P. bellii*, which was discovered by Mr John Bell (Bowburn, Co. Durham) in 1980 is given in Appendix 1.

Ginkgophytes

Of much importance are the ginkgophytes, although they are represented by only one species, *Sphenobaiera digitata* (Fig. 11). It is uncommon but is known from all the more important plant localities in Germany and England. Because of its dichotomous mode of branching, the leaf resembles some seaweeds in its habit and was initially placed in the genus *Fucoides*; later it was assigned

FIG. 3. *Sphenopteris dichotoma* Althaus. (a) Type-specimen. Kupferschiefer, Richelsdorf (after Althaus 1846); ×0.5. (b) Hilton Plant Beds, Hilton Beck, Cumbria. British Mus. (Nat. Hist.) No. V 35116; ×1. (c) Hilton Plant Beds, Hilton Beck, Cumbria. British Mus. (Nat. Hist.) No. V 35107; ×1. (d) Cadeby (Magnesian Limestone) Formation, Lower Marl Facies; Darlton Borehole, Nottinghamshire, depth 486.16 m. Inst. Geol. Sci. (London) Collection No. Bm 2910; ×1.

FIG. 5. *Sphenopteris patens* (Althaus) Geinitz. (a) Type-specimen. Kupferschiefer, Richelsdorf (orig. to Althaus 1846, pl. 4, fig. 3); Geol. Inst. Marburg Collection; × 1.2. (b) ? Hilton Plant Beds, Vale of Eden, Cumbria (exact site not recorded); Inst. Geol. Sci. (Keyworth) Collection No. 19058; × 1.2.

to the ginkgophytes. This classification had been verified by examination of the cuticular structure in the English and Rhenish specimens. The cuticular structure corresponds to that of the younger Mesozoic ginkgophytes and is shown in Fig. 10(b,c). The upper epidermis differs from the lower one in having much longer cells and considerably fewer stomata.

Cordaitales

The cordaitales persisted into early Zechstein time, and a few poorly preserved parallel-veined leaves that might belong to this group of plants have been described from both Germany and England (Weigelt 1930b, fig. 17; Stoneley 1958, pl. 38, fig. 3).

Conifers

Conifers are the plants most characteristic of the Zechstein; essentially the same species occur throughout but the rarer species are known only from certain localities.

In the phylogeny of the conifers, the monotypic genus *Pseudovoltzia* is of exceptional importance because the scales of the female cone show all the different organs in an unreduced evolutionary stage. It is therefore fortunate that almost every organ of this plant has been found with its structure preserved; only the male cone is unknown. The wood, like that of all other Zechstein conifers that have their anatomy preserved, has a structure typical of the Mesozoic Protopinaceae. This means that the pits of the

FIG. 6. *Sphenopteris kukukiana* Gothan et Nagalhard. (a) Most complete specimen known. Kupferschiefer, Hoerstgen near Kamp-Lintfort; Coll. Schweitzer, Bonn; ×0.8. (b) Better preserved specimen, locality as (a). (c) Cf. *Sphenopteris kukukiana*, Edlington Formation, Markham Moor Borehole, Nottinghamshire, depth 319.74 m; Inst. Geol. Sci. (Keyworth) Collection No. Bn 74; ×2.

tracheids of the secondary xylem are arranged in one row only and are so close together that their pit chambers flatten each other.

The branchlets are so extremely heterophyllous that one would suppose they belong to different species (Fig. 12a–c). The helically-arranged leaves have a decurrent base and branch at angles of about 45°. Abaxially they are rounded, adaxially flattened; the apex is pointed (Fig. 12d). The anatomy corresponds to that of modern conifers. The central bundle is surrounded by an aliformous transfusion tissue around which are the spongy parenchyma and the palisade tissue (Fig. 12e). The hypodermis consists of two or three

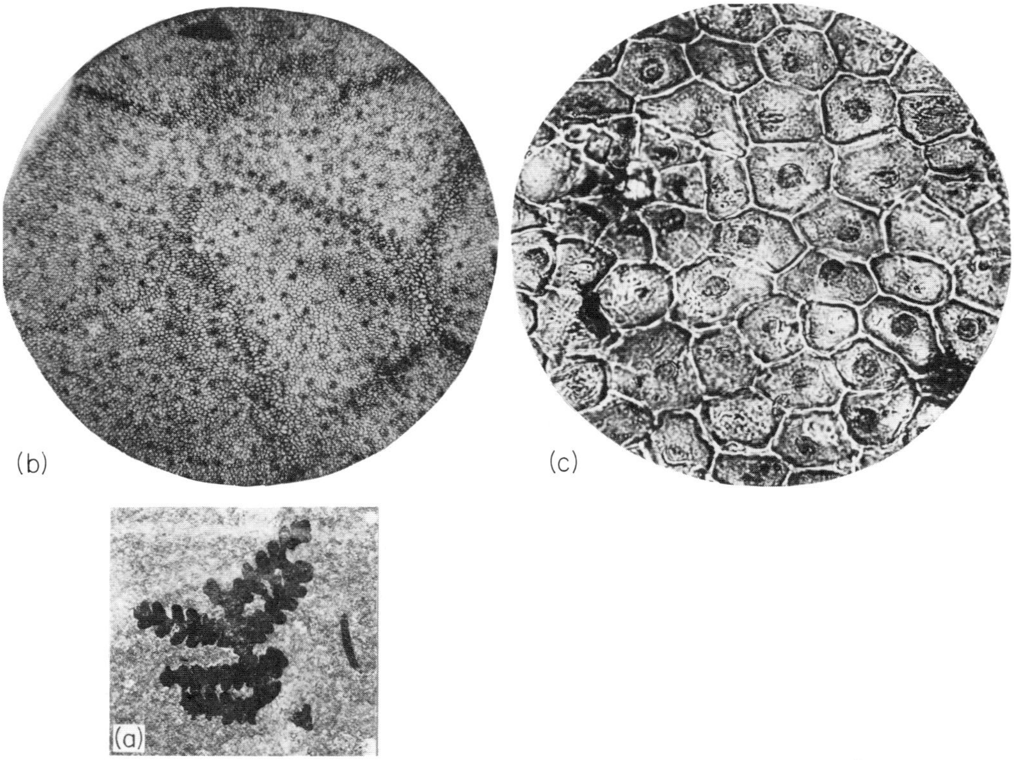

FIG. 7. *Callipteris martinsi* (Kurtze) Zeiller. (a) Kupferschiefer, Hoerstgen near Kamp-Lintfort; Coll. Schweitzer, Bonn; ×0.7. (b) Lower epidermis showing the distribution of the stomata and the network formed by the veins. (c) Enlarged section of the lower epidermis. Every cell bears a papilla.

FIG. 8. *Taeniopteris eckardti* Kurtze. Kupferschiefer, Mansfeld; Coll. Geol. Inst. Halle; ×0.7.

layers of small fibres that are arranged in separate groups. The epidermis is highly cutinized. The stomata, which form parallel rows on the upper and lower surfaces of the leaves, are deeply immersed and their vestibules are covered by papillae as in *Callipteris martinsi* (Fig. 12f). The subsidiary cells of the stomata are much smaller than the surrounding cells.

A reconstruction of the female cone of *Pseudovoltzia* is shown in Fig. 12(g). The axis bears helically arranged scales that consist of two different organs, a bract and an axillary dwarf shoot (Fig. 12h). The latter represents a flower and the whole cone consequently is an inflorescence. Basically the bract and the stalk of the dwarf shoot are fused; more distally, however, they part. The free part of the bract corresponds completely in its anatomy and morphology to a sterile leaf. The stalk of the dwarf shoot, which is rather thin, extends considerably in its distal region and ends in three large and two small flattened, rounded leaves. These leaves are spread in one plane and not lignified as in the Recent conifer scales but formed like normal leaves; this indicates that they photosynthesized. On the adaxial surface of the larger leaves there runs a median ridge ending at the base of each leaf, where an ovule is inserted. The ovules (Fig. 12i) consist of a rather thick integument with an inner solid sclerotesta and an outer parenchymatous sarcotesta. The micropyle is surrounded by very long hair-like papillae. In the interior of the ovule is a relatively large free-standing nucellus that

TABLE 1. Distribution of the species of *Sphenopteris*. See also Schaumberg (1977) and Schweitzer (1960, 1962, 1968).

	Gera	Mansfeld	Ilmenau	Richelsdorf	Niederrheim	England
Sphenopteris dichotoma	1(?)	2		3		3
Sphenopteris geinitzi	2		1			
Sphenopteris patens	4			3 or 4		1
Sphenopteris kukukiana					~30	1(?)

FIG. 9. *Pseudoctenis middridgensis* Stoneley. (a) Type-specimen. Nearly complete leaves. Marl Slate, Middridge, Co. Durham (orig. to Stoneley 1958, fig. 12); Hancock Mus. Collection, Newcastle, No. G 03.32; ×0.7. (b). Enlarged portion of (a), showing four to five veins traversing a whole pinna; ×2.

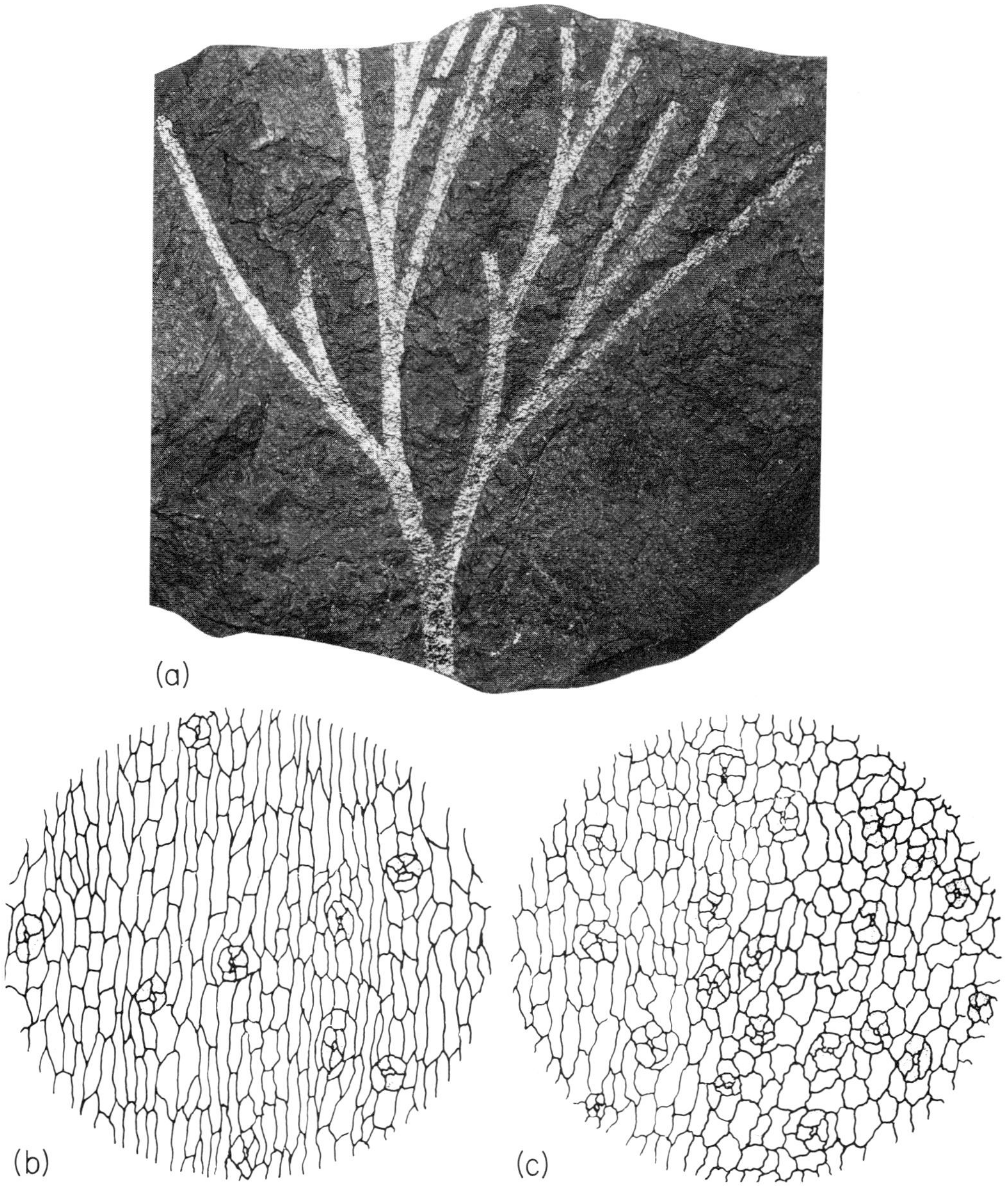

FIG. 10. *Sphenobaiera digitata* (Brongniart) Florin. (a) Nearly complete leaf. Kupferschiefer, Mansfeld (orig. to Weigelt 1928, pl. 12, Fig. 1), Coll. Geol. Inst. Halle; ×0.7. (b) Structure of the upper epidermis (Rhenish material). (c) Structure of the lower epidermis (Rhenish material).

most probably has a pollen chamber at its tip. From the structure and the insertion of the ovules it can be supposed that pollination took place as in many Recent conifers; the pollen grains, transported by wind, would have been caught by the tuft formed by the long papillae where they contacted the pollination drop; as the drop dried, the grains would have been pulled through the micropyle to the tip of the nucellus or into the pollen chamber. The function of the hair-like papillae is not clear. Either they served as catching organs for the pollen grains or as distribution organs for the seed.

Even though *Pseudovoltzia* represents the most

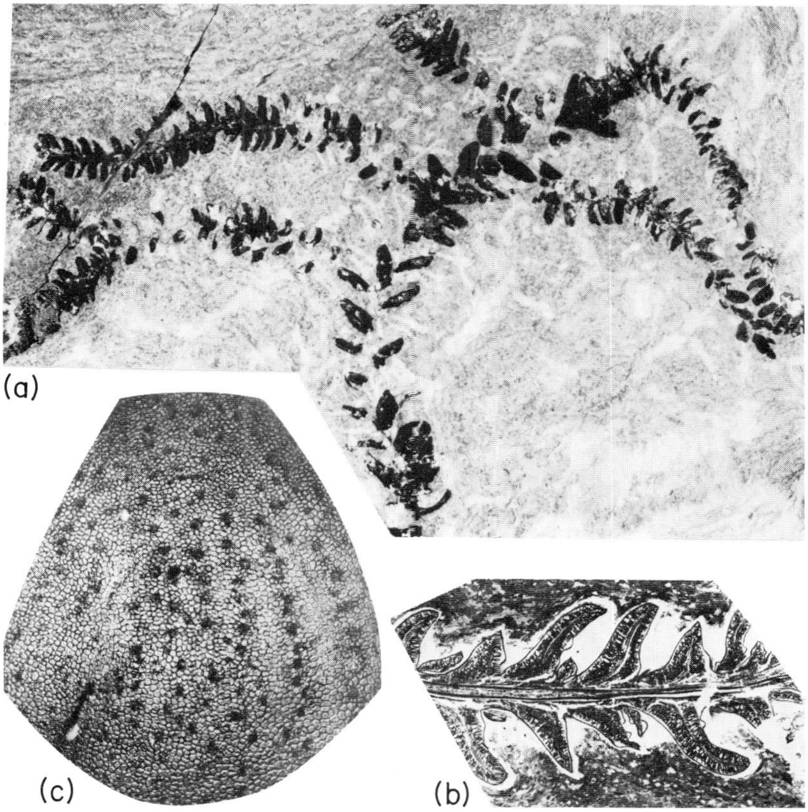

FIG. 13* *Ullmannia bronni* Göppert. (a) Largest known specimen. Kupferschiefer, Rossenray, near KampLintfort; Coll. Schweitzer, Bonn; × o.c. (b) Longitudinal thin-section of a branchlet. (c) Epidermal structure.

interesting conifer phylogenetically, the two species of *Ullmannia*, *U. bronni* and *U. frumentaria*, are much more popular among geologists. For several centuries they have been found by miners in the Kupferschiefer, where they are common. The two species might belong to different genera, but unfortunately the fertile organs of *U. bronni* are unknown; the classification has therefore to be based on the anatomical structures of the leaves. Since these features are nearly identical in both species, it is preliminarily more useful to unite them in one genus. As in *Pseudovoltzia*, the leaves have a remarkably large aliformous transfusion tissue and the hypodermis is not coherent but split into separate bundles of fibres. The epidermis, too, shows typical xeromorphic structures, with a very thick cuticle and deeply immersed stomata covered by long papillae of the auxiliary cells. Both species are amphistomatic, with their stomata arranged in parallel rows. In *U. bronni*, however, these rows are more commonly interrupted or otherwise irregular than in *U. frumentaria*.

The abundance of the two species differs geographically as well as stratigraphically. Generally, and in contrast to *U. bronni*, *U. frumentaria* is much more abundant in the Kupferschiefer of Middle Germany than in the Kupferschiefer of the Rhineland and the Marl Slate in England; also in contrast to *U. bronni*, it decreases upwards in abundance. The upwards increase in the abundance of *U. bronni* continues into the lowest part of the overlying Zechsteinkalk and it and *U. frumentaria* occur in about equal abundance a few metres above the base; above this *U. bronni* predominates. In the Hilton Plant Beds of Cumbria, NW England, which are probably generally equivalent to the German Zechsteinkalk, and especially in the Kupferletten of Frankenberg, *U. bronni* is much the most abundant

* As Figs 11 and 12 each occupy two pages, they follow Fig. 13 and appear on pp. 42–45.

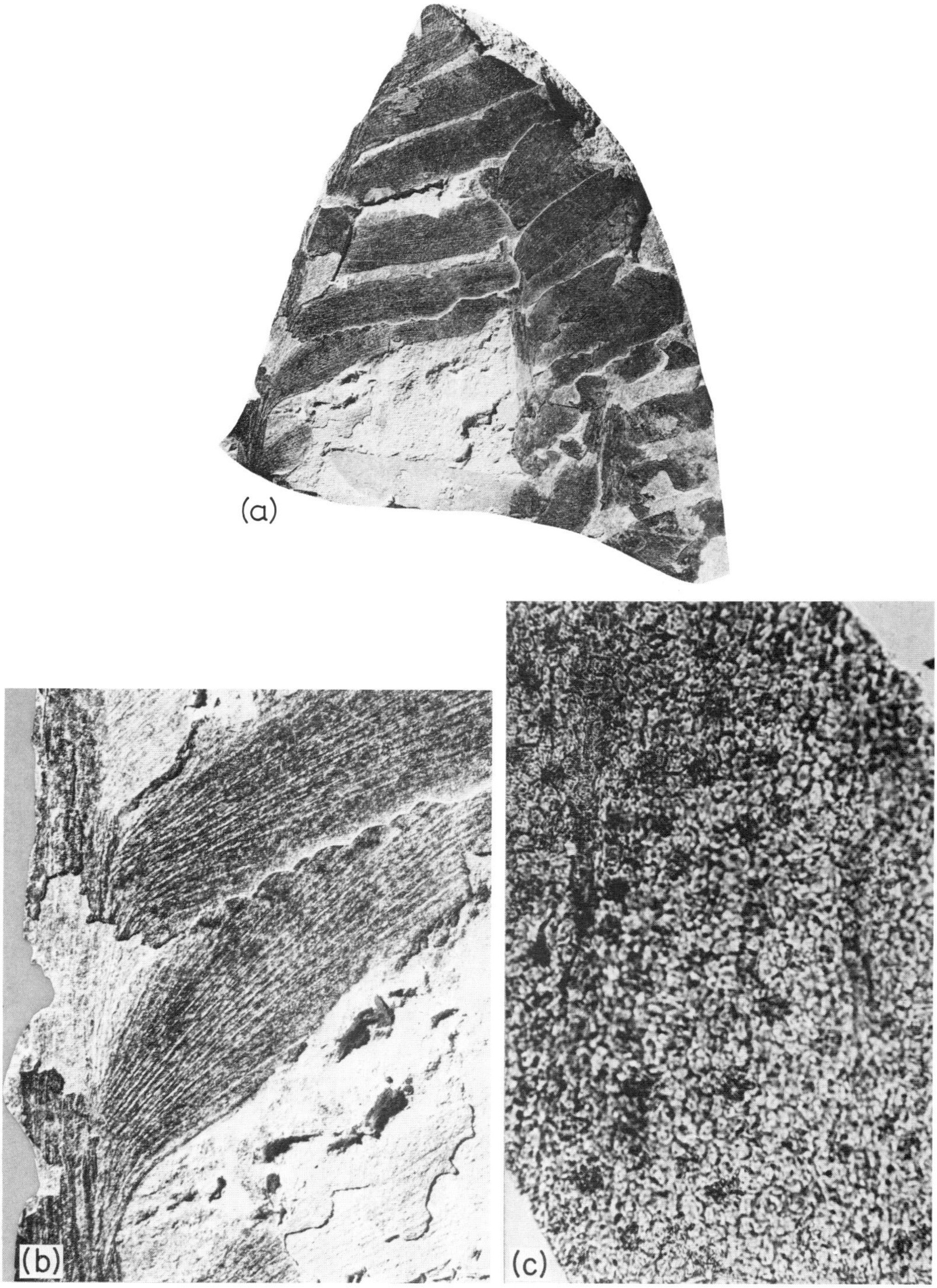

FIG. 11. *Plagiozamites bellii.* sp. nov. (a) Holotype. Marl Slate, Middridge, Co. Durham; Sunderland Mus. Collection No. TWCMS G 5756; ×0.5. (b) Enlarged section of (a) showing the oblique insertion of the leaves and the fan-shaped venation; ×1.5. (c) Epidermal structure. (d) Stomata. (e) Reconstruction of a branch. (f) Reconstruction of a leaf.

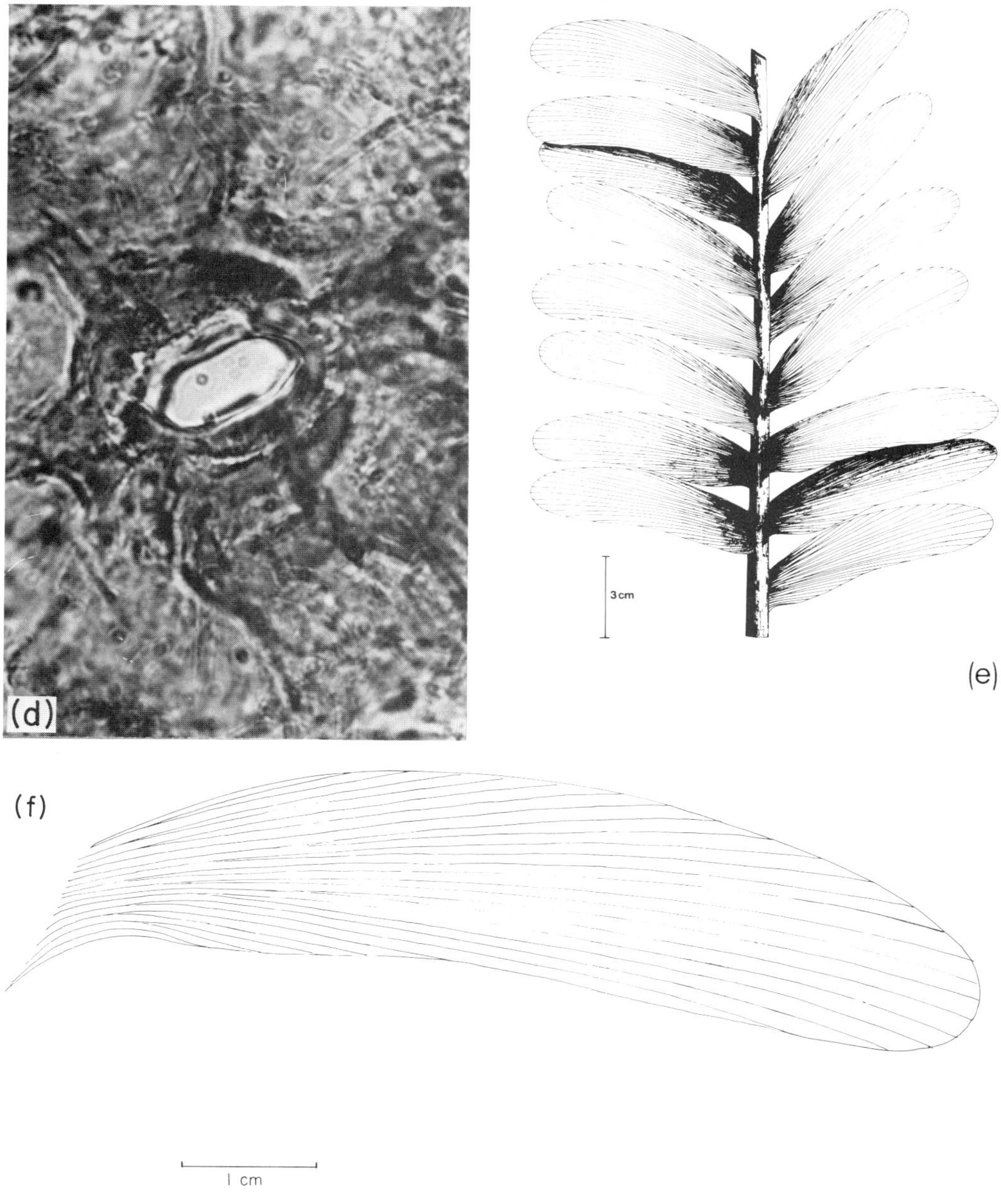

plant fossil. The Kupferletten of Frankenberg, according to the most recent reports, belongs to the upper part of Zechstein 1 (J. Kulick, pers. comm. Wiesbaden). Stoneley (1956, p. 713) described a new species, *Hiltonia rivuli*, from the Hilton Plant Beds but I consider that these remains belong to *U. bronni*.

U. bronni (Fig. 13) is characterized by relatively broad leaves that are similar to those of the Recent *Araucaria araucana* (Fig. 13a). On the older branches the leaves are much larger and usually more pointed than on the younger ones, which have a rounded apex. Isolated leaves from the older branches can therefore readily be confused with those of *Culmitzschia florini* if they are only preserved as impressions. The cuticular structures of these two species, however, are sufficiently different for them to be distinguished.

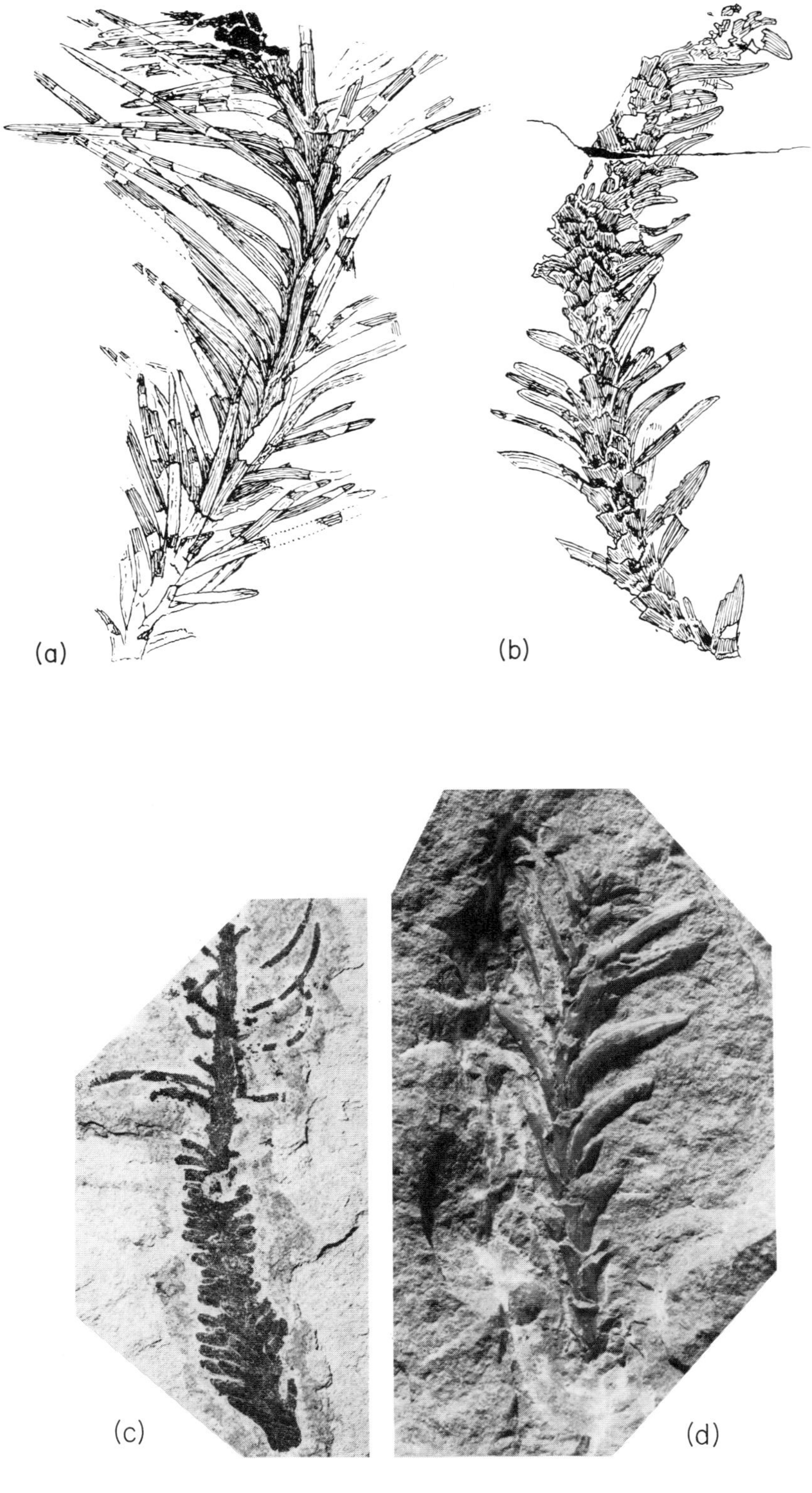
(a)
(b)
(c)
(d)

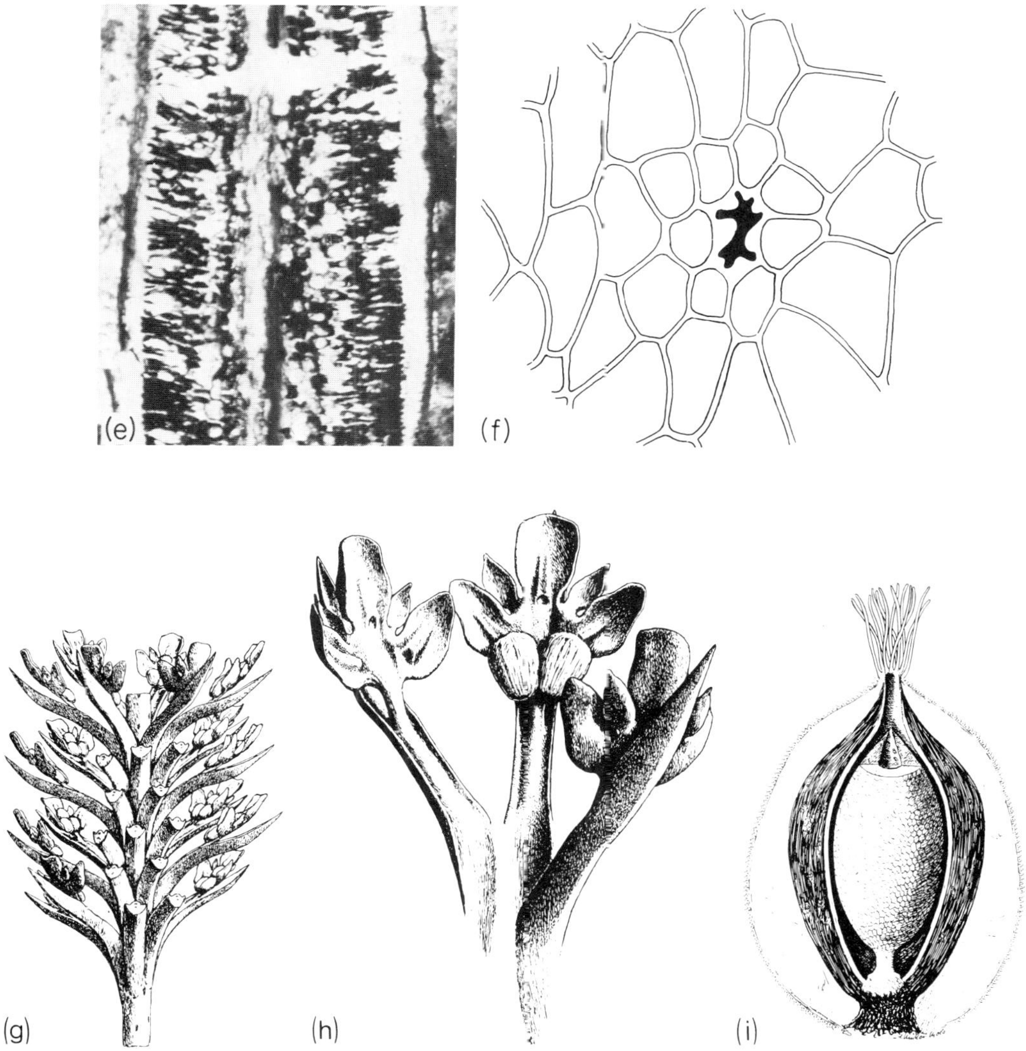

FIG. 12. *Pseudovoltzia liebeana* (Geinitz) Florin. (a–c). Three heterophyllous branchlets. (a,b) Kupferschiefer, Hoerstgen near Kamp-Lintfort; Coll. Schweitzer, Bonn; ×0.5. (c) Marl Slate, Middridge, Co. Durham; Sunderland Mus. Collection TWCMS. B 4542; ×0.5. (d) Structurally preserved branchlet; ×0.7. (e) Longitudinal thin-section of a leaf, showing the central bundle flanked by the transfusion tissue which is followed by the spongy parenchyma and palisade tissue. (f) Stomata. (g) Reconstruction of a female cone. (h). Reconstruction of the scales in different views. Each scale bears three ovules. They are partly or completely omitted to demonstrate their insertion. (i) Reconstruction of the ovule.

Many such leaves occur in the English Zechstein rocks, but examination of the epidermal structure reveals that they all belong to *U. bronni*; this is also true of the Rhenish material.

U. frumentaria is the most common Upper Permian conifer. To it belong more than half of all Zechstein plant remains. It is therefore not surprising that the plant is known in nearly all its details: wood, leaves, male cone, pollen grains and female cones including the ovule (Fig. 14). The leaves are much more needle-shaped than those of *U. bronni*, and are helically and very densely arranged. They are similar to those of *Pseudovoltzia*, but the blades are closer to the axis

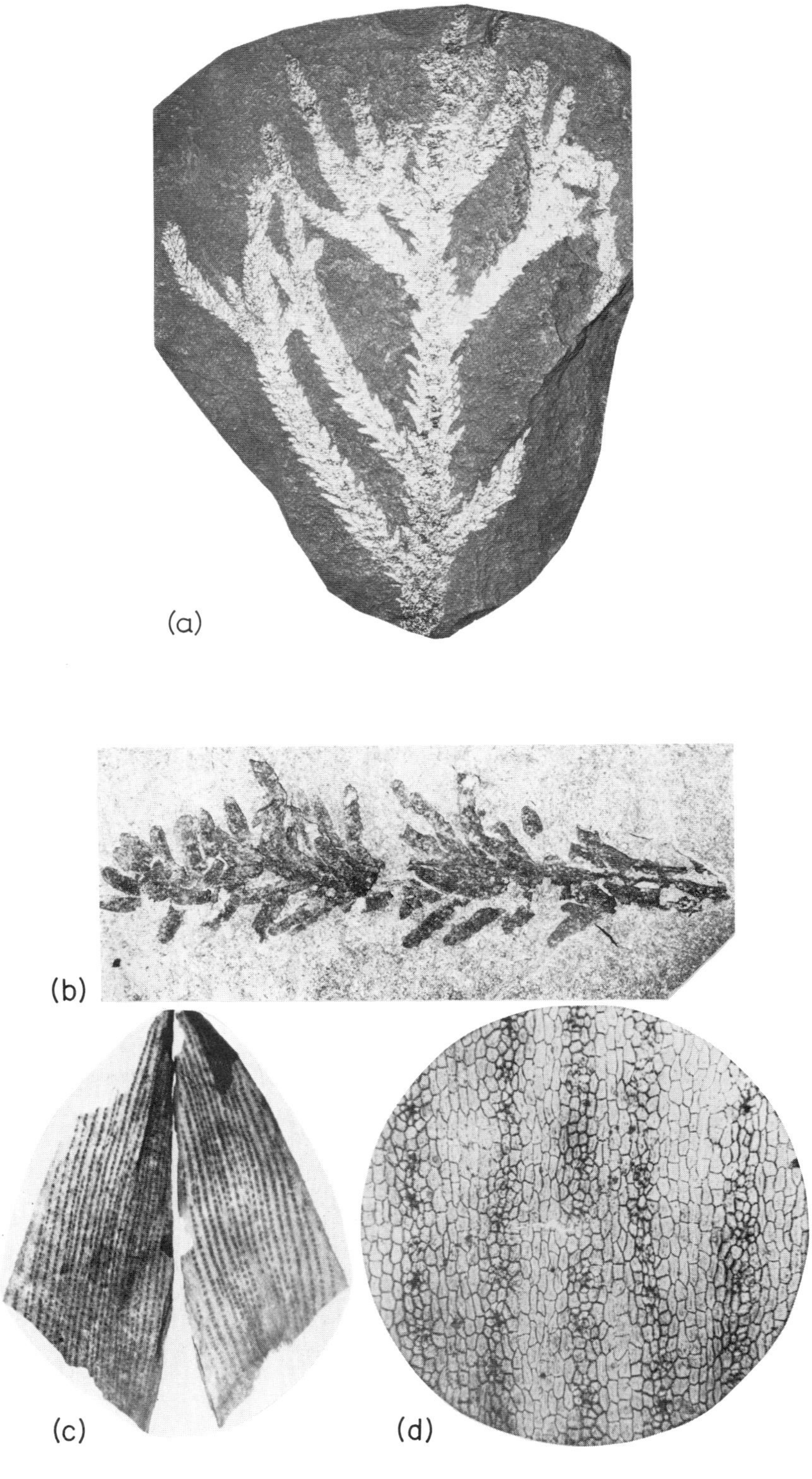
(a)
(b)
(c)
(d)

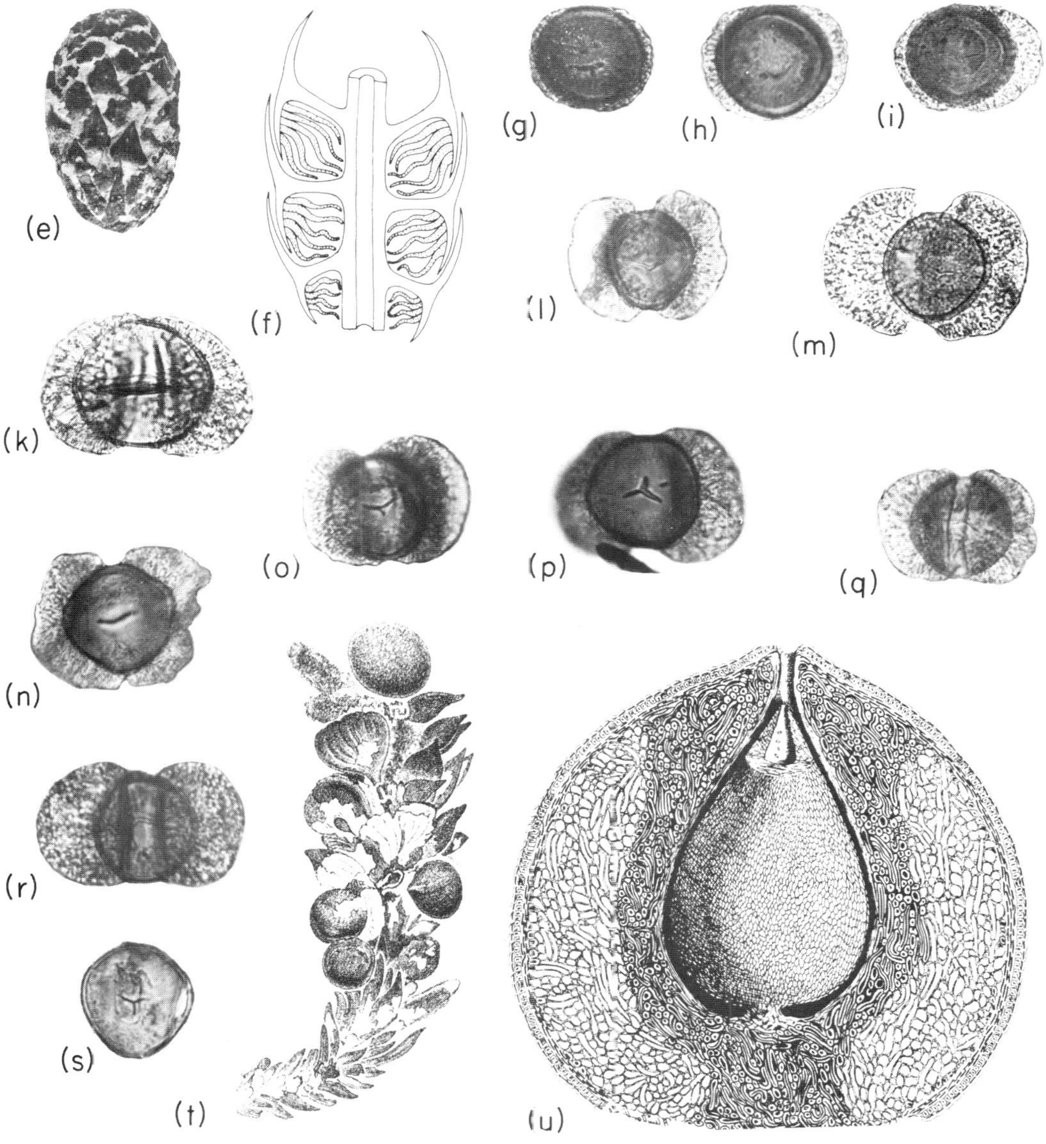

FIG. 14. *Ullmannia frumentaria* (Schlotheim) Göppert. (a) Large branch. Kupferschiefer, Mansfeld (orig. to Weigelt 1928, pl. 24, fig. 3); Coll. Geol. Inst. Halle; ×0.4. (b) Branchlet. Marl Slate, East Thickley, Co. Durham; C.T. Trechmann Collection, British Mus. (Nat. Hist.) No. V 20553; ×0.7. (c) Upper and lower surface of a macerated leaf (Rhineland). (d) Epidermal structure (e) Male cone, Zechsteinkalk, Hoerstgen near Kamp-Lintfort; Coll. Schweitzer, Bonn; ×0.8. (f) Male core, longitudinal section. (g–s) Pollen grains isolated from a structure-bearing cone. Kupferschiefer, Rossenray, near Kamp-Lintfort. (g–m: different ontogenetic stages, n–p: adult stages with differently developed marks, q–r: adult stages with different distances between the sacci, s: central body without sacci, looking like trilete spore.) (t) Female cone. Kupferschiefer, Gera (after Geinitz 1880). (u) Reconstruction of the ovule; ×8.

and less rigid. The length and density of the leaves vary considerably and because of this the branchlets were formerly described under many different names. Weigelt (1928), however, recognized by intermediate material that all the branchlets belong to the same species and that the length of the needles and their density depend on their exposure to the sun (cf. Fig. 14a,b).

Fossils of *U. frumentaria* from the Rhenish Zechstein strata include structures bearing coalified male cones. Their helically arranged peltate microphylls are similar to those of *Araucaria* with, in their abaxial part, at least eight pollen sacs that contain hundreds of bisaccate grains (Fig. 14e,f). These showed the different ontogenetic stages of the sacci and also adult stages exhibiting considerable variation (Fig. 14g–s). If the grains had been dispersed in the sediment they would have been assigned to more than 10 species of four different genera.

Some of the short-needled shoots terminate in upright female cones of rather strange appearance. As in *Pseudovoltzia*, the cone scale consists of a bract and an auxiliary dwarf shoot with five distal leaves. The axes of the dwarf shoots, however, are almost completely reduced and the leaves are not free but totally united to form one homogeneous scale (Fig. 14t). This is similar to Recent conifers. There is one difference however: the scales were not lignified but apparently fleshy. At the base of the median leaf a very large spherical ovule is inserted. Its size was the reason for Weigelt's conclusion that the ovule must have been embedded in a special excrescence, an epimatium as it exists in Recent Podocarpaceae. Therefore Weigelt considered *U. frumentaria* as an ancestor of this family, and named it *Archaeopodocarpus*. This assumption, however, could be clearly disproved by a coprolite from the Rhineland which contained 12 ovules of *U. frumentaria*; these ovules had passed undamaged through the digestive system of the animal whereas the other plant remains in the coprolite were almost completely digested. The animal probably was a *Protorosaurus*, for cones of *U. frumentaria* were found in a skeleton of this rhynchocephalid and were assumed to represent the stomach contents (Weigelt 1930a). The ovules show essentially the same structure as those of *Pseudovoltzia*, but have an extremely thick integument (Fig. 14u). There is, however, no epimatium and *U. frumentaria* cannot therefore be compared with the podocarps. Tissue of the endosperm is still preserved in the nucelli of some ovules, but an embryo is never visible. Embryos have been observed, however, in Permian conifer seeds from Texas by Miller & Brown (1973).

Compared with *Pseudovoltzia* and especially with the species of *Ullmannia*, the other genera of Zechstein conifers are of much less importance. None have yet been found in the English Zechstein, but probably some will be discovered eventually. The systematic classification of these genera is based almost solely on the antomy and on the epidermal structure of the leaves.

Quadrocladus includes two rather rare species, *Q. solmsi* and *Q. orobiformis*, which occur in Middle Germany as well as in the Rhineland. In the genus *Quadrocladus* the transfusion tissue of the leaves is aliformous as in the coniferous genera listed above, but is not so significant. Other than this the genus is characterized by the absence of a hypodermis and by stomata that usually possess 4 auxiliary cells. In *Q. solmsi*, the needles are close to the axis, round in cross-section and thickened at the apex to produce a club-shaped form (Fig. 15a). However, in *Q. orobiformis*, the needles usually spread horizontally and have a constant thickness from the base to the tip (Fig. 16a).

More abundant branchlets and isolated needles of the genus *Culmitzschia* occur, but until recently have been found only in the Kupferschiefer around Gera. A few remains have recently been discovered in Richelsdorf. Clement-Westerhof (1984) has shown that *Culmitzschia* is synonymous with the nearly contemporaneously published genus *Ortiseia* Florin. Originally both genera were founded on sterile leaves showing a very characteristic epidermal structure with bicyclic stomata (Fig. 17b). *Ortiseia* has priority and therefore the name *Culmitzschia* is unnecessary. However, because Clement-Westerhof found for the first time fertile material—male and female cones as well—she considered *Ortiseia* as a 'natural' genus containing species of which the fertile organs are known while *Culmitzschia* remained as a form genus for sterile branchlets, needles and cuticules. As in the German Zechstein only sterile material has yet been found, the name *Culmitzschia* is appropriate for this. Ullrich (1964) described only one species, *C. florini*. According to Clement-Westerhof, however, there may perhaps be more than one species united under this epithet. To the same species, or at least to the same genus, also belong the remains named *Strobilifer frumentarius* by Weigelt (1928). *Culmitzschia florini* (Fig. 17a) has rather large leaves, which resemble those of *Ullmanni bronni* as mentioned above; there are considerable differences, however, in the epidermal structure (cf. Figs 13c and 17b).

Unlike all other Zechstein conifers, *Rhenania reichelti* has minute scale-shaped leaves that are similar to those of certain *Cupressaceae*, but these are helically arranged instead of oppositely. Until

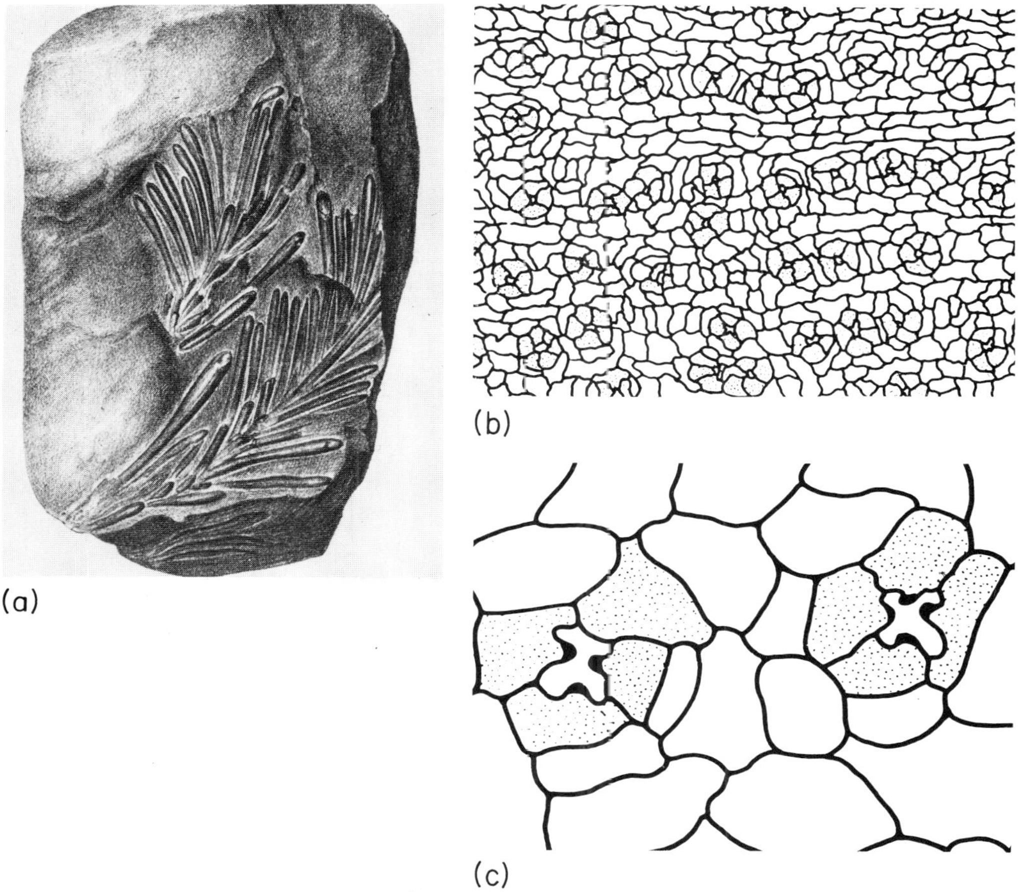

Fig. 15. *Quadrocladus solmsi* (Gothanet Nagalhard) Schweitzer. (a) Kupferschiefer, Ilmenau (after Solms-Laubach 1884); ×0.8. (b) Epidermal structure. (c) Stomata enlarged.

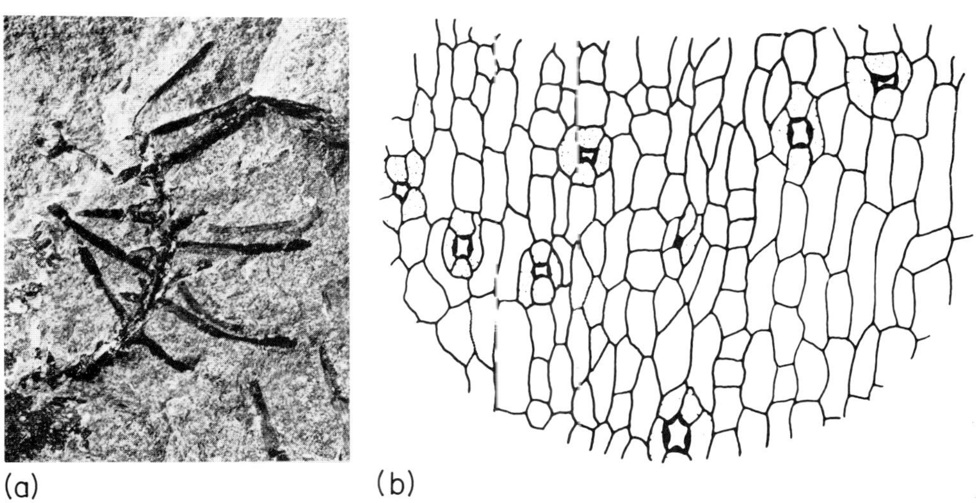

Fig. 16. *Quadrocladus orobiformis* (Schlotheim) Schweitzer. (a) Kupferschiefer, Rossenray, near Kamp-Lintfort; Coll. Schweitzer/Bonn; ×0.8. (b) Epidermal structure.

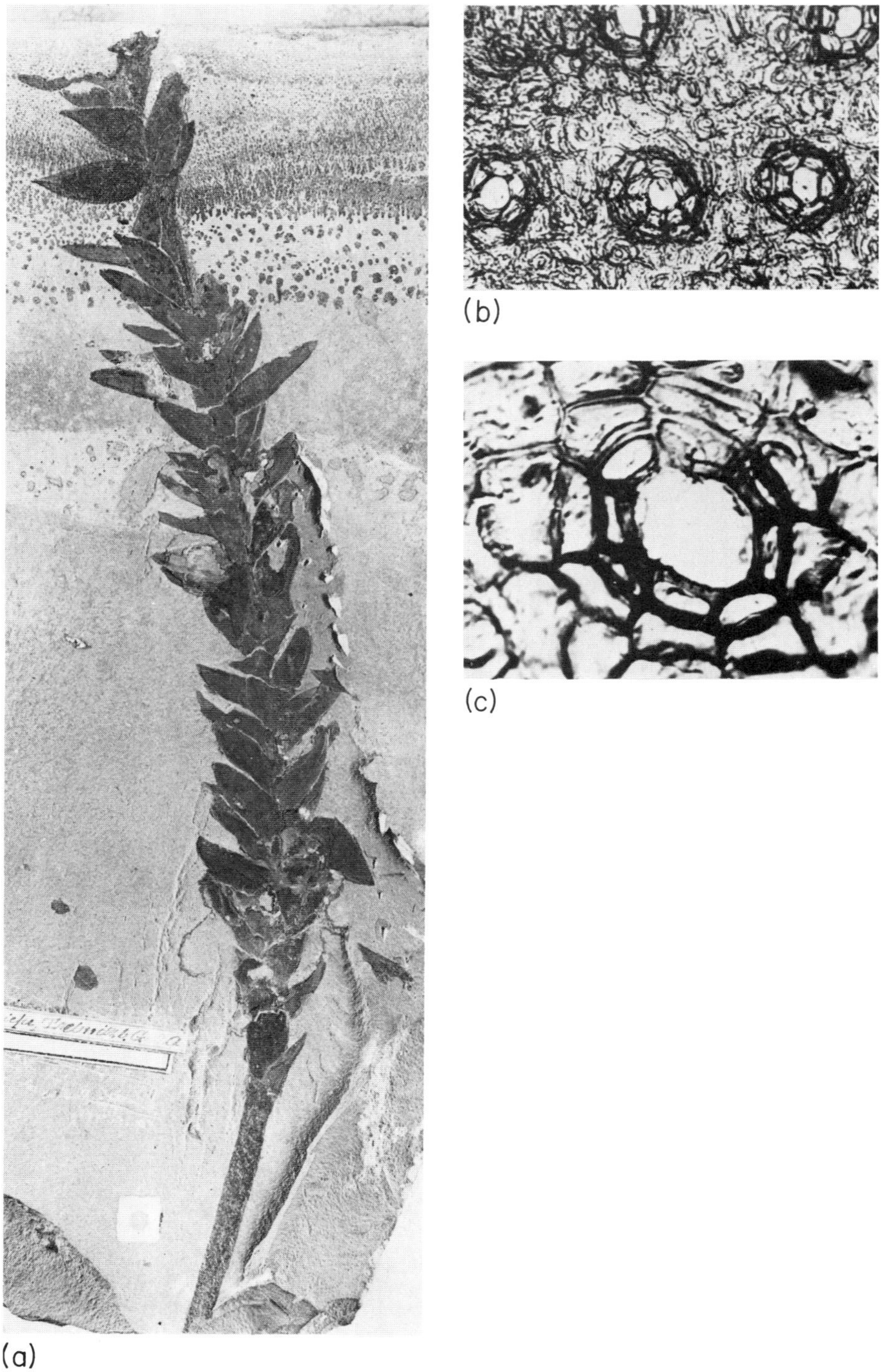

FIG. 17. *Culmitzschia florini* Ullrich. (a) Kupferschiefer, Gera (orig. to Geinitz 1880, pl. 3, fig. 4) ×1. (b) Epidermal structure (after Ullrich 1964). (c) Stomata (after Ullrich 1964).

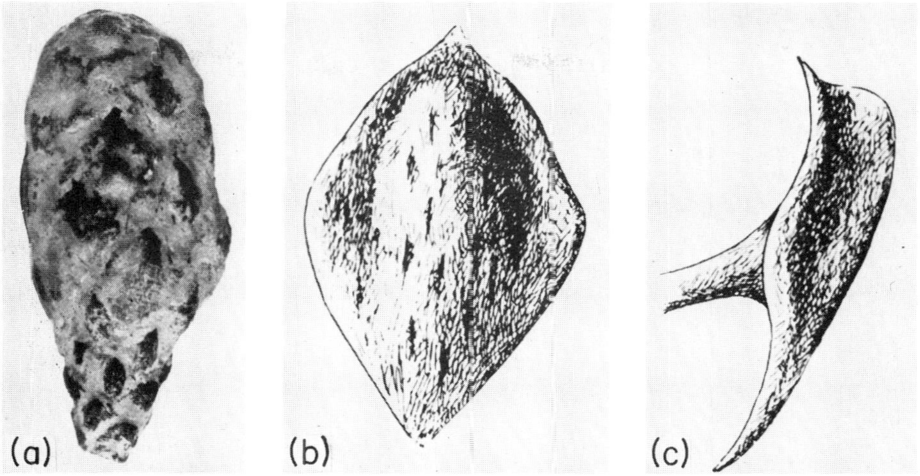

FIG. 18. *Rhenania reichelti* Schweitzer. (a) Male cone (Holotype). Zechstein kalk, Hoerstgen near Kamp-Lintfort; Coll. Schweitzer, Bonn; ×7. (b, c) Cone scale. Plan view and side-view.

now only one example has been found. This is a male cone with peltate microsporophylls (Fig. 18) that have a distinct bump in the middle of the abaxial side.

The palaeoenvironment of Zechstein land floras

Interpretation of the palaeoenvironment of the Zechstein flora is difficult because there are no determinable autochthonous plant remains: the climate can only be inferred from the anatomical structures of the plants and from a comparison with similar floras from other formations.

Of the Zechstein plants, only *Neocalamites* and the sphenopterids lack a xeromorphic habit. Probably we have to consider these plants as inhabitants of humid biotopes; *Neocalamites* perhaps lived in the shallow water of coastal bogs or lakes and the mouths of rivers and *Sphenopteris* probably preferred somewhat dryer habitats. *Pseudoctenis*, *Taeniopteris* and *Sphenobaiera* possibly grew in dryer or more or less protected valleys whereas *Callipteris* and the conifers may have grown on the more exposed slopes of hills and dunes. Thus the postulated hygrophylic *Neocalamites*-Sphenopterid association and the more xeromorphic *Callipteris*-conifer association could correspond to the *Calamites*-pecopterid and the *Callipteris-Walchia* association described by Gothan & Gimm (1930) from the Lower Permian. Of course only the extremes can be characterized by this classification; there must also have been all intermediate stages.

FIG. 19. Conifer wood with distinct growth rings. Kupferschiefer, Rossenray, near Kamp-Lintfort; Coll. Schweitzer, Bonn.

The distinct growth rings in the conifer woods (Fig. 19) are of great importance and indicate rhythmic climatic changes (Chaloner & Creber 1974). The kind of changes, however, cannot be determined from these growth rings, as either the humidity or the temperatures could have changed. Consideration of the fine lamination of the Kupferschiefer suggests that there must have been a rhythmic alternation of dryer and more humid periods. That these changes were not as abrupt as those of today in Middle Europe is shown by the differences of the radial diameters of the smallest and widest tracheids of the growth rings, which are less than in Recent Middle European conifers. However, the distances between the growth rings in these, as in the

TABLE 2. Distribution and abundance of species of land plants in Zechstein marine strata based on references cited in the text and also Schaumberg (1972) and Schweitzer (1960, 1962, 1968). x—very rare; xx—rare; xxx—dispersed; xxxx—common; xxxxx—abundant.

	Germany						England		
	Gera	Mansfeld	Ilmenau	Richelsdorf	Frankenberg	Niederrhein	Co. Durham	Hilton Beck	Other localities
Neocalamites		x				x	x	x	
Sphenopteris	x	x	x	x	x	xx	x	x	x
Callipteris	x	xx		xx	xx	x	x	x	x
Taeniopteris		xx		x			x		
Pseudoctenis							x		
Sphenobaiera	x	xx		x		xx	x	xx	
Cordaites		x(?)							x(?)
Plagiozamites							x		
Pseudovoltzia	xxx	xxx	x		xxx	xxx	xx	xx	x
Ullmannia bronni	xxx	xx	x		xxxxx	xxxx	xxx	xxxx	x
U. frumentaria	xxx	xxxx	xxx	xxxxx	xx	xxxxx	xx	xx	x
Quadrocladus solmsi	x	x	xx			x			
Q. orobiformis	x	x(?)	x			x			
Culmitzschia	xxx			x					
Rhenania						x			

Zechstein conifers, are equal. This does not prove that the rhythmic climatic changes of the Late Permian are similar to modern annual climatic cycles but nothing contradicts such a similarity. Thus the growth rings may support the interpretation of the lamination ('varves') in the Upper Permian evaporites as an indicator of annual climatic changes (see Richter-Bernburg 1957).

The geographical distribution of all Zechstein plants mentioned in the text is shown in Table 2.

ACKNOWLEDGMENTS: The author is greatly indebted to all those who supported the present paper. These are the collectors of the fossils, Mr T. Pettigrew and his colleagues from Sunderland, Mr J. Bell, Bowburn, Co. Durham, and others. Additional specimens were borrowed from the British Museum (Natural History), London; the Palaeontology Department of the Institute of Geological Sciences, London and Leeds, and the Hancock Museum, Newcastle upon Tyne. Mr A. Piel and Mrs I. Suchanek, both of the Palaeontological Institute of Bonn, prepared the illustrations and the photographs respectively.

References

ALTHAUS, J. 1846. Über einige neue Pflanzen aus dem Kupferschiefer von Riechelsdorf. *Palaeontogr.* **1**, 30–3.

BELL, J., HOLDEN, J., PETTIGREW, T.H. & SEDMAN, K.W. 1979. The Marl Slate and Basal Permian Breccia at Middridge, Co. Durham. *Proc. Yorks. geol. Soc.* **42,** 439–60.

CLEMENT-WESTERHOF, J.A. 1984. Aspects of Permian palaeobotany and palynology. IV. The conifer *Ortiseia* Florin from the Val Gardena Formation of the Dolomites and the Vicentian Alps (Italy) with special reference to a revised concept of the Walchiaceae (Göppert) Schimper. *Rev. Palaeobot. Palynol*, **41**, 51–166.

GEINITZ, H.B. 1848. Die Versteinerungen des deutschen Zechsteingebirges. In: GEINITZ, H.B. & GUTBIER, A.V. (eds) *Die Versteinerungen des Zechsteingebirges und des Rothliegenden oder des permischen Systems in Sachsen*, **1**, 1–27. Dresden.

——1880. Nachträge zur Dyas, I. *Mitt, K. mineral. Mus. Dresden*, **3**, 1–43.

GOTHAN, W. & GIMM, O. 1930. Neue Beobachtungen und Betrachtungen über die Flora des Rotliegenden in Thüringen. *Arb. Inst. Paläobot. Preuss. geol. Landesanst.* **2**, 39–74.

—— & NAGALHARD, K. 1921. Kupferschieferpflanzen aus dem niederrheinischen Zechstein. *Jb. K. Preuss. geol. Landesanst Bergakad., Berlin*, **42**, 440–60.

HALLE, T.G. 1927. Palaeozoic plants from Central Shansi. *Palaeont. sin.* **2**, 1–316.

KERP, J.H.F. 1982. Aspects of Permian palaeobotany and palynology II. *Acta Bot. Neerl.* **31**, 417–27.

MILLER, C.N. & BROWN, J.T. 1973. Paleozoic seed with embryos. *Science, New York*, **179**, 184–5.

RICHTER-BERNBURG, G. 1957. Isochrone Warven im Anhydrit des Zechstein 2. *Geol. Jb. Hannover*, **74**, 601–10.

SCHAUMBERG, G. 1977. Der Richelsdorfer Kupferschiefer und seine Fossilien. IV. Die pflanzlichen Fossilien des Kupferschiefers, sonstige Lebensspuren, Aufrechnung der Fundhäuigkeit. *Aufschluss*, **28**, 427–42.

SCHUSTER, A. 1933. Oberrotliegendes und Zechstein in Sachsen. *Abh. sächs geol. Landesamts.* **13**, 1–85.

SCHWEITZER, J.-J. 1960. Die Sphenopteriden des Zechsteins. *Senckenberg. Lethaea*, **41**, 37–57.

——1962. Die Makroflora des niederrheinischen Zechsteins. *Fortschr. Geol. Rheinld Westf.* **6**, 331–76.

——1968. Die Flora des Oberen Perms in Mitteleuropa. *Naturw. Rdsch. Stuttg.* **21**, 93–102.

SMITH, D.B. 1980. The evolution of the English Zechstein. In: FÜCHTBAUER, H. & PERYT, T.M. (eds) *The Zechstein Basin with Emphasis on Carbonate Sequences. Contr. Sediment.* **9**, 7–34.

SOLMS-LAUBACH, H. Graf zu. 1884. Die Coniferenformen des deutschen Kupferschiefers und Zechsteins. *Palaeont. Abh.* **2**, H.2, 81–118.

STONELEY, H.M.M. 1956. *Hiltonia*, a new plant genus from the Upper Permian of England. *Ann. Mag. nat. Hist.* **9**, 713–20.

——1958. The Upper Permian flora of England. *Bull. Br. Mus. nat. Hist. Geol* **3**, 295–337.

TOWNROW, J.A. 1960. The Peltaspermaceae, a pteridosperm family of Permian and Triassic age. *Palaeontology*, **3**, 332–61.

ULLRICH, H. 1964. Zur Stratigraphie und Paläontologie der marin beeinflussten Randfazies des Zechsteinbeckens in Ostthüringen und Sachsen. *Freiberger Forschhft.* 1–163.

WEIGELT, J. 1928. Die Pflanzenreste des mitteldeutschen Kupferschiefers und ihre Einschaltung ins Sediment. *Fortschr. Geol. Palaeont. Berlin*, **6**, 395–592.

——1930a. Uber die vermutliche Nahrung von *Protosaurus* und über einen körperlich erhaltenen Fruchtstand von *Archaeopodocarpus germanicus* aut. *Leopold.* **6**, 269–80.

——1930b. Neue Pflanzenfunde aus dem Mansfelder Kupferschiefer. *Leopold.* **6**, 643–68.

ZALESSKY, M.D. 1927. Flore permienne des limites ouraliennes de l'Angaride. *Mem. Com. geol. St. Petersbourg.* (N.S.) **176**, 1–52.

ZEILLER M.R. 1894. Notes sur la flora des Couches Permiennes de Trienbach (Alsace). *Bull. Soc. geol. Fr.* **3**, Ser. 12, 163–82.

H.-J. SCHWEITZER, Rhein. Friedr, Wilhelms-Univ., Paläontolog. Inst., Nußallee 8, 53 Bonn 1, Federal Republic of Germany.

Appendix 1

Diagnosis: *Plagiozamites bellii* sp.nov.
Genus: *Plagiozamites* (Zeiller 1894, p. 174).
Types species: *Plagiozamites planchardi* (Renault) Zeiller.
Species: *Plagiozamites bellii* sp.nov.
Name: After Mr John Bell (Bowburn, Co. Durham) who discovered the plant in 1980.
Species diagnosis: Branchlets with spatulate leaves up to 9 cm long and 1.5 cm broad. These were originally helically arranged, apparently in a more or less 2/5 phyllotaxis. However due to the plagiotropic direction of the ultimate shoots they are planated, thus forming two irregular rows one on each side of the axis. Some leaves are attached a little more on the upper side, others a little more on the lower. Leaves arise with broad but not decurrent bases, which are oblique, forming an angle of about 85° with the axis. Above the bases the leaves broaden slightly but then the margins are sub-parallel up to the rounded apex. Venation on both surfaces is clearly visible, each vein dichotomizing mostly only once in the basal third. Then the veins spread in an approximately fan-shaped manner towards the leaf margins.

The leaves are amphistomatic with no visible differences between lower and upper surfaces. Epidermal cells are polygonalic, isodiametric and above the veins slightly elongated. Stomata are mostly irregularly arranged, only rarely forming rows, separated by 3–4 (rarely 1–2) cells and surrounded by 5–7 auxilliary cells. These are more intensively cutinized and

are mostly somewhat smaller than the neighbouring cells. They are without papillae. Vestibules of the stomata are rounded or elliptical.
Holotype: Figured in Fig. 10(a). Registered in Sunderland Museum: TWCMS: G5746.
Locality: Middridge Quarry, Co. Durham N (GR NZ 249252).
Stratum typicum: Marl Slate, Upper Permian.

Discussion

This most important recent find represents a new species. Unfortunately only one sterile specimen has as yet been discovered. According to the habit and the insertion of the leaves the plant must be put in the genus *Plagiozamites*, which is considered to be heterogenus. The new species does not belong to the ferns but is a gymnosperm, due to its epidermal structures. Until additional, more complete material is found the systematic classification cannot be completed. A detailed description and a comparison with other species will be given in a future monograph on the English Zechstein flora (Schweitzer, in preparation).

The conodont *Merrillina divergens* (Bender & Stoppel) from the Upper Permian of England

A. Swift

SUMMARY: Conodont elements have been recovered from English Zechstein Cycle 1 carbonate rocks (EZ1 Ca) from three localities in NE England. Most of the elements belong to the apparatus of a single species, *Merrillina divergens* (Bender & Stoppel). The systematic concept of this species is revised.

In a report on the first discoveries of identifiable indigenous conodonts from the British Upper Permian, Swift & Aldridge (1982) noted that the resolution of taxonomic and biostratigraphic problems was hindered by the low numbers and low diversity of specimens. Conodonts are moderately abundant from the three localities reported here, and are mostly well preserved. This increase in the availability of identifiable material now permits a systematic revision of the conodont *Merrillina divergens* (Bender & Stoppel).

Samples

The conodont remains described in this paper were extracted from specimens of (1) the Raisby Formation from National Coal Board B2 Offshore Borehole, Blyth, Northumberland and from Claxheugh Rock, Sunderland and (2) the Ford Formation from Downhill Quarry, West Boldon, near Sunderland and from Claxheugh Rock, Sunderland. Sample details are given in the Appendix and their localities and those of sites from which indigenous Upper Permian conodonts have been recorded before are shown on Fig. 1. Locality numbers accord with the system in use in the Micropalaeontology Unit of the Department of Geology, University of Nottingham.

Details of the faunas

Preservation of individual elements is generally very good, but the collections from the reef of the Ford Formation at Claxheugh Rock are fragmentary. The colour of the conodonts from all localities is straw or pale orange, approximating to a colour alteration index (CAI) of 1.0–1.5 (Epstein *et al.* 1977). This indicates very little post-burial heating.

The collections from each locality and each stratigraphic level are remarkably uniform, and all the identifiable specimens, with the exception of the fauna of four individuals from 287/1, can be assigned to just six elements (Table 1). This low diversity is also typical of the previously reported British, German and Polish Zechstein Cycle I conodont faunas (Bender & Stoppel 1965; Jordan 1969; Szaniawski 1969). Six-element conodont apparatuses are known from the Ordovician to the Triassic, almost the entire known range of conodonts (e.g. Zeigler 1977, 1981). The six elements recognized in Zechstein collections conform with this well-established pattern and, furthermore, show close interelement similarities in denticulation and basal cavity morphology. The recognition of a symmetry transition between the hibbardelliform, angulodiform and hindeodelliform elements is also an important factor in regarding these elements as the apparatus of a single multielement species.

Sample 287/1, collected from just above the base of the Raisby Formation at Claxheugh Rock, yielded a fauna of just four specimens (Fig. 3 (14–17)) which do not fit the above reconstruction. Unless they are reworked, which seems unlikely in view of their delicacy and fair preservation, they indicate the presence of one or more additional conodont species in the earlier carbonate beds of EZ1. More material is needed before any interpretation can be attempted.

The dominance of a single species of conodont in EZ1 carbonate rocks follows the pattern of low diversity recognized in the Zechstein macrofauna, reflecting the difficult living conditions in the wholly or partially enclosed basin, where evaporation, sea-level and salinity changes frequently severely affected the fauna.

Systematic palaeontology

Genus *Merrillina* Kozur 1975

Type species: Spathognathodus divergens (Bender & Stoppel 1965).

Amended diagnosis: The skeletal apparatus consists of Pa, Pb, M, Sa, Sb and Sc elements. The

From Harwood, G.M. & Smith, D.B. (eds), 1986, *The English Zechstein and Related Topics*, Geological Society Special Publication No. 22, pp. 55–62.

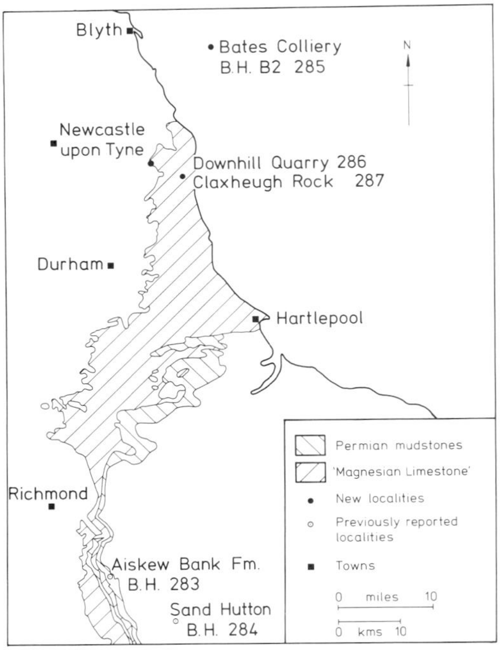

FIG. 1. Map showing sample localities and previously reported English localities that have yielded indigenous Upper Permian conodonts. The outcrop of the 'Magnesian Limestone' and Permian mudstones is from the *1:625 000 Geological Map of the U.K.*, 3rd edition, 1979, Institute of Geological Sciences.

TABLE 1. Frequency and distribution of element types of *Merrillina divergens*. Multielement notation follows Sweet & Schönlaub (1975).

	Frequency of element types					
Sample number	Pa	Pb	M	Sa	Sb	Sc
285/1	16	12	11	7	13	28
286/1	10	6	18	6	8	49
287/2		2	3	1	1	6
287/3	1		1			1
287/4	2	1	2			2

characteristic Pa element is short, with denticles which tend to increase in size and inclination posteriorly. The basal cavity is widely flared, broadest and deepest beneath the posteriorly positioned main denticle, and tapering anteriorly.

All other elements possess a more or less well-developed basal cavity, broadest and deepest in the area beneath the main denticle, tending to extend along all processes as a groove. The Sa element has a well-developed posterior bar.

White matter is present in the denticles of all elements.

Remarks. The genus *Merrillina* was proposed by Kozur (1975, p. 21) to accommodate some apparatuses whose Pa elements had previously been referred to as *Neospathodus*. His grounds for the separation were differences in basal cavity and carina morphology of the Pa elements and differences in his interpretation of multielement structure. The Pa elements from Zechstein faunas closely resemble the type species of *Neospathodus*, *N. cristagalli* Huckreide, but there is no evidence from the associated conodonts (Huckreide 1958, p. 167) that the apparatus of *Neospathodus* is comparable to that recognized herein for *Merrillina*. If *Neospathodus* proves to have a seximembrate apparatus comparable with that of *Merrillina* then a synonomy between *Neospathodus* and *Merrillina* would have to be considered.

The apparatus of *Ellisonia* as currently conceived (Sweet 1970, p. 224) has a suite of ramiform elements similar to those of *Merrillina*, but the holotype of the type species, *E. triassica* Müller, is an Sa element having 'no pit or groove beneath the main cusp' (Müller 1956, p. 822). Sa elements of *Merrillina* have a clearly developed basal cavity.

Stepanovites is reconstructed at present (Kozur 1975, p. 22) to have a quadrimembrate non-platform apparatus. The Pb, M, Sa and Sb elements of the type species *S. meyeni* Kozur and Movšovič, are very similar to the corresponding ramiforms in *Merrillina*; if other elements are found which indicate that the apparatuses of *S. meyeni* and *M. divergens* are sufficiently similar to be congeneric, then *Merrillina* would have page priority over *Stepanovites* as the valid generic designation.

Kozur (1975, p. 23) noted an absence or rarity of white matter in elements of *Stepanovites*; all elements of the apparatus of *M. divergens* show conspicuous distribution of white matter in their denticles (Fig. 2).

Merrillina divergens (Bender & Stoppel 1965), Figs 2 and 3 (1–13).

1965 *Hindeodella triassica* Müller; Bender & Stoppel, p. 343, pl. 14, fig. 12; pl. 15, figs 3, 4 (Sc element).

1965 *Hindeodella triassica* Müller; Bender & Stoppel, p. 343, pl. 15, figs 1, 2 (Sb element).

1965 *Lonchodina inflata* Bender & Stoppel, p. 346, pl. 15, fig. 8(a–c); pl. 16, fig. 18(a–c) (M element).

1965 *Lonchodina* cf. *inflata* Bender & Stoppel,

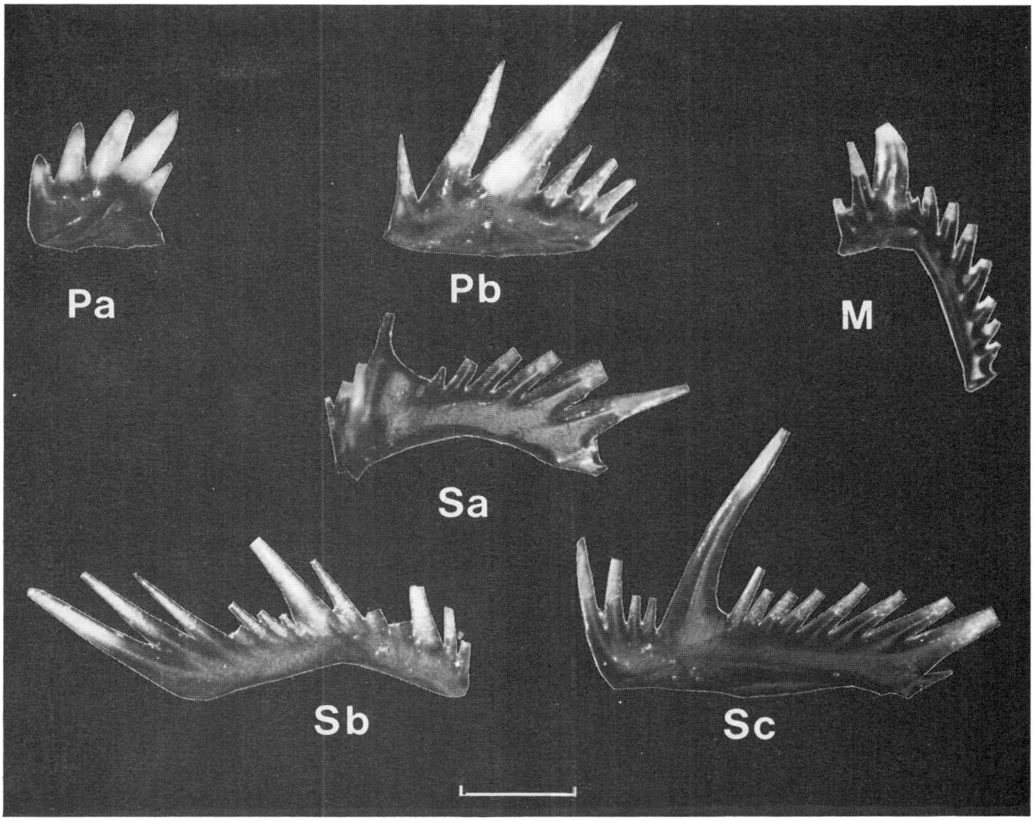

FIG. 2. The apparatus of *Merrillina divergens* showing distribution of white matter. Element numbers 4513/29A (Pa), 4512/28A (Pb), 4508/4A (M), 4508/5A (Sa), 4508/6A (Sb), 4509/9A (Sc). All elements × 60. The scale bar represents 0.25 mm.

p. 346, pl. 15, fig. 11; pl. 16, fig. 24 (M element).

1965 *Lonchodina mülleri* Tatge; Bender & Stoppel, p. 347, pl. 15, fig. 12(a,b) (M element).

1965 *Roundya* sp.A.; Bender & Stoppel, p. 350, pl. 15, fig. 19 (Sa element).

1965 *Roundya* sp. B; Bender & Stoppel, p. 350, pl. 15, fig. 20(a–c) (Sa element).

1965 *Spathognathodus divergens* Bender & Stoppel, p. 350, pl. 16, figs 1–3, 21 (Pa element).

1969 *Spathognathodus divergens* Bender & Stoppel; Jordan, p. 217, pl. 1, figs 1, 2; pl. 2, fig. 11(a,b) (Pa element).

1969 *Lonchodina mülleri* Tatge; Jordan, p. 217, pl. 1, fig. 8(a,b); pl. 2, fig. 9(a,b) (M element).

1969 *Hindeodella triassica* Müller; Jordan, p. 217, pl. 1, fig. 7 (a,b); pl. 2, fig. 8 (Sb element).

1969 *Hindeodella triassica* Müller; Jordan, p. 217, pl. 1, figs 9(a,b), 10; pl. 2, figs 1, 4, 7(a,b), 10, 13 (Sc element).

1969 *Hindeodella triassica* Müller; Szaniawski, p. 329, pl. 1, figs 7(a–c), 8, 9, 11 (Sc element).

1969 *Hindeodella triassica* Müller; Szaniawski, p. 329, pl. 1, fig. 10 (Sb element).

1969 *Lonchodina inflata* Bender & Stoppel; Szaniawski p. 330, pl. 1, figs 5, 6 (M element).

1969 *Lonchodina vistulensis* Szaniawski, p. 331, pl. 1, figs 1–4 (M element).

1969 *Prioniodina lindstroemi* Szaniawski, p. 332, pl. 2, figs 6–9 (Pb element).

1969 *Hibbardella baltica* Szaniawski, p. 332, pl. 2, figs 11, 12 (Sa element).

1969 *Spathognathodus divergens* Bender & Stoppel; Szaniawski, p. 333, pl. 2, figs 1–5 (Pa element).

1972 *Plectospathodus simuelleri* Kozur &

Mostler, p. 27, pl. 2, figs 2, 3 (Sc element).
1972 *Plectospathodus simuelleri* Kozur & Mostler, p. 27, pl. 2, fig. 1 (Sb element).
1975 *Merrillina divergens* (Bender & Stoppel); Kozur, p. 21 (Pa, M elements described, but not figured).
1975 *Stepanovites inflatus* (Bender & Stoppel); Kozur, p. 25 (Pb, M, Sa, Sc elements described, but not figured).
1978 *Merrillina divergens* (Bender & Stoppel); Kozur, pl. 5, figs 27, 28; kop. Bender & Stoppel (1965), pl. 16, figs. 2c, 3 (Pa element).
1978 *Stepanovites inflatus* (Bender & Stoppel); Kozur, pl. 7, figs 1, 2; kop. Bender & Stoppel (1965), pl. 15, figs 1, 2 (Sb element).

Type material

Holotype BS 7/12(H), Pa element figured as *Spathognathodus divergens* by Bender & Stoppel (1965), pl. 16, fig. 2 (a–c); from Schacht Hoerstgen shaft, depth 280–285 m. Paratypes, BS 7/11 (H), BS 7/13 (H), Pa elements from Schacht Hoerstgen shaft, depth 280–285 m.

Material, localities and horizons

Pa element, 29 specimens; Pb element, 21 specimens; M element, 35 specimens; Sa element, 14 specimens; Sb element, 22 specimens; Sc element, 86 specimens. Found in the Raisby Formation near Sunderland and offshore Northumberland, and in reef facies of the Ford Formation in Tyne and Wear. All specimens from strata of English Zechstein I age (Upper Permian). Also found in EZ1 strata in North Yorkshire (Swift & Aldridge 1982), Germany (Bender & Stoppel 1965; Jordan 1969) and Poland (Szaniawski 1969).

Diagnosis

Pa element short, slightly laterally bowed, bearing between three and seven denticles fused to mid-height in large specimens. Basal cavity lachryform, occupying almost the entire aboral surface, with the exception of a very short unexcavated blade at the anterior tip of the unit. Sa elements have a prominent posterior bar which is approximately twice the length of each of the lateral bars. The denticles in the distal fifth of the longer process of S elements are conspicuously large, the largest positioned above the point where there is commonly a sharp downward deflection of the process, tending towards a geniculation. Basal cavity broad and deep on Pa, Pb and M elements, but subject to a large degree of variability on S elements, its extent varying from a narrow slit to a broad slash. The anterior tip is the only unexcavated area on all elements. White matter is present in all elements, but is confined to denticles.

Description

Pa element: A short, paired, slightly bowed unit with three to seven subrounded denticles, discrete in small specimens, fused to mid-height in larger ones. Denticles increase in size and inclination posteriorly, with a larger main denticle terminal or succeeded by a single very small denticle. Basal cavity laterally expanded, broadly flaring beneath the main denticle, tapering anteriorly. A very short unexcavated blade is normally present in the anterior-most tenth of the unit. Striae are generally present on the flanks of the denticles (Fig. 3(2)). All Pa elements have denticles which are typically deeply filled with white matter, but on some there are one or two clear denticles.

Pb element: A gently arched, paired element with moderate lateral curvature. Denticles discrete, subrounded in transverse section. Cusp posteriorly inclined, stout and prominent, equalling the length of the whole element in some specimens (Fig. 3(4)). Anterior process short, bearing two or three denticles; posterior process longer, carrying up to six straighter more peg-like denticles which trend parallel to the cusp. Basal cavity is of similar size and shape to that of Pa element but is less laterally extended. White matter patchily distributed in most denticles.

FIG. 3. (1–13) *Merrillina divergens* (Bender & Stoppel): (1, 3), inner lateral views of dextral and sinistral Pa elements 4315/29A, ×80, 286/1; 4511/23A, ×150, 287/4; (2) inner lateral view of Pa element 4513/29A, ×250, 286/1, showing striations; (4, 5) inner lateral views of dextral and sinistral Pb elements 4512/28A, ×60, 286/1; 4510/15A, ×60, 287/2; (6, 7) inner lateral views of dextral and sinistral M elements 4508/4A, ×60, 285/1; 4513/34A, ×60, 286/1; (8, 9) lateral views of Sa elements 4508/5A, ×60, 285/1; 4513/31A, ×60, 286/1; (10, 11) inner lateral views of dextral and sinistral Sb elements 4513/33A, ×60, 286/1; 4508/6A, ×60, 285/1; (12, 13) inner lateral views of sinistral and dextral Sc elements 4508/8A, ×40, 285/1; 4509/9A, ×40, 285/1. (14–17) Unassigned elements from 287/1: (14), lateral view of M? element 4509/11A, ×150; (15) lateral view of P? element 4509/10A, ×80; (16) lateral view of Sb? element 4509/13A, ×80; (17) lateral view of M? element 4509/12A, ×150. Repository of specimens illustrated: Conodont Reference Collection, Department of Geology, University of Nottingham.

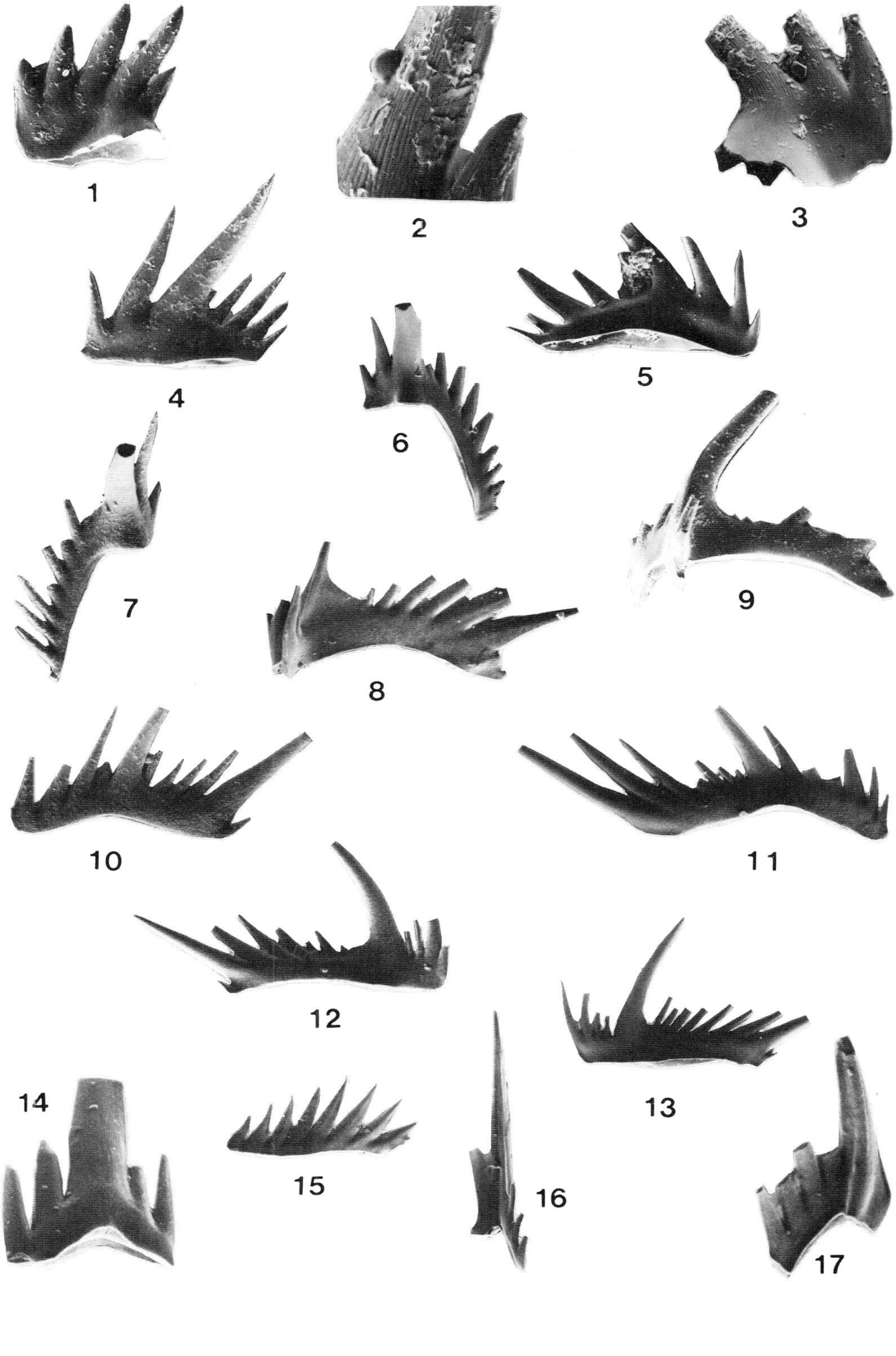

M element: Cusp recurved, stout, slightly twisted in some specimens. From its point of maximum depth and lateral extension below the cusp, the basal cavity extends along both processes as a narrow groove. On larger specimens the lateral extension of the basal cavity tends to form a short undenticulate third process. Posterior process long, strongly downcurved, bearing up to 10 denticles fused variably to a maximum of one-third of their height. Anterior process short, directed downwards, with up to five denticles which decrease in height away from the cusp. The denticle adjacent to the cusp on the anterior bar is commonly large and twisted. All M elements show a patchy distribution of white matter in some or all denticles. Elements are paired.

Sa element: A symmetrical unit with a well-developed gently arched posterior process, bearing up to seven subrounded denticles variably fused up to mid-height, increasing in size and inclination posteriorly, the ones in the posterior-most fifth being conspicuously large. Lateral processes directed downward, enclosing an angle of approximately 140° on the few unbroken specimens; both carry three to six subrounded denticles which are sub-erect and semi-fused. Basal cavity narrow, deepest beneath the large posteriorly curved cusp, extending along all processes as a groove. White matter present in some or all of the denticles.

Sb element: An arched, paired, laterally curved element; anterior and posterior processes of equal length, each bearing five to eight sharp subrounded denticles which are fused variably up to a maximum of one-third of their length. Denticle size increases posteriorly, the ones in the posterior-most fifth being large and prominent, the largest exceeding the length of the conspicuous cusp. A downward deflection on the posterior process is present in some specimens at the point where this largest denticle lies, and a small denticle may be present on this downcurved section (Fig. 3(10)). Basal cavity slightly laterally extended beneath the cusp, continuing almost to the end of both processes as a groove. A variable filling of white matter is present in most of the denticles.

Sc element: Elements are paired, gently arched or almost straight, and show a variable degree of lateral curvature. The posterior-most fifth of the blade shows a strong tendency towards geniculation, beginning at the site of a large prominent denticle. One or two small denticles are present on this downcurved section (Fig. 3(12)). Anterior process short with four or five subrounded denticles which are larger distally from the tall posteriorly curved cusp. Posterior process longer, with eight to 10 subrounded denticles which gradually increase in size and inclination posteriorly. Basal cavity varies from a narrow groove to a broad slash, occupying all but the anterior-most tip of the aboral surface. Denticles are variably fused up to mid-height and most contain white matter to varying depths.

Remarks. In the earliest report of conodont species from the Zechstein, Sc elements assigned to *Hindeodella triassica* Müller were the first to be described (Bender & Stoppel 1965, p. 343), but *triassica* is not adopted as a species name because the Zechstein Sc elements possess a well-defined basal cavity; this is at variance with the original description of *H. triassica* which states that 'there is no pit or groove beneath the main cusp' (Müller 1956, p. 826). Of the other elements in the first report *Lonchodina inflata* Bender & Stoppel has page priority over *Spathognathodus divergens* Bender & Stoppel (Bender & Stoppel 1965, pp. 346, 350) but *divergens* is preferred as a species name here because the Pa element is the more characteristic. The morphology of the M element is more conservative and the use of *inflata* as a species name could lead to difficulties in a study of closely related species.

Bender & Stoppel (1965) did not record Pb elements from their German faunas; this is notable because their collections were fairly large and included all the other elements of *M. divergens*. A re-examination of their collections and a re-sampling of the relevant German strata may help to resolve this apparent anomaly.

In the smaller faunas, e.g. 287/2, there is not a full representation of all the elements of the apparatus of *M. divergens*, but this would not be expected where such small numbers are involved.

The M and Sa elements here recognized as belonging to the apparatus of *M. divergens* have previously been described from Zechstein strata as a number of form species; the Sb element, although figured, has not previously been recognized as a separate element. Kozur (1975) combined some of these elements into two multielement species, *M. divergens*, in which he included Pa and M elements, and *Stepanovites inflatus* (Bender & Stoppel) whose apparatus consisted of Pb, M, Sa and Sc elements. He noted that *Hindeodella triassica* Müller and *Hibbardella baltica* Szaniawski, the Sc and Sa elements of *S. inflatus*, were invariably found in association with the Pa element of *M. divergens* in Zechstein collections; he further stated that there was a possibility that these elements could belong to both *Merrillina* and *Stepanovites*, and that the taxonomic problem could only be resolved by the recovery of faunas containing solely *Merrillina*. The Zechstein conodont faunas are shown here to meet that criterion.

The ancestry of *M. divergens* is difficult to trace in Europe because of the long break in marine sedimentation during the Lower Permian, but a suite of elements figured from the lower part of the Anisian stage (Triassic) in the Thuringian basin of Germany (Kozur & Mostler 1972, pl. 1, figs 4, 7, 8, 18–27; pl. 2, figs 4, 6, 9, 12) apparently form the apparatus of a descendant. This apparatus differs only slightly from *M. divergens* in the morphology of the Pa element, which has a large cusp.

There are a number of records of conodonts referred to *M. divergens* from outside the Zechstein basin, but evaluation is difficult. Clark & Behnken (1971) figured a Pa element as *M. divergens* from the upper part of the Gerster Formation (higher Capitanian) at Phalen Butte, Nevada, North America, but the illustrated specimen (Clark & Behnken, pl. 2, fig. 6) appears to have a longer anterior unexcavated blade than the Zechstein Pa elements. In addition, the figured ramiforms from the same formation (Clark & Behnken pl. 2, figs 3, 7–9, 13, 18) either range throughout or are from the lower part only, so the rest of the apparatus is unknown.

Clark & Behnken (1971) used *M. divergens* as a biostratigraphic index fossil for a zone in the higher Capitanian, as did Behnken (1975), but they did not figure any specimens.

Wardlaw & Collinson (1979) also reported *M. divergens* from the same stratigraphic level, in sections in the Great Basin–Rocky Mountain region of North America, but again no specimens were figured.

Two Pa elements identified as *M. divergens* (Igo 1981, pl. 12, figs 18, 19) have been reported from the late Middle Sakamotozawan (Lower Permian) of Japan. These specimens are not well preserved but the cusp appears to be larger and less inclined than on Zechstein Pa elements, although basal cavity morphology is very similar. The elements are found in beds which also contain species of the genus *Neostreptognathodus*, including *N. pequopensis* Behnken and *N. sulcoplicatus* (Youngquist, Hawley and Miller), each of which wholly or partly defines a zone in the Leonardian stage (Lower Permian) of North America (Behnken 1975). If the Japanese identifications are correct, then either the range of *M. divergens* will have to be extended into the Lower Permian or the possibility of a condensed deposit or reworking would have to be considered.

ACKNOWLEDGMENTS: Dr R.J. Aldridge of the Geology Department at Nottingham University is thanked for his valuable guidance and assistance in the preparation of this paper. The North-East Area office of the National Coal Board gave permission to publish information from the Bates Colliery bore. Scanning electron micrographs were taken by Mr L. Green at the Wolfson Institute for Interfacial Technology, and prepared with the assistance of Mr D. Jones. Dr G.F. Aplin is thanked for field guidance, and Mrs J. Wilkinson and Mrs J. Pearson are thanked for cartographic assistance and typing respectively.

References

BEHNKEN, F.H. 1975. Leonardian and Guadalupian (Permian) conodont biostratigraphy in western and southwestern United States. *J. Paleont.* **49,** 284–315.

BENDER, H. & STOPPEL, D. 1965. Perm-Conodonten. *Geol. Jb. Hannover*, **82,** 331–64.

CLARK, D.L. & BEHNKEN, F.H. 1971. Conodonts and biostratigraphy of the Permian. In: SWEET, W.C. & BERGSTROM, S.M. (eds) *Symposium on Conodont Biostratigraphy. Mem. geol. Soc. Am.* **127,** 415–39.

EPSTEIN, A.G., EPSTEIN, J.B. & HARRIS, L.D. 1977. Conodont color alteration—an index to organic metamorphism. *Prof. Pap. U.S. geol. Surv.* **995,** 1–27.

HUCKREIDE, R. 1958. Die Conodonten der mediterranen Trias und ihr stratigraphischer Wert. *Palaeont. Z. Berlin*, **32,** 141–75.

IGO, H. 1981. Permian conodont biostratigraphy of Japan. *Spec. Pap. palaeont. Soc. Japan. No. 24*, 50 pp.

JORDAN, H.P. 1969. Conodonten aus dem Zechsteinkarbonat (Ca 1) des Thüringer Beckens. *Geologie, DDR*. **18,** 216–21.

KOZUR, H. 1975. Beiträge zur Conodontenfauna des Perm. *Geol. Paläont. Mitt. Innsbruck*, **54,** 1–41.

—— 1978. Beiträge zur Stratigraphie des Perms. Teil II. Die Conodontenchronologie des Perms. *Freiberger Forschhft.* **334,** 85–161.

KOZUR, H. & MOSTLER, W. 1972. Die Conodonten der Trias und ihr stratigraphischer Wert. Teil I. Die 'Zahnreihen-Conodonten' der Mittel- und Obertrias. *Abh. geol. Bundesanst. Wien*, **28,** 1–36.

MAGRAW, D. 1978. New boreholes into Permian beds off Northumberland and Durham. *Proc. Yorks geol. Soc.* **42,** 157–83.

MÜLLER, K.J. 1956. Triassic conodonts from Nevada. *J. Paleont.* **30,** 818–30.

SWEET, W.C. 1970. Uppermost Permian and Lower Triassic conodonts of the Salt Range and Trans-Indus Range, West Pakistan. In: KUMMEL, B. & TEICHERT, C. (eds) *Stratigraphic Boundary Problems; Permian and Triassic of West Pakistan. Univ. Kans. Dept Geol. Publ.* **4,** 207–75.

SWEET, W.C. & SCHÖNLAUB, H.P. 1975. Conodonts of the Genus *Oulodus* Branson & Mehl, 1933. *Geol. Palaeont.* **9,** 41–59.

SWIFT, A. & ALDRIDGE, R.J. 1982. Conodonts from Upper Permian strata of Nottinghamshire and North Yorkshire. *Palaeontology*, **25,** 845–56.

SZANIAWSKI, H. 1969. Conodonts of the Upper Permian of Poland. *Acta palaeont. Pol.* **14,** 325–41.

WARDLAW, B.R. & COLLINSON, J.W. 1979. Youngest Permian conodont faunas from the Great Basin and Rocky Mountain region. In: SANDBERG, C.A. & CLARK, D.L. (eds) *Conodont Biostratigraphy of the Great Basin and Rocky Mountains. Geol. Stud. Brigham Young Univ.* **26,** 151–9.

ZIEGLER, W. (ed.) 1977. *Catalogue of Conodonts III.* E. Schweizerbart'sche Verlagsbuchhandlung, Stuttgart.

—— 1981. *Catalogue of Conodonts IV.* E. Schweizerbart'sche Verlagsbuchhandlung, Stuttgart.

A. SWIFT, Department of Geology, University of Nottingham, Nottingham NG7 2RD.

Appendix: locality and sample details

Offshore Northumberland

Loc. 285. Raisby Formation, Bates Colliery, National Coal Board borehole B2 (NZ 4391 7858). One 1500 g sample (285/1) of grey laminated dolomite with darker grey argillaceous intercalations, depth of 154.23–154.53 m. Base of Raisby Formation at 164.77 m.

Sunderland area

Loc. 286. Ford Formation, basal coquina. Downhill Quarry, West Boldon, near Sunderland (NZ 348 601). One 4000 g sample (286/1) of bioclastic dolostone.

Loc. 287. Raisby Formation and Ford Formation. Claxheugh Rock and the adjacent railway cutting to the south (NZ 362 575), Sunderland. Two samples, 287/1 (2000 g) and 287/2 (2000 g), from the atypically thin Raisby Formation and two samples, 287/3 (2000 g) and 287/4 (2000 g), from the topographic base and top respectively of the unconformably succeeding reef of the Ford Formation. 287/4 is from the railway cutting.

Sedimentology of sandwaves in an oolite shoal complex in the Cadeby (Magnesian Limestone) Formation (Upper Permian) of eastern England

J. Kaldi

SUMMARY: Four depositional facies are recognizable in the Sprotbrough Member of the Cadeby Formation (formerly the Lower Magnesian Limestone) in eastern England; they are sabkha, semi-restricted lagoon, oolite shoal complex and open marine shelf. Of these, the oolite shoal complex occupies most of the present outcrop in the Yorkshire Province, i.e. south of the Cleveland High. It comprises large-scale cross-stratified oolite bedforms that are interpreted as sandwaves from their size, type of sediment and internal features. The sandwaves probably originated from the concentration of wave energy by eminences on the floor of the Zechstein Sea. Biogenic structures in the sandwaves show that they were inhabited by marine organisms.

A complex pattern of palaeocurrent styles and azimuths is present within the sandwaves. At many exposures, the long axes of the sandwaves trend NW–SE, suggesting predominant currents normal to this direction. A predominant current from the NE is further indicated by the preferential abutment of SW-dipping beds on the NE flanks of bedforms; this was caused by sandwaves that climbed the backs of others. Occasional storms from the SE produced spillover lobes oriented towards the NW. Storms were also probably responsible for the formation of internal structures such as hummocky cross-stratification, fan-shaped bedding and contemporaneous erosion surfaces.

Introduction

The importance of a detailed understanding of depositional facies within carbonate rock systems has been emphasized recently by the oil industry's recognition that hydrocarbon produced from carbonate reservoirs comes mainly from non-reefal rocks. For this reason, the sedimentology of modern carbonate environments, especially the dynamics of carbonate sand bodies of the Bahamas and Yucatan, has received considerable study (see, e.g. Ball 1967; Hine 1977; Harms *et al.* 1978; Dravis 1979; Halley & Harris 1979; Harris 1979; Hine *et al.* 1981).

Research on modern carbonate sand bodies is facilitated by excellent lateral and regional sand body continuity, well-defined shoreline-basin relationships and a fairly well-documented understanding of local variations in level, duration and magnitude of the physical forces acting on the system. In contrast, the study of ancient analogues is hindered by limited exposures, vaguely defined regional palaeogeography and only the structures themselves as indicators of possible past energies; furthermore, many of the primary clues in carbonate rocks have been destroyed by diagenesis. This may be the reason that there has been only limited work (Klein 1965; Anderson 1967; Laporte 1969; Purser 1975) on the detailed depositional dynamics of ancient carbonate sand bodies.

It is the aim of this paper to demonstrate that ancient carbonate sand bodies reveal a plethora of useful depositional and diagenetic information. Sections through the sandwaves of the Permian oolite shoals in England allow access to and analysis of detailed internal structures and bedforms that would be impossible in modern sandwaves; the description, classification and interpretation of these structures provide valuable insight into palaeoenvironmental processes within this ancient carbonate sand system.

Geological setting

The Upper Permian Cadeby Magnesian Limestone Formation (formerly the Lower Magnesian) of Yorkshire, Derbyshire and Nottinghamshire is the first cycle marine carbonate (EZl Ca) laid down on the west margins of the Zechstein Sea. It crops out in a narrow belt extending for 170 km in east England. The formation consists of a lower (Wetherby) and an upper (Sprotbrough) member, separated in the Yorkshire Province (i.e. south of the Cleveland High of Smith 1970) by the Hampole Discontinuity and Hampole Beds (Smith 1968). Rocks of the Cadeby Formation have received extensive study by Smith (1968, 1970, 1974a,b, 1980); this paper deals exclusively with the Sprotbrough Member.

Four depositional facies are recognizable in outcrops and boreholes of the Sprotbrough Member in the Yorkshire Province: arid supratidal (sabkha); semi-restricted lagoon; oolite shoal

From Harwood, G.M. & Smith, D.B. (eds), 1986, *The English Zechstein and Related Topics*, Geological Society Special Publication No. 22, pp. 63–74.

complex; and open marine shelf (Fuzesy 1970; Smith 1974a,b, 1976; Kaldi 1980a,b).

The oolite shoal complex occupies most of the present outcrop. The shoals consist of large-scale (sets up to 12 m) cross-stratified bedforms which at many localities are linear, ridge-like and asymmetric; they are interpreted as oolite sandwaves (Kaldi 1980a). The 12 m cross-strata within these structures set this figure as a minimum water depth (Smith 1974a, p. 129).

Petrography

The sediment of the sandwaves is partly or wholly of dolomite, as are most of the rocks of the Cadeby Formation at outcrop. In the central and northern sectors of the study area (Fig. 1) the shoals consist mostly of grainstone that has been replaced by dolomicrospar or dolomicrite; although many of the micro-textures have been obliterated by diagenesis, the gross morphology of many allochems has been preserved. Ooids and peloids occur on palaeotopographic highs, and packstones mostly of ooids, peloids, pisoids, skeletal fragments and grapestones (composite grains) are found in hollows and areas of low relief. Non-carbonate grains are almost absent in these central and northern sectors.

In the southern sector, the rocks contain coarsely crystalline (200–250 μm) mosaics of anhedral dolomite in which many crystals poikilotopically enclose vague ghosts of former allochems. Lozenge-shape non-concentric allochems were probably peloids and those with concentric internal laminae were probably ooids. Subrounded to angular quartz grains (75–250 μm) are commonly enclosed within dolomite crystals and some form the nuclei of ooids. The non-carbonate content of these sediments is as high as 25% near Mansfield, where the Mansfield River is inferred to have drained uplands to the west (Smith *et al.* 1967).

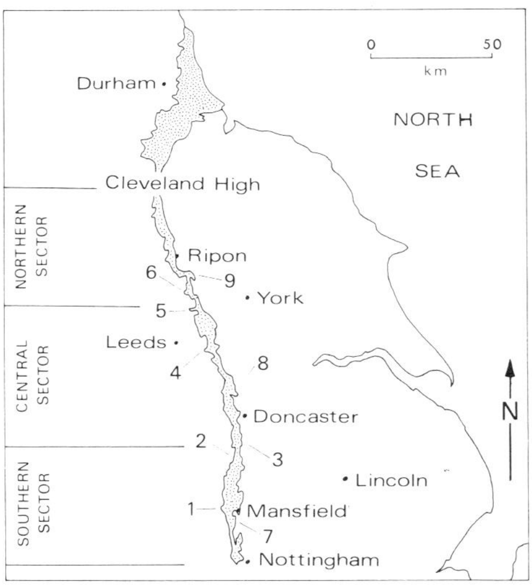

FIG. 1. Outcrop of the Cadeby Formation, showing the three sectors referred to in the text: (1) Pleasley–Shirebrook railway cutting; (2) Creswell Crags; (3) Maltby railway cutting; (4) New Micklefield Quarry; (5) River Wharfe cliffs, Boston Spa; (6) Knaresborough Gorge; (7) Mansfield Quarry; (8) Hampole Grange Quarry; (9) Farnham Quarry.

Field relationships and details

Exposures and major structures

Nine localities where oolite sandwaves in the Sprotbrough Member are particularly well exposed were selected for special study; the locations and other details of these localities are summarized in Table 1.

The exposure with the greatest number of bedforms is a disused railway cutting between Pleasley and Shirebrook where there is 1200 m of continuous vertical section 3–12 m high; Fig. 2 is a sketch of the southernmost 300 m of this section. The cutting is 10–12 m wide and trends SW–NE. The orientation is fortunate in that many of the sandwaves have long axes oriented normal to the cutting; in most places there it is possible to measure true dips. Eighteen sandwaves, several of which are composite structures of two or more bedforms, were located, numbered and their internal structures examined. The bedforms at this locality typify many of the others of the Sprotbrough Member and the internal features seen here also occur at many other exposures.

The size-range of the sandwaves is similar at all localities where they are present: 2–50 m across, and 0.6–15 m high. In the Pleasley–Shirebrook railway cutting and the Knaresborough gorge there is a progressive increase in maximum size from the SW to the NE. The steepest dips (19°–35°) within the sandwaves at all localities are on the flanks; near the centres, where the crests or highest parts are located, all laminae and the bounding surfaces have dips of 3°–15°.

Internal structures and their origins

Bedform contacts

The nature of the contact between sandwaves varies, but most adjoining waves have sharp

TABLE 1. Bedforms of the Sprotbrough Member of the Cadeby Formation

Locality	*Grid ref.*	*No. of bedforms*	*Internal structures*
Old quarry, Mansfield	SK 535601	2	CUC, CE, MF
Railway cutting, Pleasley–Shirebrook	SK 507652–SK 521672	18+	D, CUC, FB, HCS, CE, B, LB, MF
Gorge, Creswell Crags	SK 534743	5+	CUC, FB, CE, B, LB, MF
Railway cutting, Maltby	SK 549918	2	D, CE, B
Railway cutting, Hampole Grange	SE 516097	2	B, FB
Old quarry, New Micklefield	SE 445325	2+	FB, CUC, CE, B
Gorge, River Wharfe, Boston Spa	SE 426468		CE, HCS, CUC
Old quarry, Knaresborough	SE 351566	2	FB, CE, B
Old quarry, Farnham	SE 353602		CE, CUC, B, MF

Key to internal structures: B = Burrows, CE = evidence of contemporaneous erosion, CUC = convex-up cross-stratification, D = drape structures, FB = fan bedding, HCS = hummocky cross-stratification, LB = lobate bedform, MF = marine fossils.

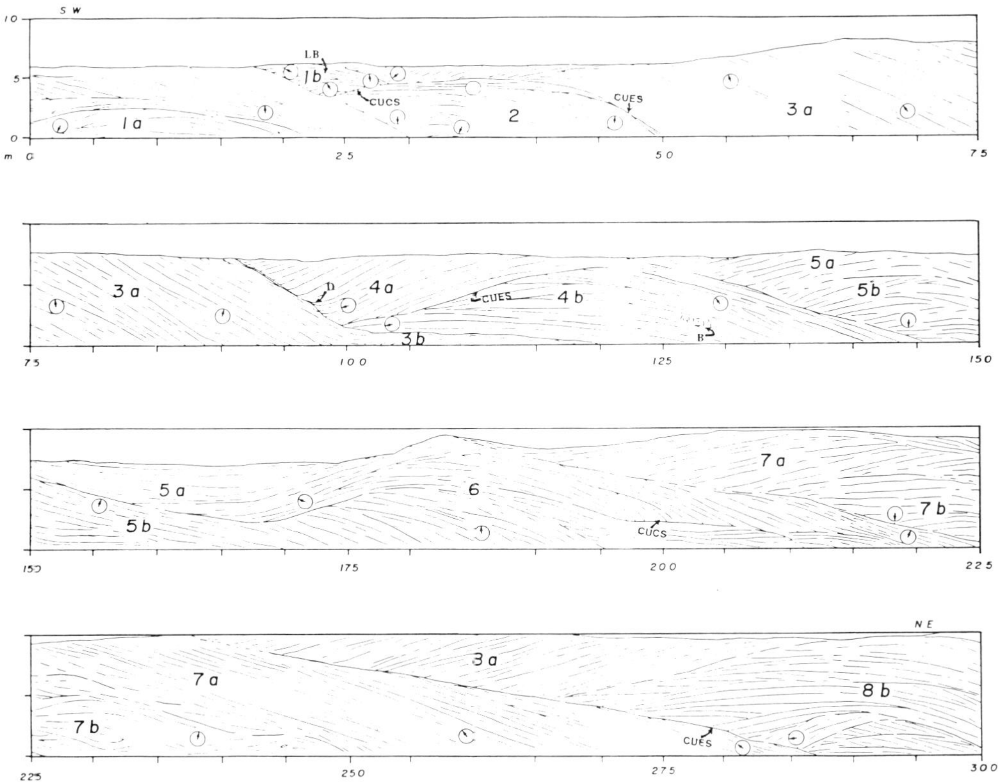

FIG. 2. Sketch of 300 m section in the Pleasley–Shirebrook railway cutting.

mutual contacts in which the bedding planes of one sandwave terminate abruptly on the upper bounding surface of the other. South-west dipping beds preferentially abut against the NE flanks of other bedforms. Sharp angular junctions between bedforms and the surfaces over which they migrate were also reported by Harms *et al.* (1978) from the subaqueous carbonate dunes of Isla Mujeres. Examples of abrupt contacts occur between bedforms 1 and 2, 4 and 5, 6 and 7, 7 and 8 (Fig. 2). In several places, the upper bounding surface truncates the internal laminae of the underlying set, a feature indicative of contemporaneous erosion. The uniformity of depositional styles suggests that uniform conditions existed during the formation of the sandwaves; the time interval between subsequent depositional episodes was probably short, perhaps of the order of several seasons or years. Contemporaneous erosion (CE) surfaces are present between bedforms 3 and 4 (Fig. 2) and are also shown in Fig. 3(a). The lowest deposits of a sandwave commonly drape the irregularities of that below ('D' in Figs 2 and 3a). Exceptionally, adjacent bedforms are laterally linked and the bounding surfaces may be followed, without break, across inter-wave areas. Examples of this type of contact occur between bedforms 5 and 6 (Fig. 2) and at the exposure depicted in Fig. 5(a,b).

The abrupt contacts, contemporaneous erosion surfaces and drape structures in these bedforms suggest that the sandwaves underwent some degree of stabilization or hardening of their top surfaces between episodes of sediment transport. Stabilization of carbonate grains may be accomplished by subtidal algal mats (Bathurst 1967) and early diagenetic hardening of submarine carbonate 'mounds' was noted in the Straits of Florida (Neumann *et al.* 1977). The rapid and widespread formation of Recent oolitic hardgrounds up to 30 cm thick in high-energy shoal environments has been documented by Harris (1978) and Dravis (1979) in different areas in the Bahamas. In places, organisms or currents have excavated the crusts, exposing unlithified sediment; where organisms were responsible for the excavation, original lamination has been destroyed by bioturbation. The patchy distribution of bioturbated zones within the sandwaves of the Sprotbrough Member (Fig. 3a) may have a similar explanation.

Lobate bedforms

Along the Pleasley–Shirebrook railway cutting it is possible to reconstruct the three-dimensional shape of some sandwaves by matching the two exposed faces. Bedforms 1b (Fig. 2, top) and other unfigured sandwaves can be projected from one side of the cutting to the other. In these examples the structure exposed on the SE face is larger and higher up the face than that on the NW face. Bedding plane azimuths converge and dips

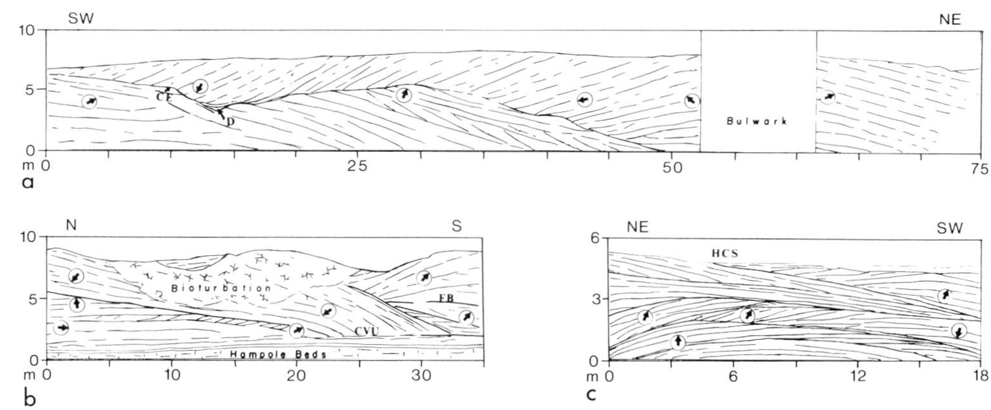

FIG. 3. Sketches of field sections: (a) Maltby railway cutting; (b) New Micklefield Quarry; (c) River Wharfe cliffs, Boston Spa. Symbols used are as in Fig. 2.
Keys to symbols for internal structures used in Figs 2 and 3:

B—bioturbation,
CE—evidence of contemporaneous erosion,
CUCS, CVU—convex-up cross-stratification,
CUES—convex-up erosion surface,
D—drape structures,
FB—fan bedding,
HCS—hummocky cross-stratification,
LB—lobate bedform,
MF—marine fossils.

Encircled arrows show the azimuths of the dips of cross beds. Numbers denote individual sandwaves (see Table 1).

increase towards the NW; this suggests that the structures are lobate, with noses projecting towards the NW. Another lobate bedform is almost wholly exposed in a small quarry at Creswell Crags; its geometry is similar to that of the structures in the railway cutting, though its nose projects towards the SW. The lobate bedforms probably originated as washover lobes such as those described by Ball (1967), Deery & Howard (1977), Leatherman *et al.* (1977) and Fisher & Stauble (1978). Commonly associated with the lobate structures are cross-strata with convex-up profiles. These unusual structures originate nowadays when storm surges supercharged with sediment break over a barrier system, rapidly depositing sediment from both suspension and traction into quieter waters (Kaldi & Allen 1986).

Hummocky cross-stratification

Hummocky cross-stratification is seen in the Sprotbrough Member (Figs 3c and 4b) at several exposures. This structure was first described and illustrated by Harms (1975, p. 87) from the Upper Cretaceous Gallup Sandstone of the Rocky Mountains (Fig. 4a). The essential characteristics of hummocky cross-stratification, both in the Sprotbrough Member and the Gallup Sandstone are: the lower bounding surfaces of sets are erosional and commonly planar, dipping at angles of up to 15°, or are curved reactivation surfaces (*sensu* Collinson 1970); laminae above the erosional set boundaries are parallel or subparallel to that surface; bedding can systematically thicken or thin laterally within a set so that the traces on a vertical section are wedge-shaped; the dip directions of erosional set boundaries and of the overlying laminae are scattered (see dip azimuths of Fig. 3c).

These characteristics suggest that the depositional surface was scoured into low hummocks and shallow swales with no regular orientation; this topography was then mantled by laminae of material swept over the hummocks and into the swales. Hummocky cross-stratification has not been reproduced in flume tanks nor its formation observed directly in recent environments, so that its hydrodynamic interpretation is speculative. However, Harms (1975) suggested that it results from strong surges of varied direction that are

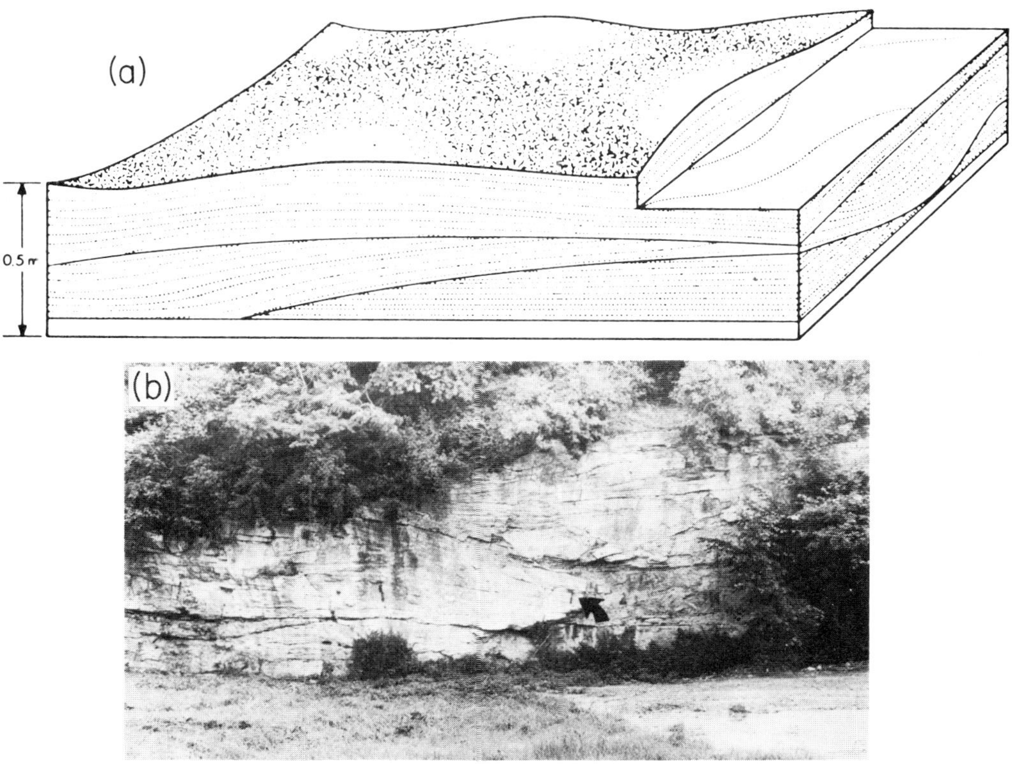

FIG. 4. (a) Illustration of hummocky cross-stratification from Harms (1975). (b) Hummocky cross-stratification in the Sprotbrough Member, North Deighton (NGR SE 391503). Hammer (arrow) is 30 cm.

generated by relatively large storm waves. Subsequently, Hamblin & Walker (1979), Bourgeois (1980), Kreisa (1981) and Wright & Walker (1981) have used hummocky cross-stratification as an indicator of storm-generated nearshore marine conditions.

Storms producing huge waves from different directions can occur in a shallow sea such as the Zechstein, especially in areas that have subaqueous bedforms (sandwaves) producing submarine relief (Johnson 1978). The hydrodynamic conditions necessary for the formation of hummocky cross-stratification are envisaged as analogous to those encountered in the Western Approaches to the English Channel during the Fastnet sailing race of the summer of 1979 (Page *et al.* 1979). Investigations have shown that the force-ten gales that swept the area caused a build-up of high seas on the shallow continental shelf over which the race was sailed; the presence of the Labadie Bank, a topographic high on the shelf floor, caused vicious waves to crash together from different directions, capsizing many yachts and claiming several lives.

Fan bedding

Fan bedding is common in the sandwaves. This style of cross-bedding is characterized by sets whose bounding surfaces change in dip and, in places, azimuth, across the structure (Figs 3b and 5c). The steepest dips of fan-shaped sets occur within the lowermost beds; higher beds have progressively shallower dips. The uppermost beds in a fan-shaped set commonly dip in a direction opposite to those in the lowest beds. Individual beds contain no internal laminae; they thin systematically from the toeset toward the topset of a unit. This is similar to the wedge-shaped beds found in hummocky cross-strata, and indeed, the structures may be genetically related. The major difference in the two types of bedforms is that hummocky cross-strata lie on erosional, sloping, or curved lower bounding surfaces, while many fan-bedded sets rest on surfaces that are non-erosional sub-horizontal planes.

The origin of fan bedding may be related to the lateral migration of sandwaves. As the bedform migrates, areas of negative relief on its edges tend to be filled first; this is accomplished by aggradation off the leeside of the bedform and filling of the depression until an equilibrium base level is reached. Each differently oriented bedding surface marks the position of the leading edge of the shifting bedform during a depositional hiatus. A process which may be compared with this is the formation of longitudinal cross-bedding on the point bars of meandering channels; in this setting, deposition takes place on the convex sides of the point bars, so producing packets of inclined beds (Reineck & Singh 1975, p. 89). In places, because of the shifting direction of meandering channels, several packets of inclined beds dipping in different directions may be deposited above one another. A problem with comparing this process with that responsible for producing fan bedding in the sea is that longitudinal cross-bedding requires channelized flows, which are ordinarily absent on the seafloor. However, the presence of substantial relief on the seafloor could locally control or channel currents and Harms *et al.* (1978) reported currents between sandwaves flowing parallel to the avalanche faces. A major part of a point bar sequence is deposited during floods, especially when water starts to subside (Leopold *et al.* 1964). Similarly, in the marine environment, the waning stages of a storm might produce fan bedding; sediment moved in a storm on the shallow seafloor is carried in suspension, but is rapidly dropped as the surge diminishes (Harms 1975). Each depositional episode would produce one entire bed of a set, thus accounting for the lack of distinct laminae within beds of fan-shaped sets.

Bioturbation

Many of the sandwaves contain trace fossils, which are commonly preserved as discrete structures within the rock. These ichnofossils are found at parting surfaces (Fig. 6a) or within beds, passing through and destroying primary sedimentary features (Fig. 6b). The traces found on bedding planes appear to be systematically constructed, convex epirelief structures (Hallam 1975) that meander across the sediment surface. They are probably *Pascichnia* (feeding trails) or *Repichnia* (locomotion traces) (terms from Seilacher 1964a,b); the former interpretation is preferred because the traces cover a considerable area of the sediment without reworking previously formed structures; this probably reflects the organism's grazing or its search for food. Such traces are made today mainly by mobile deposit-feeding epibenthos that feed at or near the sediment surface. Thin sections through this type of ichnofossil in the sandwaves reveal no obvious petrographic difference between the structure of the fossil and that of the host sediment.

The other common biogenic structure found within the sandwaves comprises horizontal and vertical branching and interconnected three-dimensional networks of cylindrical tunnels or shafts of constant diameter. The walls are preferentially dolomitized and on weathered surfaces

FIG. 5. (a,b) Mansfield, old quarry; continuous (laterally linked) contact between two adjacent bedforms. (c,d) Fan bedding; scale is 15 cm. (e) Pleasley–Shirebrook railway cutting; drape contact between two bedforms. (f) Pleasley–Shirebrook railway cutting; abrupt contact between two bedforms. (g) Section through sandwave, Farnham (SE 353602); note contemporaneous erosion surface (arrow). Fence posts are 1.3 m high.

stand out in relief from the surrounding rock (Fig. 6b). This preferential dolomitization may have resulted from the former presence of mucus-impregnated sediment around the burrow (Kaldi, *this volume*); such secretions are commonly used by organisms to line their burrows. These structures are probably *Domichnia* (dwelling burrows) or *Fodichnia* (feeding burrows); the former is more likely because of the preferentially dolomitized linings. Burrow linings are used as 'retaining

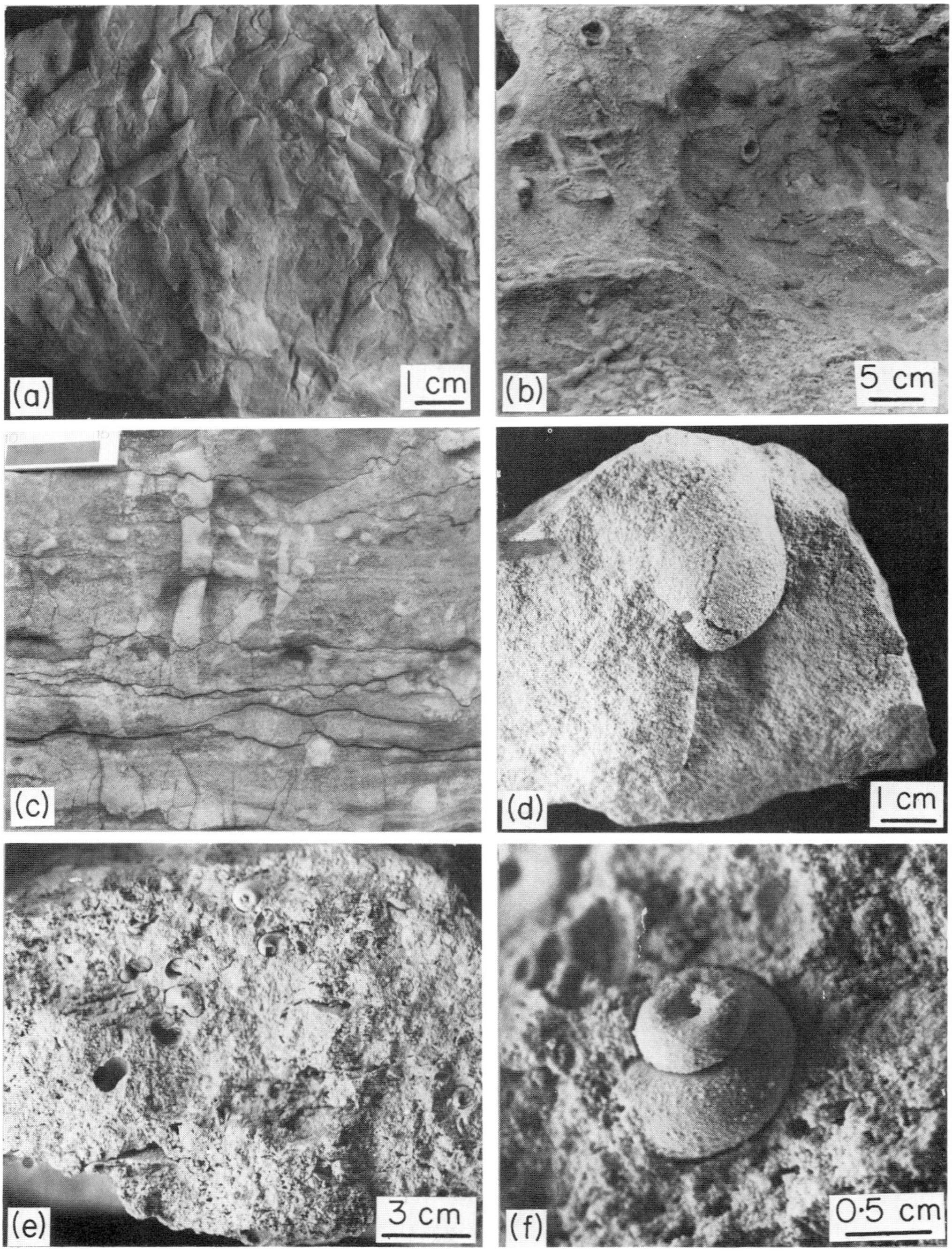

FIG. 6. (a) Trace fossils on bottom surface of bedding plane; probably Pascichnia (feeding trails). Markland Grips (SK 509750). (b) Horizontal and vertical traces; probably Domichnia (dwelling burrows). Note preferential preservation of burrow walls related to dolomitization and reticulate burrow pattern on left. New Micklefield Quarry (SE 445325). (c) Vertical trace fossils, without preferentially dolomitized walls; probably Fodichnia (feeding burrows). Hazelmere Farm (SK 513747). (d) Internal mould of bivalve, probably *Schizodus obscurus* (Sowerby). Bulwell Quarry (SK 533453). (e) Internal moulds of indeterminate gastropods from the Sprotbrough Member. (f) Details of (e) J. Pattison (pers. comm.) suggests that this is *Yunnania? helcine* (Schlotheim). Toulston Grange Quarry (SE 466427).

walls' in recent dwelling structures (Simpson 1975, p. 44). Recent *Domichnia* are produced chiefly by sessile or semi-sessile endobenthos such as suspension feeders, though some are made by predators such as the tube-dwelling polycheate *Diopatra* (Howard & Frey 1975). The animals may spend most or all of their life cycle within the burrow, as do burrowing shrimps and some other crustaceans (Frey 1978). The distribution of this type of ichnofossil in the Sprotbrough Member is patchy; zones of intense bioturbation are surrounded by, and within a few decimetres pass laterally into, undisturbed sediment (Fig. 6c).

The possibility that these trace fossils are the casts of plant roots is rejected for several reasons: (1) the networks formed by the traces are interconnected in grid patterns, whereas root traces are generally crowded and irregular (Boyd 1975) and commonly assume either a distributary pattern with progressively smaller branches or are made up of a thick, central rhizopore with many root hairs; (2) nearly every trace examined in this study is of constant thickness, whereas most plant roots taper; (3) plant roots in aeolian dune environments tend to be very extensive either laterally or vertically (Glennie & Evamy 1968) whereas the trace fossils in the Sprotbrough Member commonly have equally well developed horizontal and vertical components; (4) some traces of organic carbon would be expected from roots but none has been detected in these structures.

Marine fossils

Several fossils of marine invertebrates were found in the sandwaves (Fig. 6d–f). These include the bivalves *Bakevellia binneyi* (Brown) and *Schizodus obscurus* (Sowerby), the brachiopod *Horridonia* and gastropods. Dolomitization has obliterated all internal structures and the organisms are preserved only as interior moulds; most of them are complete and not in mutual contact, suggesting minimal transport.

Palaeocurrents

Data for palaeocurrent interpretation were analysed by Kaldi (1980a) and include more than 3800 measurements from more than 50 localities. Current roses of cross-bedding dip azimuths were produced using a program described by Parks (1974). The outcrop area of the Sprotbrough Member was divided into northern, central and southern sectors (Fig. 1). Palaeocurrent roses from Knaresborough Gorge in the north (Fig. 7) and Creswell Crags in the south (Fig. 8) are included here because they typify the range of cross-bedding styles in their respective sectors.

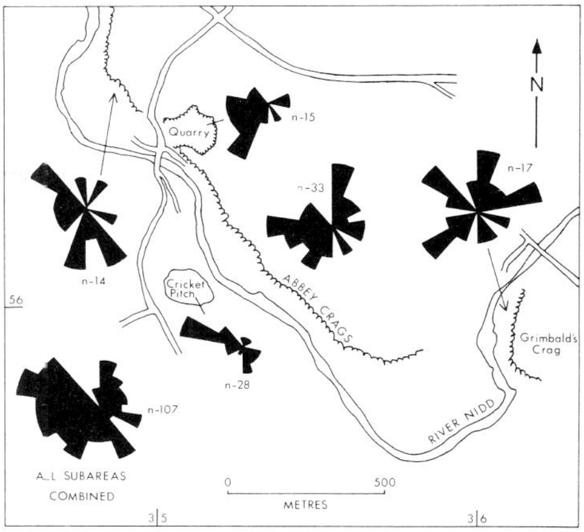

FIG. 7. Palaeocurrent roses, Knaresborough.

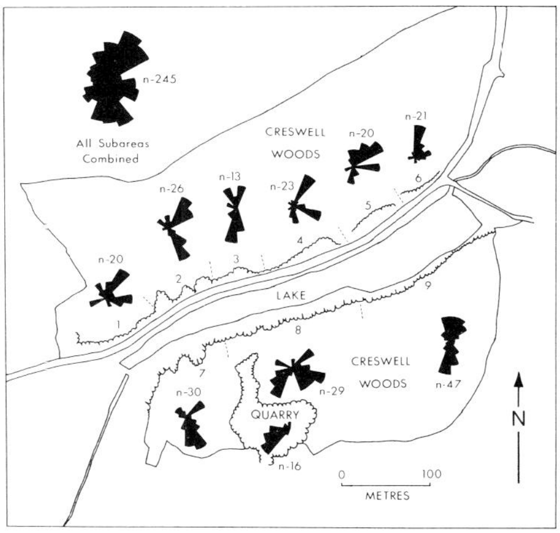

FIG. 8. Palaeocurrent roses, Creswell Crags.

General observations

Azimuths of cross-bedding vary greatly throughout the study area. In the central and northern sectors they are generally more randomly orientated than in the southern sector. In the central and northern sectors, exposures near each other commonly have cross-bedding azimuths that dip in widely different directions; in contrast, several adjacent localities in the southern sector have similar palaeocurrent orientations. In the north, local palaeocurrent indicators at Knaresborough (Fig. 7) display different orientations within a few tens of metres. In the south, the combined local dip readings at Creswell Crags are roughly aligned towards the NE and SE (Fig. 8). Current roses from most northern and central sectors are predominantly polymodal; those from the south are commonly unimodal or bipolar with east and E–W dipping modes. In addition, small-scale oscillation ripples with roughly NNE–SSW trending crests are present in the southern sector; such ripples have not been found in the central or northern sectors.

Discussion

The randomness of palaeocurrents of the central and northern sectors of the study area may be indicative of a particular depositional environment. An open shelf or broad shallow sea is one of the few environments where such randomness could be expected. This setting is characterized by a range of factors that control the dispersal of sediments in non-confined flows (Klein 1965). These factors are likely to change periodically, causing differences in the current orientations. Currents driven both by predominant winds and waves and by storm surges sometimes flow independently of palaeoslope of shoreline. For this reason, marine shelf sandstones have regional cross-beds that variously dip offshore and downslope (Michelson & Dott 1973), oblique to slope (Adams 1970) or parallel to slope (Potter & Pettijohn 1977, p. 143).

Current rose modal patterns vary in shallow marine sands, especially in areas with a sandwave bottom topography. Sandwaves in the St Peter's Sandstone, an Ordovician shallow sea deposit, have polymodal azimuths suggesting that they were formed without tidal or other flow-constraining influences (Pryor & Amaral 1971; Dott & Roshardt 1972).

The predominant eastward-dipping unimodal and E–W dipping bipolar palaeocurrents of the southern sector suggest that it was influenced by a combination of tidal and deltaic processes because: (1) tidal influence produces bipolar cross-bedding in sandwaves of the North Sea (Reineck 1963); (2) tidal flows become bipolar as they approach shore-lines (McCave 1971); and (3) the only shallow marine setting where seaward-dipping unidirectional current indicators are characteristic is the deltaic environment (Klein 1965; Potter & Pettijohn 1977, p. 142). South of Nottingham the shoreline of the Zechstein Sea swings abruptly eastward (Edwards 1951; Smith 1971). The palaeocurrent azimuths in the south are consistent with this shoreline trend if they are interpreted as coast-parallel patterns produced by nearshore tidal systems. Tides affected the southern sectors preferentially because of the shape of the shoreline. By analogy, in the North Sea of western Europe, the shoreline off Denmark swings westward at the Elbe estuary and tidal energies near this bend in the coast increase sharply as a result of wave concentration (Elliot 1978, p. 144). Similarly, even if only minimal tides were active in the Zechstein, if funnelled into a coastal re-entrant, they would have been strong enough to transport sediment. The sediment ripples, oriented NNE–SSW, probably were formed by wind-generated waves travelling normal to this trend.

Origin of sandwaves

The origin of sandwaves may be related to seafloor topography. Various sources of energy, such as winds, tides, storms and earthquakes, give rise to waves in the deeper parts of a basin. Trains of these waves move in all directions; those that move into shoaling waters are increasingly distorted until they break or translate to form currents (Ball 1967, p. 557). The constriction caused by shallowing or bottom topography, initiates and directs currents that transport sediments. Once sandwaves are established in response to control by regional bottom topography, they become part of the local relief and exert an influence on both their own features and those of the sediments around them. Thus, the evolution of sandwaves and their internal structures is part of a continuing cause-and-effect process. The problem is of initiating the structures.

Residual hills on the sub-Zechstein unconformity (Smith 1974b) probably provided the topographic relief on which sandwaves were initiated in the north of the study area. In the south, pre-Permian topographic control has not been observed. However, it is possible that reefs in the underlying Wetherby Member, though draped by the Hampole Beds, in places provided sufficient relief to initiate sandwave formation.

Conclusions

Large-scale sandwaves characterize the oolite shoal facies of the Sprotbrough Member of the Cadeby Formation in the Yorkshire Province. Sandwaves at many exposures in the study area are oriented with their long axes NW–SE, indicative of predominant currents normal to this direction (Harms *et al.* 1978). A predominant current from the NE is suggested by the preferential abutment of SW-dipping beds on the NE flanks of bedforms; this was caused by migrating sandwaves that climbed up the backs of others. Occasional storms from the SE produced spillover lobes oriented towards the NW. Storms were also responsible for the formation of internal structures such as hummocky cross-stratification, fan bedding and contemporaneous erosion surfaces. The sandwaves were inhabited by mainly soft-bodied organisms that caused locally intensive bioturbation.

The style and orientation of bedding in the sandwaves was controlled primarily by their migration (generally shoreward) near the margins of the Zechstein Sea. Cross-strata in the south of the study area are more consistently oriented than their northern counterparts, suggesting that these may have been essentially tide-dominated structures. Longshore currents and deltaic processes also influenced sediment dispersal in the south. Bedforms in the central and northern sectors are extremely diverse in form and orientation, suggesting that sediment was distributed by a number of shallow marine processes; storm-driven currents and circulating currents that crossed the Zechstein from many directions could all have been contributing factors. In many areas, sandwave topography controlled local currents by channelling or refracting flows. The exact manner in which the sandwaves originated is not clear, but one possibility is that they were initiated where existing eminences on the Zechstein Sea floor caused a concentration of wave energy.

ACKNOWLEDGMENTS: This paper is based on the results of postgraduate research at the University of Cambridge; many thanks go to Peter Friend, Ian Fairchild and Phil Allen for constructive criticism and advice. Jack Pattison and Colin Forbes aided in fossil identification; Nick Winkworth helped with computing; Dave Bursill did the darkroom magic. Special thanks go to Denys Smith of the Institute of Geological Sciences for introducing me to this project and guiding me through it.

A substantial part of this work was financed by research grants from Steetley Minerals (UK) Ltd, Clare Hall, the Cambridge Philosophical Society, and Sigma Xi.

References

ADAMS, R.W. 1970. Loyalhanna Limestone—cross-bedding and provenance. In: FISHER, G.W., PETTIJOHN, F.J., REED, J.C. (JR) & WEAVER, K.N. (eds) *Studies of Appalachian Geology*, pp. 83–100. Wiley (Interscience), New York.

ANDERSON, E.J. 1967. *Paleoenvironments of the Coeymans Formation (Lower Devonian) of New York State*. Unpublished PhD dissertion, Brown University.

BALL, M.M. 1967. Carbonate sand bodies of Florida and the Bahamas. *J. sedim. Petrol.* **37,** 556–91.

BATHURST, R.G.C. 1967. Subtidal gelatinous mats, sand, stabilizer and food, Great Bahama Bank. *J Geol., Chicago*, **75,** 736–8.

BOURGEOIS, J. 1980. A transgressive shelf sequence exhibiting hummocky stratification: the Cape Sebastian Sandstone (Upper Cretaceous), southwestern Oregon. *J. sedim. Petrol.* **50,** 681–702.

BOYD, D.W. 1975. False or misleading traces. In: FREY, R.W. (ed.) *The Study of Trace Fossils*, 65–83. Springer-Verlag, New York.

COLLINSON, J.D. 1970. Bedforms of the Tana River, Norway. *Geogr. Annlr.* **52A,** 31–55.

DEERY, J.R. & HOWARD, J.D. 1977. Origin and character of washover fans on the Georgia coast, U.S.A. *Trans. Gulf-Cst Ass. geol. Socs* **28,** 259–71.

DOTT, R.H. (JR) & ROSHARDT, M.A. 1972. Analysis of cross-stratification orientation in the St. Peter Sandstone in southwestern Wisconsin. *Bull. geol. Soc. Am.* **83,** 2589–96.

DRAVIS, J. 1979. Rapid and widespread generation of Recent oolitic hardgrounds on a high energy Bahamian platform, Eleuthera Bank, Bahamas. *J. sedim. Petrol.* **49,** 195–207.

EDWARDS, W. 1951. *The Concealed Coalfield of Yorkshire and Nottinghamshire*, 3rd edn. *Mem. geol. Surv. G.B.* London.

ELLIOT, T. 1978. Clastic shorelines. In: READING, H.G. (ed.) *Sedimentary Environments and Facies*, 143–77. Blackwell Scientific Publications, Oxford.

FISHER, J.S. & STAUBLE, D.K. 1978. Impact of Hurricane Belle on Assateague Island washover. *Geology*, **5,** 765–8.

FREY, R.W. 1978. Behavioural and ecological implications of trace fossils. In: BASAN, P.B. (ed.) *Trace Fossil Concepts, Soc. econ. Paleont. Mineral. Short Course No. 5, Oklahoma*, 43–66.

FUZESY, L.M. 1970. *Petrology of the Lower Magnesian Limestone in the neighbourhood of Selby, Yorkshire.* Unpublished PhD thesis, University of Cambridge.

GLENNIE, K.W. & EVAMY, B.D. 1968. Dikaka: plants and plant root structures associated with eolian sand. *Palaeogeogr. Palaeclim. Palaeoecol.* **4,** 77–87.

HALLAM, A. 1975. Preservation of trace fossils. In: FREY, R.W. (ed.) *The Study of Trace Fossils*, 55–65. Springer-Verlag, Berlin.

HALLEY, R.B. & HARRIS, P.M. 1979. Fresh-water cementation of a 1,000-year-old oolite. *J. sedim. Petrol.* **49,** 969–88.

HAMBLIN, A.P. & WALKER, R.G. 1979. Storm-dominated shallow marine deposits: the Ferni-Kootenay (Jurassic) transition, southern Rocky Mountains. *Can. J. Earth Sci.* **16,** 1673–90.

HARMS, J.C. 1975. Stratification and sequence in prograding shoreline deposits. In: *Depositional Environments as Interpreted from Primary Sedimentary Structures and Stratification Sequences. Soc. econ. Paleont. Mineral. Short Course No. 2, Tulsa*, 81–102.

HARMS, J.C., CHOQUETTE, P.W. & BRADY, M.J. 1978. Carbonate sand waves, Isla Mujeres, Yucatan. In: WARD, W.C. & WEIDIE, A.E. (eds) *Geology and Hydrogeology of Northeastern Yucatan*: pp. 60–84. Field Trip Guide, New Orleans Geological Society, Field Trip 2, 1978 Annual Convention GCAGS/SEPM.

HARRIS, P.M. 1978. Holocene marine cemented sands, Joulters Ooid Shoal, Bahamas. *Trans. Gulf-Cst Ass. geol. Socs* **28,** 175–84.

HARRIS, P.M. 1979. Facies anatomy and diagenesis of a Bahamian ooid shoal. *Sedimenta VII*. Comparative Sedimentology Laboratory, University of Miami.

HINE, A.C. 1977. Lily Bank, Bahamas: history of an active oolite sand shoal. *J. sedim. Petrol.* **47,** 1554–83.

HINE, A.C., WILBER, R.J. & NEUMANN, A.C. 1981. Carbonate sand bodies along contrasting shallow bank margins facing open seaways in northern Bahamas. *Bull. Am. Ass. Petrol. Geol.* **65,** 261–90.

HOWARD, J.D. & FREY, R.W. 1975. Characteristic physical and biogenic sedimentary structures in Georgia estuaries. *Bull. Am. Ass. Petrol. Geol.* **57,** 1169–84.

JOHNSON, H.D. 1978. Shallow siliciclastic seas. In: READING, H.G. (ed.) *Sedimentary Environments and Facies*, 207–59. Blackwell Scientific Publications, Oxford.

KALDI, J. 1980a. *Aspects of the sedimentology of the Lower Magnesian Limestone (Permian) of eastern England.* Unpublished PhD thesis, University of Cambridge.

—— 1980b. The origin of nodular structures in the Lower Magnesian Limestone (Permian) of Yorkshire, England. In: FUCHTBAUER H. & PERYT, T.M. (eds) *The Zechstein Basin with Emphasis on Carbonate Sequences. Contr. Sedim.* **9,** 45–60.

KALDI, J. & ALLEN, P.A. 1986. The origin of convex-up cross-stratification: examples from the Permian Lower Magnesian Limestone of eastern England. *Sedimentology*, submitted.

KLEIN, G. DE V. 1965. Dynamic significance of primary structures in the Middle Jurassic Great Oolitle Series, southern England. In: MIDDLETON, G.V. (ed.) *Primary Sedimentary Structures and their Hydrodynamic Interpretation. Spec. Publ. Soc. econ. Paleont. Mineral., Tulsa*, **12,** 173–91.

KREISA, R.D. 1981. Storm-generated sedimentary structures in subtidal marine facies with examples from the Middle and Upper Ordovician of southwestern Virginia. *J. sedim. Petrol.* **51,** 823–48.

LAPORTE, L.F. 1969. Recognition of a transgressive carbonate sequence within an epeiric sea: Helderberg Group (Lower Devonian) of New York State. In: FRIEDMAN, G.M. (ed.) *Depositional Environments in Carbonate Rocks. Spec. Publ. Soc. econ. Paleont. Mineral., Tulsa*, **14,** 98–118.

LEATHERMAN, S.P., WILLIAMS, A.T. & FISHER, J.S. 1977. Overwash sedimentation associated with a large-scale northeaster. *Mar. Geol.* **24,** 109–21.

LEOPOLD, L.B., WOLMAN, M.G. & MILLER, J.P. 1964. *Fluvial Processes in Geomorphology*. Freeman, San Francisco.

MCCAVE, I.N. 1971. Sandwaves in the North Sea off the coast of Holland. *Mar. Geol.* **10,** 199–222.

MICHELSON, P.C. & DOTT, R.H. (JR) 1973. Orientation analysis of trough-cross-stratification in Upper Cambrian sandstones of western Wisconsin. *J. sedim. Petrol.* **43,** 784–94.

NEUMANN, A.C., KOFOED, J.W. & KELLER, G.H. 1977. Lithotherms in the Straits of Florida. *Geology*, **5,** 4–10.

PAGE, F., NICHOLLS, J. & CROOKSTON, P. 1979. How 34 fathoms doomed the Fastnet. *Observer Mag.* 25 November, 26–45.

PARKS, J.M. 1974. Paleocurrent analysis of sedimentary crossbed data with graphic output using three integrated computer programs. *J. int. Ass. math. Geol.* **6,** 353–62.

POTTER, P.E. & PETTIJOHN, F.J. 1977. *Paleocurrents and Basin Analysis*, 2nd edn. Academic Press, New York.

PRYOR, W.A. & AMARAL, E.J. 1971. Large-scale cross-stratification in the St. Peter Sandstone. *Bull. geol. Soc. Am.* **82,** 229–44.

PURSER, B.H. 1975. *Sedimentation et diagenese précoce des series carbonatées du Jurassique moyen de Bourgogne.* Unpublished DSc thesis, l'Université de Paris-sud.

REINECK, H.E. 1963. Sedimentgefuge im bereich der sudlichen Nordsee. *Abh. senckenb. naturforsch. Ges. Frankfurt*, **505,** 1–38.

REINECK, H.E. & SINGH, I.B. 1975. *Depositional Sedimentary Environments.* Springer-Verlag, New York.

SEILACHER, A. 1964a. Sedimentological classification and nomenclature of trace fossils. *Sedimentology*, **3,** 253–6.

—— 1964b. Biogenic sedimentary structures. In: IMBRIE, J. & NEWELL, N.D. (eds) *Approaches to Paleocology*, 296–313. Wiley, New York.

SIMPSON, S. 1975. Classification of trace fossils. In: FREY, R.W. (ed.) *The Study of Trace Fossils*, pp. 39–54. Springer-Verlag, Berlin.

SMITH, D.B. 1968. The Hampole Beds—a significant marker in the Lower Magnesian Limestone of Yorkshire, Derbyshire and Nottinghamshire. *Proc. Yorks. geol. Soc.* **36,** 463–77.

—— 1970. The palaeogeography of the English Zechstein. In: RAU, J.L. & DELLWIG, L.F. (eds) *Third Symposium on Salt*, **I.** 20–3. Northern Ohio Geological Society, Cleveland.

—— 1971. The Lower Permian in the British Isles. In: FALKE, H. (ed.) *The Continental Lower Permian of Western Europe*, 1–33. Brill, Leiden.

—— 1974a. Permian. In: RAYNER, D.H. & HEMINGWAY, J.E. (eds) *The Geology and Mineral Resources of Yorkshire*, 115–44. Yorkshire Geological Society, Leeds.

—— 1974b. The stratigraphy and sedimentology of Permian rocks at outcrop in north Yorkshire. *J. Earth Sci. Leeds Geol. Ass.* **8,** 365–86.

—— 1976. The Upper Permian sabkha sequence at Quarry Moor, Ripon, Yorkshire. *Proc. Yorks. geol. Soc.* **40,** 639–52.

—— 1980. The evolution of the English Zechstein Basin. In: FUCHTBAUER, H. & PERYT, T.M. (eds) *The Zechstein Basin with Emphasis on Carbonate Sequences. Contr. Sedim.* **9,** 7–34.

SMITH, E.G., RHYS, G.H. & EDEN, R.A. 1967. Geology of the country around Chesterfield, Matlock and Mansfield. *Mem. geol. Surv. G.B.* 430 pp.

WRIGHT, M.E. & WALKER, R.G. 1981. Cardium Formation (U. Cretaceous) at Seebe, Alberta—storm-transported sandstones and conglomerates in shallow marine depositional environments below fair-weather wave base. *Can. J. Earth Sci.* **18,** 795–809.

J. KALDI, Shell Canada Limited, Calgary Research Centre, PO Box 2506, Calgary, Alberta T2P 3S6, Canada.

The diagenetic history of Cadeby Formation carbonate (EZ1 Ca), Upper Permian, eastern England

G.M. Harwood

SUMMARY: The Upper Permian Cadeby Formation (formerly the Lower Magnesian Limestone) crops out from North Yorkshire to Nottinghamshire in eastern England. The Cadeby Formation comprises shallow water carbonates within which penecontemporaneous, eogenetic, mesogenetic and teleogenetic diagenetic events can be defined. Penecontemporaneous events included facies-controlled dolomitization and evaporite calcitization. Pervasive dolomitization, largely independent of facies, was eogenetic. Late eogenetic to early mesogenetic events were dominated by replacive and pore-filling anhydrite crystallization. Both pervasive dolomitization and subsequent anhydrite formation occluded porosity. Other mesogenetic events were of local effect only. Anhydrite to gypsum transformation, together with evaporite dissolution and carbonate leaching, dominate the later near-surface, or telogenetic, events.

The diagenetic history of the EZ1 carbonate, the Cadeby Formation (formerly the Lower Magnesian Limestone), at outcrop is dominated by dolomitization which has destroyed much of the original sediment texture. This is in contrast to the texture-preserving dolomitization reported by Clark (1980) in Dutch and Danish Zechstein carbonates. Depositional carbonate sediment characteristics are readily overprinted by different diagenetic regimes, or diagenetic environments (Choquette & Pray 1970; Longman 1980). Dolomitization is thus only one of a series of diagenetic changes which took place, and may still be taking place, in the Cadeby Formation at outcrop. This paper attempts to define the major processes that occurred within each diagenetic regime and to link these processes with the evolution of porosity and permeability within the carbonates. Terms used for major diagenetic environments are based on those of Choquette & Pray (1970) except that a distinction is made here between processes that act immediately after deposition, i.e. during *penecontemporaneous* diagenesis, and those that act slightly later in the sediment history but still within the zone of 'influence of processes that operate from the surface, or depend for their effectiveness on proximity to the surface' (Choquette & Pray 1970, p. 219), i.e. during *eogenetic* diagenesis (Fig. 1).

Carbonate facies

The Cadeby Formation (Lower Magnesian Limestone; EZ1 Ca) (Fig. 2) crops out from Catterick in North Yorkshire to south of Nottingham (Sedgwick 1829) within the Yorkshire province of the English Zechstein sequence (Smith 1970); the formation averages 30–40 m in thickness at crop. Two major shallowing-upwards sequences are recognized within near-shore, intertidal and supratidal carbonates (Smith 1974a,b, 1981; Harwood 1981). The lower, Wetherby Member (Lower Subdivision, EZ1a Ca) contains varied depositional environments. Mudstones and wackestones with some thin beds of displacive evaporites (Harwood 1980) are interpreted to have formed in restricted, or semi-restricted, lagoons (Harwood 1980, 1981). Islands and peninsulas of resistant pre-Permian strata, together with peloid grainstone shoals with large-scale cross-stratification formed the barriers to these lagoons. Seaward of these barriers, and in near-shore and shoreline environments where barriers and lagoons were absent, peloid and skeletal grainstones and packstones were deposited (Smith 1974a,b; Harwood 1981). These sediments contain small-scale cut-and-fill structures and ripple cross-laminations, with early lithification and storm reworking, characterizing deposition in shallow water environments. Away from the shoreline, bryozoan-algal patch reefs (Smith 1981), with flanking peloid-pisoid grainstones and packstones, formed on the open carbonate shelf. In the south of the outcrop, quartz-silt- and clay-rich carbonate sandstones evidence continental, possibly partially fluvial, input.

A minor discontinuity, the Hampole Discontinuity (Smith 1968), followed by a bed of peloid packstones with fenstrae (the Lower Hampole Dolomite), together mark the culmination of the shallowing-upwards sequence within the Wetherby Member (Smith 1968). Further evidence of shallowing is from stromatolitic cappings on some bryozoan patch reefs (Smith 1981). The two clay beds within the Hampole Beds (Smith 1968) are thought to be of terrestrial derivation.

From HARWOOD, G.M. & SMITH, D.B. (eds), 1986, *The English Zechstein and Related Topics*, Geological Society Special Publication No. 22, pp. 75–86.

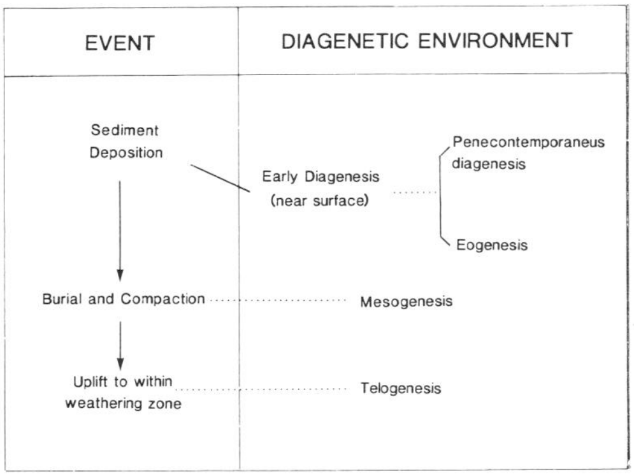

FIG. 1. Major diagenetic environments (modified from Choquette & Pray 1970).

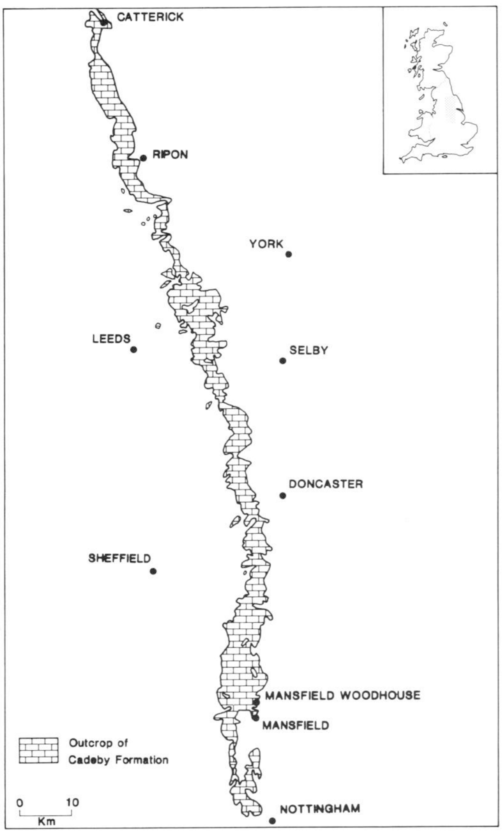

FIG. 2. Outcrop map of the Cadeby Formation showing localities mentioned in the text.

The upper, Sprotbrough Member (Upper Subdivision; EZ1b Ca) contains, at its base, evidence of rapid inundation of the carbonate platform. Sediments within the Sprotbrough member are dominated by peloid grainstone shoals throughout most of the outcrop area (Smith 1974a,b; Kaldi 1980 and *this volume*; Harwood 1981). The relationship between these shoals and highly bioturbated wackestones present in the north of the area (Ripon to Catterick, North Yorkshire) has yet to be ascertained. Inter-shoal skeletal peloid grainstones/packstones show dispersed bioturbation, and dense bioturbated patches also occur within shoal grainstones. In the south, lower energy bioturbated carbonates contain silt- and fine sand-grade quartz grains. Shallowing upwards near the top of the Sprotbrough Member is evidenced by isolated areas of cryptalgal laminates, subtidal through supratidal (Smith 1974a,b; Kaldi 1980; Harwood 1981); one such sequence was interpreted as a fossil sabkha (Smith 1976). In the Doncaster area the reworking of lithified crests of the grainstone shoals resulted in a thin veneer of breccia being deposited on the unlithified surfaces between shoal crests.

The Cadeby Formation is overlain by the Edlington Formation (the former Permian Middle Marls), which comprises poorly exposed reddened sediments and sediments with lenticular sandstones and some thin beds of dolostone. Evidence of the former existence of thick evaporites within this formation comes from foundering of the overlying carbonate, the Brotherton Formation (the Upper Magnesian Limestone; EZ3 Ca) and from numerous subsidence pits and hollows throughout much of the outcrop belt (Tute 1870; Fox-Strangways 1908; Cooper, *this volume*). Gypsum remains present at outcrop north of Ripon, North Yorkshire where it directly overlies the Cadeby Formation (James *et al.* 1981).

The abrupt change of lithology from carbonates to interbedded clastics and evaporites at the top of the Cadeby Formation indicates a marked increase in evaporation of the marine Zechstein waters which continued to extend over the carbonate platform (Harwood 1981). Eastwards progradation of continental-derived clastics commenced at the same time.

Penecontemporaneous diagenetic events

Penecontemporaneous diagenetic events are only recognizable in a few areas. The events include early calcite cementation, dolomitization and evaporite calcitization. In the Mansfield area (Fig. 2) lagoon carbonates near the base of the Wetherby Member contain dolomitized clay-rich (5–15%) carbonates intimately interbedded with clay-poor (< 3%) primary limestone wackestones (location; SK 532648). The primary limestone is preserved both within individual beds and/or within burrows. Burrow margins have sharp (Fig. 3a) or gradational contacts with the surrounding dolomite. Burrow morphology exhibits little evidence of compaction (Fig. 3a) whereas the surrounding dolomitized, clay-rich carbonates contain microstylolites showing some degree of compaction around the cemented burrows after dolomitization. Calcite cementation thus preceded dolomitization. This early calcite cementation probably restricted migration of the dolomitizing fluids, resulting in the preferential dolomitization of the clay-rich carbonate matrix. The small (< 20 μm) subhedral to euhedral dolomite rhombs (Fig. 3a) are distinct from most other dolomite crystal morphologies within the Cadeby Formation.

Micrite or lime muds are commonly selectively dolomitized (Davies 1979; Choquette & Steinen 1980). These authors infer that this is the result of three major factors; sediment composition (by possible analogy to modern carbonates, a mix-

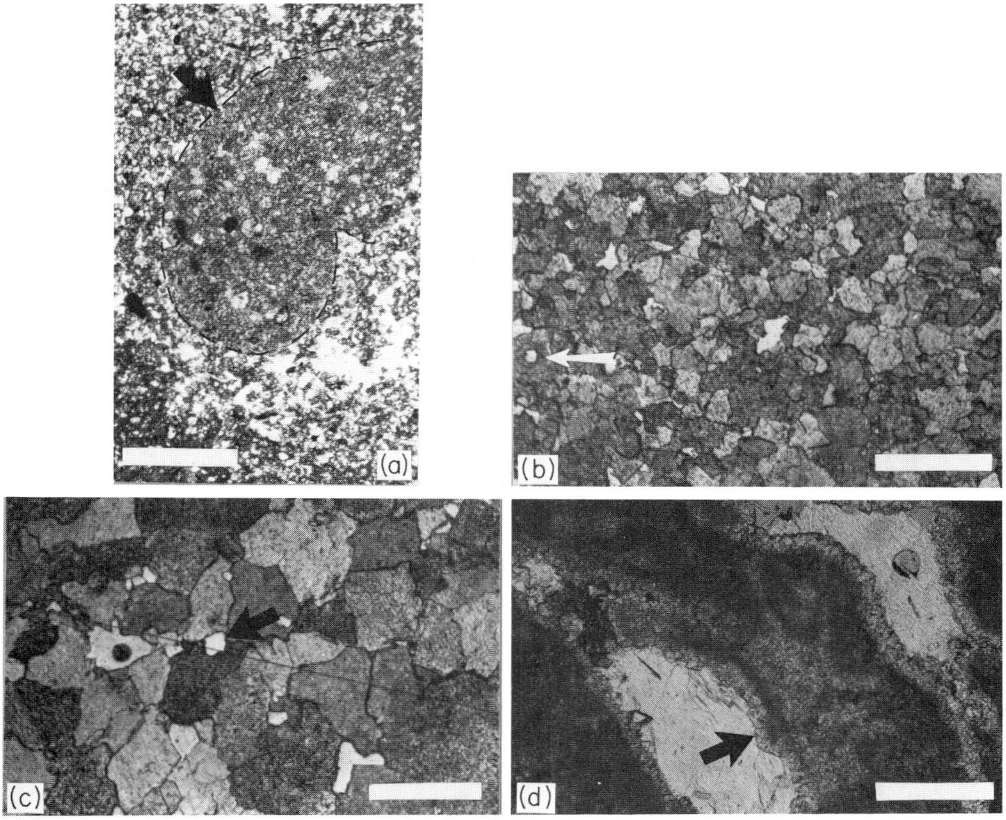

FIG. 3. Penecontemporaneous and eogenetic dolomites.
(a) Limestone-dolomite contact (arrowed) at burrow margin (location: Mansfield Woodhouse, SK 532648). Burrow is of calcite (stained with Alizarin Red S); surrounding sediment consists of clay-rich penecontemporaneous dolomite. Burrow outline indicates lack of compaction before cementation by calcite. Bar is 500 μm.
(b) Anhedral eogenetic dolomite crystals from near top of peloid grainstone shoal (location: Shirebrook, SK 530687). Dolomitization was fabric-destructive. Faint ghost of former ooid (arrowed) has quartz nucleus. Bar is 500 μm.
(c) Anhedral eogenetic dolomite crystals 1 m below that in (b). Dolomitization was fabric-destructive; ghosts are rare. Photomicrograph is of same magnification as (b). Percolating groundwaters created leached porosity at crystal boundaries during telogenesis (arrowed). Bar is 500 μm. (d) Penecontemporaneous dolomite cements (arrowed) coating clasts in breccia near top of Sprotbrough Member (location: New Edlington brick works quarry, SK 532986). Clasts were lithified and fractured before inclusion in breccia. Porosity occluded by later calcite cements. Bar is 500 μm

ture of aragonite and high magnesian calcite), small particle size (and consequent large surface area to volume ratio), and the catalytic action, during dolomitization, of clay mineral and organic contents. These factors all reflect the influence on diagenesis of original sediment composition, and hence the depositional environment.

Near the base of the Wetherby Member in the Ripon area (Fig. 2) mudstones and wackestones contain beds of displacive evaporite nodules, now calcitized, and hopper halite casts (location; SE 304650 and SE 257812). The host carbonates comprise anhedral dolomite crystals (<10–40 μm). Early bacterial reduction of the evaporite sulphate, with the possible influence of meteoric-derived pore fluids, caused penecontemporaneous calcitization of the anhydrite nodules. This calcitization demonstrably post-dates dolomitization (Harwood 1980; Harwood & Coleman 1983). Penecontemporaneous dolomitization is further indicated by the small dolomite crystal size and the association with displacive evaporites.

Fauna in these dolomites comprises only scattered foraminifers whereas adjacent carbonates contain bryozoans, brachiopods and more abundant foraminifera. This decrease in fauna reflects the increased salinity evidenced by the displacive evaporites and halite casts. The host mudstones and wackestones are indicative of low energy depositional environments, in contrast with the higher energy grainstones and packstones elsewhere in the Wetherby Member (Smith 1974a,b; Harwood 1981). Both low energy environment and periodic salinity increases suggest that these carbonates were deposited within a lagoon. These sediments contain no sedimentological evidence of exposure nor of deposition in supratidal or intertidal environments.

Pore fluids in the lagoon bottom sediments therefore fluctuated between 'normal' marine during recharge periods, and hypersaline. Mixed, or meteoric/mixed, groundwater may also have had some influence, as the margins of the meteoric groundwater lens would have migrated between periods of recharge and restriction/evaporation. Evidence suggests, however, that dolomitization was linked to evaporite formation. Anhydrite growth within the unlithified lagoon bottom sediments would have depleted the pore fluids in calcium, thus enhancing their Mg/Ca ratio and promoting dolomitization (Fig. 4b). Whole rock X-ray fluorescence (XRF) analyses of these wackestones show strontium contents statistically higher (>100 ppm) than other Cadeby Formation dolomites (<80 ppm) providing further evidence for an evaporative origin.

Penecontemporaneous dolomitization in these

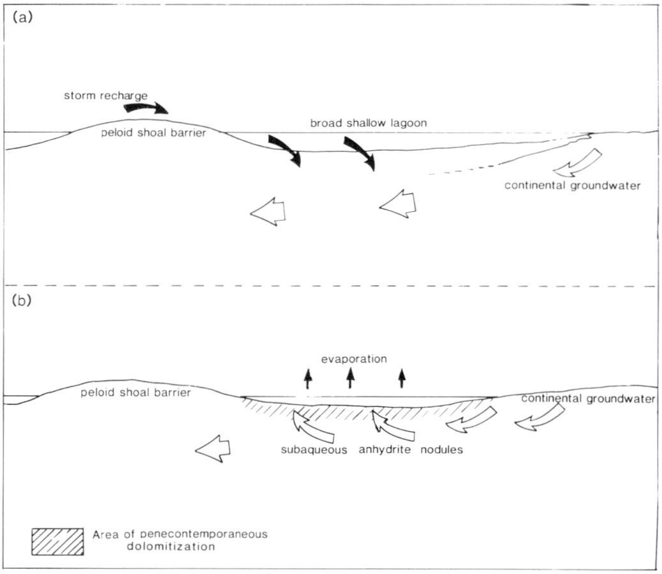

FIG. 4. Penecontemporaneous dolomitization.
(a) Storm recharge and/or temporary breaching of lagoon barrier with water of near normal salinity. Colonization of lagoon by restricted faunal species.
(b) Evaporation of lagoon waters and restriction of lagoon, leading to hypersaline waters and pore fluids. Dolomitization of carbonates in the lagoonal sediments with some growth of displacive evaporites. Limited fauna within lagoon.

sediments was, therefore, controlled by a complex hydrological system. This system was governed by three major factors; storm recharge, evaporation and meteoric groundwater penetration. However, unlike that within the hydrological systems of sabkhas (Hsü & Siegenthaler 1969; McKenzie *et al.* 1980), here penecontemporaneous dolomitization was entirely subaqueous. This dolomitization was also confined within the lagoon bottom wackestones and mudstones and thus is facies-restricted.

Eogenetic diagenetic events

Eogenetic dolomites

Medium to coarsely crystalline dolomite crystals (50–1500 μm) characterize most of the Cadeby Formation at outcrop. In grainstones and packstones preservation of pre-dolomitization textures is commonly poor; Smith (1981, fig. 1b) illustrates a rare example of good fabric retention within a patch reef. Areas of selective fabric retention in grainstones include burrow margins and some irregular zones in the shallow subsurface. Within these areas where original grains are visible, two factors indicate that dolomitization was eogenetic: (i) relic peloids have point or short-planar grain-to-grain contacts and (ii) dolomitization took place in both grains and extant intergranular cements. There is no evidence of pressure solution either between grains or between the intergranular cements visible. Dolomitization, therefore, took place before substantial burial of the carbonates. There is also no evidence for volume reduction during dolomitization; rather the anhedral interlocking dolomite crystals cut across both grains and intergranular areas (Fig. 3b,c), showing that either volume-for-volume replacement occurred or that there was a volume increase with consequent occlusion of any remaining intergranular porosity. Dolomite crystal size is independent of original grain size; on the flank of one ooid grainstone shoal crystal size ranges from averages of 120 μm (Fig. 3b) to 400 μm (Fig. 3c) within 1 m. Original grain composition cannot be determined because of poor fabric preservation. Pelmouldic porosity, perhaps an indication of meteoric water flushing of former ooid shoals, is absent except on the crests of isolated grainstone shoals. From this it can be concluded that, although the original grains were probably ooids, there was no regional meteoric influence before dolomitization.

There is also no evidence for different episodes of eogenetic dolomitization between the two members of the Cadeby Formation. Texture preservation is not better in one member than the other; dolomite crystal form does not vary between members more than it does within members. Furthermore, the clays of the Hampole Beds have been shown not to be aquicludes (Aldrick 1978), a conclusion which is supported by their rare occurrence in the subsurface and their irregularity in some exposures. This reinforces the hypothesis of one major episode of eogenetic dolomitization which, in general, occluded porosity. Differences in crystal size and variations in texture destruction were controlled by local, and as yet undetermined, factors. Areas of penecontemporaneous dolomitization appear little changed by this later dolomitization event.

Direct comparison of Holocene models of dolomitization with dolomitization of the Cadeby Formation is tenuous because of the lack of large modern evaporite basins. Three major applicable models of dolomitization have been proposed: (1) seepage reflux (Adams & Rhodes 1960; Deffeyes *et al.* 1965, (2) supratidal models, both arid (McKenzie *et al.* 1980) and humid (Shinn *et al.* 1978) and (3) mixed water models (Hanshaw *et al.* 1961). Holocene supratidal dolomites are penecontemporaneous and are facies restricted. Further, hydrological models which cause dolomitization within these facies cannot be extended to invoke large-scale dolomitization of varied facies. Thus only the mixed water and seepage reflux models, both eogenetic, are applicable to dolomitization of mixed facies as found in the Cadeby Formation. However, several points argue against mixed water dolomitization being the major causative model here:

(1) Substantial drawdown of sea-level and subsequent meteoric influx after sedimentation of the Cadeby Formation would have led to the development of leached fabrics and evidence of early meteoric water related diagenesis throughout, at least, the Sprotbrough Member. It was noted above that there is evidence for meteoric diagenesis only in isolated localities and there is not the regional evidence of early diagenetic leaching that would be expected from a large sea-level fall.

(2) Substantial drawdown would have had to precede deposition of the overlying evaporites (up to 25 m still exposed in North Yorkshire; James *et al.* 1981) or these would have been lost by dissolution. However, correlation with subsurface core data indicates that bedded evaporites were present overlying much of the Cadeby Formation and in places these are intimately interbedded with dolomitized grainstones. This can be taken as an indication that drawdown was not on a large scale and that the Cadeby Formation carbonates remained under water after their

deposition, albeit that the water was substantially more saline.

(3) Pore fluids in the Cadeby Formation were subsequently supersaturated in calcium sulphate (Harwood 1980). This supersaturation took place after porosity occlusion by dolomitization (e.g. Fig. 3b,c) but before final compaction (see below). It is perhaps unlikely that mixed meteoric/marine fluids could effectively dolomitize a carbonate formation after lithification but before subsequent supersaturation with calcium sulphate was reached during deposition of the overlying evaporites. On the other hand, the continued diagenetic effects of increasingly more saline fluids could account for both dolomitization and calcium sulphate supersaturation.

In addition, in order to dolomitize totally the original Cadeby Formation carbonates, pore fluid movement should be considered. Meteoric-derived fluids possess low Mg/Ca ratios and low total Mg^{2+} content (Back & Hanshaw 1970; Badiozamani 1973) compared to hypersaline fluids (Butler 1969; McKenzie *et al.* 1980). Dolomitization requires the provision of sufficient Mg^{2+} ions combined with the flushing of excess Ca^{2+} ions throughout the carbonates. Dolomitization by mixed meteoric/marine fluids would necessitate flow through the existing pore water volumes of about three orders of magnitude greater than for hypersaline fluids. This, combined with lack of petrographic evidence of meteoric flushing and the presence of high salinities both before and after dolomitization, suggests that Cadeby Formation dolomitization may have resulted from seepage reflux.

Figure 5 illustrates the possible situation during dolomitization of the Cadeby Formation. The Zechstein Sea extended over the carbonate platform with only isolated areas of exposure. High evaporation rates combined with low precipitation to produce hypersalinity and evaporite precipitation. The resultant brines were of high ionic strength and had high Mg/Ca ratios. Combined gravity and chemical potentials resulted in the downwards and basinwards seepage of the brines, dolomitizing the carbonates as they moved. With a constant, or slowly rising, sea-level, continued evaporation produced a constant replenishment of dense hypersaline fluids which in turn refluxed through the sediments. Thus brine quantities were not limited by a restricted lagoon (e.g. Adams & Rhodes 1960) but were continuously supplied over a period. Dolomitization, therefore, is thought to result from large-scale seepage reflux through the Cadeby Formation carbonates. Clark (1980) came to the conclusion that dolomitization was caused by seepage reflux in both Dutch and Danish Zechstein carbonates.

Sucrose dolomites

Sucrose dolomites, with peloid ghosts preserved in their cores (Fig. 6a), are found only south of Mansfield Woodhouse (Fig. 2), in near-shore carbonates with up to 30% silt- and fine sand-grade quartz. These sucrose dolomite crystals have limpid outer zones (Fig. 6a), originally partially ferroan. Later alteration, however, has meant that zones within these rims have been subject to dedolomitization and now contain opaque goethite aggregates (e.g. Harwood 1983b, fig. 2). The remaining limpid rims are of near-stochiometric dolomite. By contrast, the crystal

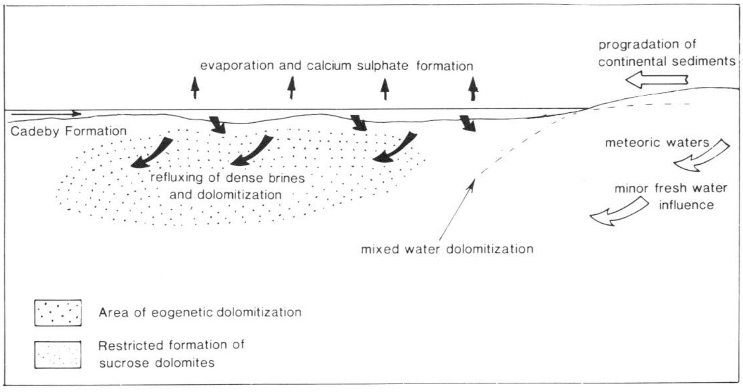

FIG. 5. Eogenetic dolomitization by large-scale seepage reflux. The hypersaline Zechstein Sea covered most of the carbonate platform, precipitating the overlying evaporites. Dense hypersaline brines with high Mg/Ca ratios flowed downwards through the carbonates, causing pervasive dolomitization of varied facies. Minor sucrose dolomites in nearshore areas were formed by interaction with meteoric groundwaters.

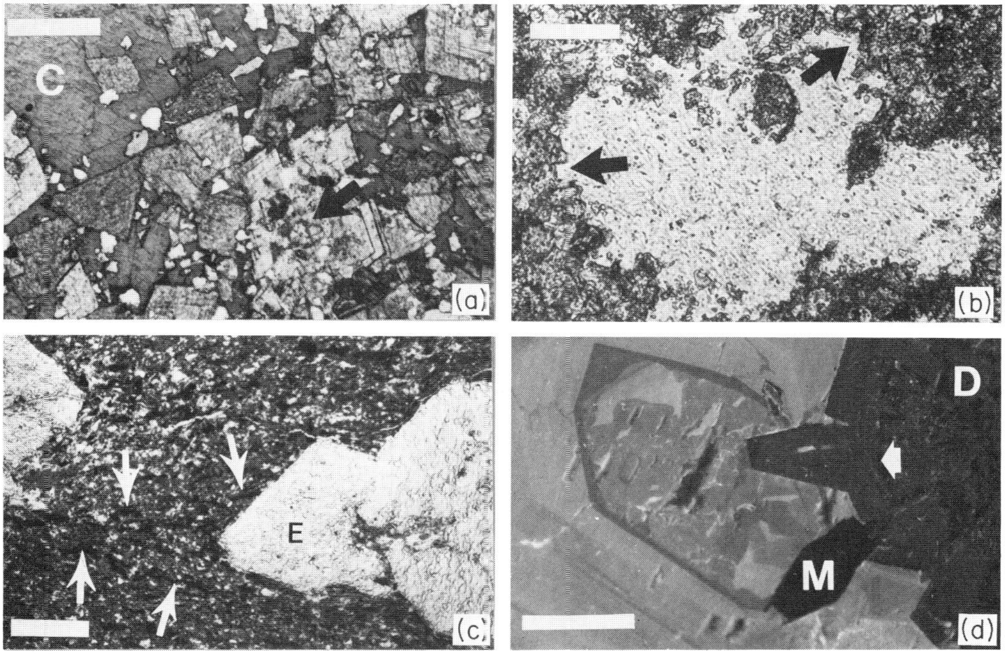

FIG. 6. Dolomites and evaporites.

(a) Sucrose dolomite rhombs with centres containing peloid ghosts (arrowed) (location: Kirkby in Ashfield, SK 570561). Outer limpid zone of crystals has no peloid ghosts. Thin section shows up to 10% silt-sized quartz grains and later calcite (C) occluding porosity (stained with Alizarin Red S). Bar is 500 μm.

(b) Corroded dolomite crystals in gypsum (location Wistow Wood Borehole, SE 567356, 201.2 m). Rectilinear outlines (arrowed) show that replacive evaporite was originally anhydrite, now replaced by gypsum during telogenesis. Bar is 100 μm.

(c) Replacive evaporite (E) in dolomite wackestone (location: Bank End Borehole, SK 706972, 474 m). Laminae (arrowed) are partially truncated by evaporite, but also show compaction after evaporite formation. Evaporite mineral is now gypsum after anhydrite. Bar is 500 μm.

(d) Cathodoluminescence photograph showing zoned dolomite crystals growing into void (top left) (location: Wistow Wood Borehole, SE 567356, 249 m). Zoned dolomites grow on original dolomicrospar (D), commencing with dark, non-luminescent zone (arrowed). Marcasite (M) also grew into void which was later partially occluded by zoned calcite crystals. Bar is 250 μm.

centres are both more calcian and contain more inclusions, both fluid and solid (Fig. 6a); their crystal chemistry is similar to the eogenetic dolomites throughout the rest of the Cadeby Formation (Harwood 1981). It is possible, therefore, that the crystal centres are of similar origins (i.e. seepage reflux). However, continued eastwards progradation of continental sediments (Fig. 5) over these nearshore carbonates curtailed the influence of hypersaline brines after limited dolomite nucleation. The later, more iron rich, outer zones of these crystals may be evidence of the presence of meteoric pore fluids during and after progradation of the continental clastics. As such, these rims may have resulted from continued growth of dolomite within mixed hypersaline-meteoric pore fluids on the landwards extremity of the carbonate shelf. The resultant dolomites had high intercrystalline porosities, now occluded by calcite (Fig. 6a).

Replacive and pore-filling evaporites

Evidence for the former existence of replacive and pore-filling evaporites is found throughout the Cadeby Formation except in the south, in the Nottingham area. In cores from the shallow subsurface of the Selby area, Yorkshire, the top few metres of Sprotbrough Member dolomitized grainstones have been almost entirely replaced by massive anhydrite, now hydrated to gypsum (Harwood 1981). At outcrop, irregular clusters of former replacive evaporite crystals extend up to several centimetres in diameter within the dolomites of differing facies. These replacive clusters cross-cut sedimentary lamination (Harwood 1980, fig. 3) and their frequency of occurrence is not facies-dependent. In the subsurface, inclusions of corroded dolomite crystals within the evaporites (Fig. 6b) demonstrate that evaporite replacement post-dated dolomitization. Rectilinear outlines (Fig. 6b) indicate that replacement was originally by anhydrite rather than gypsum. Shallow subsurface replacive evaporites now, however, comprise gypsum because of hydration of anhydrite: in exposed carbonates, evaporite dissolution has resulted in vugs. Fowler (1944), Raymond (1953) and Jones (1969) noted replacement anhydrite and/or gypsum in subsurface English Zechstein carbonates and Clark (1980) and Clark & Tallbacka (1980) have documented replacive anhydrite in subsurface Zechstein carbonates of the Netherlands and Denmark. Pore-filling evaporites in the subsurface occluded the slight intercrystalline porosity that remained after dolomitization. Pore-filling evaporites were probably more extensive than replacive evaporites, but little evidence of them now remains in the exposed carbonates.

Although replacive evaporites have been shown to cross-cut sedimentary lamination, core samples through clay- and quartz silt-rich wackestones in the Cadeby Formation show some compaction around replacive evaporites (Fig. 6c). Evaporite replacement, therefore, is concluded to have taken place after eogenetic dolomitization (Fig. 6b) but prior to final compaction (Fig. 6c), probably during late eogenesis. The calcium sulphate-supersaturated pore fluids, which were the agents of carbonate replacement by evaporites, probably were derived from the overlying evaporites because (1) replacive evaporite clusters are more numerous near the top of the Cadeby Formation and (2) there are no replacive evaporites in the southern, Nottingham, area where the Cadeby Formation is overlain by evaporite-free continental clastics of the Edlington Formation. Sulphur isotope analyses of replacive evaporites within the Cadeby Formation are indistinguishable from those of overlying bedded evaporites (Harwood & Coleman 1983), reinforcing this hypothesis. Clark (1980) suggested that, during the transition from gypsum to anhydrite, the crystal water expelled became saturated in calcium sulphate by partial dissolution of the evaporites. The resultant fluid was forced out from the evaporites in all directions. Some was forced downwards, and basinwards, through the carbonates, causing some replacement. The fluid flow direction thus paralleled that followed earlier by the brines which effected dolomitization by large-scale seepage reflux.

Mesogenetic diagenetic events

The combined cover of later Permian and Mesozoic sediments probably never exceeded 1 or 2 km in the outcrop area. The burial effects of chemical compaction are therefore slight, with perhaps 5% reduction due to pressure solution (D.B. Smith, pers. comm. 1981). The most common expression of chemical compaction is low amplitude stylolites which follow bedding surfaces in peloid grainstones and packstones. Stylolitization does increase with depth of burial eastwards into the North Sea basin, but no integrated study of stylolite density or amplitude within these carbonates has yet been made.

Replacement of carbonates by anhydrite probably continued during early mesogenesis on a regional scale. Other events during mesogenesis are much more restricted in extent but do include two examples of secondary porosity formation:
(1) Anhedral dolomite crystals within irregular, small subsurface zones were progressively replaced by euhedral ferroan dolomites with good intercrystalline porosity and permeability. In some parts of these zones, dedolomitization of the ferron dolomites resulted in a relatively soluble ferroan-magnesian-calcite. Further enhancement of porosity occurred where this calcite was leached. The delineation of these zones of ferroan alteration is a subject of current research.
(2) Near the base of the Cadeby Formation, fluid flow from the overpressured aquifers of the Basal Permian Sands caused dissolution of former displacive evaporites. Voids thus formed have since been partially occluded by a centripetal dolomite cement with later calcite and marcasite (Fig. 6d). This leaching has created a highly porous and permeable facies-controlled zone within the subsurface, which, in the Selby area, is filled with overpressured saline groundwaters.

Telogenetic diagenetic events

Telogenesis was, and is, dominated by dissolution and leaching of both carbonates and evaporites. The anhydrite to gypsum transition and subsequent gypsum dissolution influence present topography. Within the carbonates, evaporite dissolution produces vuggy (Guy 1911) and intercrystalline porosities; the vugs result from the dissolution of both displacive and replacive evaporites (Guy 1911; Harwood 1980, fig. 3). Where the evaporites immediately overlying the Cadeby Formation carbonates have been leached, the resultant groundwater had a high calcium content. This water, percolating downwards, has caused dedolomitization. The resultant dedolomites are compact, low porosity, secondary limestones which contain corroded dolomite relics (Harwood 1981); they are laterally discontinuous and have very irregular thicknesses (e.g. Sedgwick 1829, plate 7, fig. 3; Harwood 1983a). Where thin carbonate beds were originally interbedded with evaporites, leaching of the evaporites has resulted in cavernous dedolomites.

Percolation of meteoric groundwaters has also caused leaching of the carbonates, with the development of intercrystalline porosity at corroded triple crystal boundaries (Fig. 3c). Late phreatic calcite cements, formed from a combination of leached carbonates and evaporites with soil-derived carbon dioxide (Harwood & Coleman 1983) were precipitated in voids (e.g. Harwood 1980, fig. 3) and partially occlude the leached porosities. There is no evidence at outcrop of leaching and calcite precipitation preceding the present (Pleistocene and Holocene uplift) groundwater pattern.

Conclusions

A summary of the diagenesis of the Cadeby Formation is presented in Fig. 7. The major conclusions are:

(1) Three major dolomite types are present in the Cadeby Formation; (a) penecontemporaneous dolomites; (b) eogenetic anhedral dolomites and (c) eogenetic sucrose dolomites.
(2) Primary limestones exist only where early calcite cementation protected the sediments from dolomitizing fluids.
(3) Penecontemporaneous dolomitization was facies-restricted; it took place in lagoonal sediments and was effected by alternating periods of lagoon recharge and evaporation.
(4) Eogenetic dolomitization, the major diagenetic event, was fabric-destructive in most places; it occluded porosity throughout the formation and was pervasive, not facies controlled. Eogenetic dolomitization was probably caused by large-scale seepage reflux of brines from the overlying hypersaline Zechstein Sea.
(5) Sucrose dolomites, with high intercrystalline porosities, were a product of dolomitization by mixed marine-meteoric waters in nearshore areas.
(6) Replacive and pore-filling evaporites were formed before final burial compaction. These evaporites efficiently occluded any remaining porosity. Replacement of carbonates by evaporites may have continued during mesogenesis.
(7) Mesogenetic diagenetic events included some stylolitization and localized secondary porosity

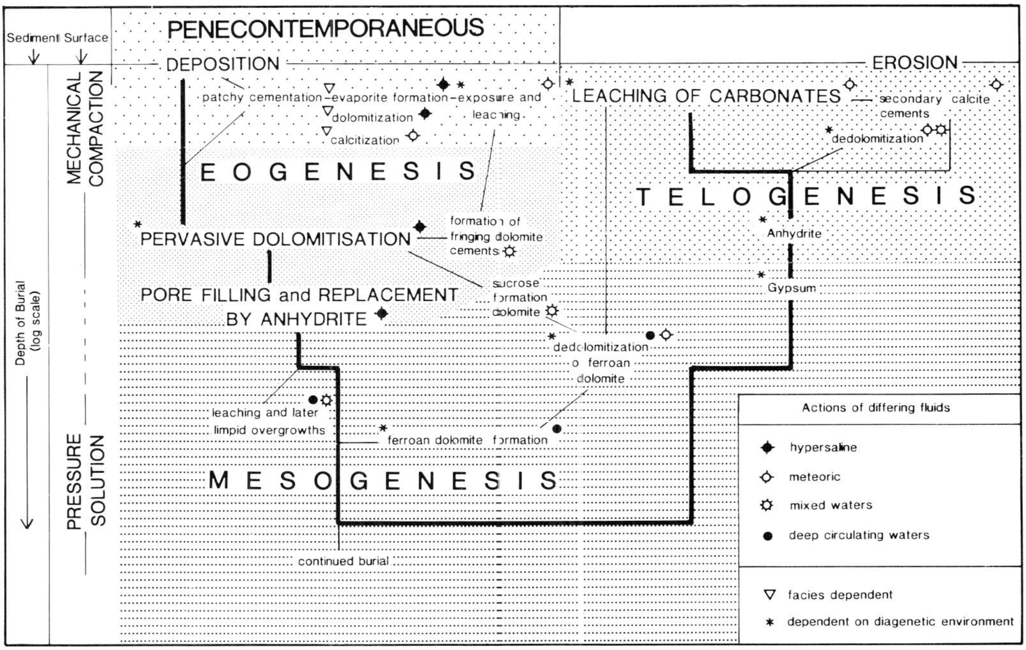

FIG. 7. Summary diagram of the diagenetic history of the Cadeby Formation.

creation through subsurface evaporite leaching, ferroan dolomite formation and some dedolomitization of ferroan dolomite.
(8) Uplift and weathering with percolation of meteoric groundwater caused the hydration of anhydrite to gypsum. Subsequent dissolution of gypsum, together with leaching of carbonates, formed vuggy and leached intercrystalline porosities. Late meteoric phreatic calcite cements partially occlude these porosities.

ACKNOWLEDGMENTS: This paper results from my PhD dissertation research at the Open University supervised and encouraged by Professor I.G. Gass and Dr G. Hornung. Dr D.B. Smith was a constant aid to field research and Dr J.A.D. Dickson helped with cathodoluminescence techniques and discussion on carbonate diagenesis. The Department of Earth Sciences at Leeds University is to be thanked for their many facilities used throughout my research. T.F. Johnstone, J.A. Pearce, J. Todd and H. Clemmy provided encouragement throughout my studies. Finally I should like to thank Barbara Delaville, Paula Reece and Liz Walton who typed various drafts of this manuscript and Dupe Duplechin, Mary Lee Eggart and Christine Jeans for their help with the diagrams.

References

ADAMS, J.E. & RHODES, M.L. 1960. Dolomitisation by seepage reflux. *Bull. Am. Ass. Petrol. Geol.* **44,** 1912–20.

ALDRICK, R.J. 1978. The hydrogeology of the Magnesian Limestones in Yorkshire between the River Wharfe and the River Aire. *Q. J. eng. Geol.* **11,** 193–201.

BACK, W. & HANSHAW, B.B. 1970. Comparisons of chemical hydrogeology of the carbonate peninsulas of Florida and Yucatan. *J. Hydrol.* **10,** 330–58.

BADIOZAMANI, K. 1973. The Dorag dolomitisation model—application to the Middle Ordovician of Wisconsin. *J. sedim. Petrol.* **43,** 965–84.

BUTLER, G.P. 1969. Modern evaporite deposition and the geochemistry of co-existing brines, the Sabkha, Trucial Coast, Arabian Gulf. *J. sedim. Petrol.* **39,** 71–9.

CHOQUETTE, P.W. & PRAY, L.C. 1970. Geologic nomenclature and classification of porosity in sedimentary carbonates. *Bull. Am. Ass. Petrol. Geol.* **54,** 207–50.

CHOQUETTE, P.W. & STEINEN, R.P. 1980. Mississippian non-supratidal dolomite, St Genevieve Limestone, Illinois Basin: Evidence for mixed-water dolomitization. In: ZENGER, D.H., DUNHAM, J.B. & ETHINGTON, R.L. (eds) *Concepts and Models of Dolomitization. Spec. Publ. Soc. econ. Paleont. Mineral., Tulsa,* **28,** 163–96.

CLARK, D.N. 1980. The diagenesis of Zechstein carbonate sediments. In: FUCHTBAUER, H. & PERYT, T.M. (eds) *The Zechstein Basin with Emphasis on Carbonate Sequences. Contr. Sedim.* **9,** 167–203.

CLARK, D.N. & TALLBACKA, L. 1980. The Zechstein of southern Denmark. In: FUCHTBAUER, H. & PERYT, T.M. (eds) *The Zechstein Basin with Emphasis on Carbonate Sequences. Contr. Sedim.* **9,** 205–31.

COOPER, A.H. (*this volume*). Subsidence and foundering of strata caused by the dissolution of Permian gypsum in the Ripon and Bedale areas, North Yorkshire.

DAVIES, G.R. 1979. Dolomite reservoir rocks: processes, controls, porosity development. In: *Geology of Carbonate Porosity. Am. Ass. Petrol. Geol. Short Course Notes,* **11,** C1–17.

DEFFEYES, K.S., LUCIA, S.J. & WEYL, P.K. 1965. Dolomitization of Recent and Plio-Pleistocene sediments by marine evaporite waters on Bonaire, Netherlands Antilles. In: PRAY, L.C. & MURRAY, R.C. (eds) *Dolomitization and Limestone Diagenesis; a Symposium. Spec. Publ. Soc. econ. Paleont. Mineral., Tulsa,* **13,** 71–88.

FOWLER, A. 1944. A deep borehole in the Cleveland Hills. *Geol. Mag.* **81,** 193–206, 254–65.

FOX-STRANGWAYS, L. 1908. Geology of the country north and east of Harrogate. *Mem. geol. Surv. G.B.* HMSO, London.

GUY, E. 1911. On the formation of cavities in the Magnesian Limestone. *Trans. Leeds geol. Ass.* **4,** 10–5.

HANSHAW, B.B., BACK, W. & DEIKE, R.G. 1971. A geochemical hypothesis for dolomitisation by ground water. *Econ. Geol.* **66,** 710–24.

HARWOOD, G.M. 1980. Calcitized anhydrite and associated sulphides in the English Zechstein First Cycle Carbonate (EZ1 Ca). In: FUCHTBAUER, H. & PERYT, T.M. (eds) *The Zechstein Basin with Emphasis on Carbonate Sequences. Contr. Sedim.* **9,** 61–72.

—— 1981. *Controls of mineralization in the Cadeby Formation (Lower Magnesian Limestone).* Unpublished PhD thesis, Open University.

—— 1983a. Former evaporites within English Zechstein First Cycle Carbonates (Abstract). *Bull. Am. Ass. Petrol. Geol.* **66,** 480.

—— 1983b. The application of cathodoluminescence in relative dating of barite mineralization in the Lower Magnesian Limestone (Upper Permian), United Kingdom. *Econ. Geol.* **78,** 224–8.

—— (*this volume*). The diagenetic history of the Cadeby Formation.

—— & COLEMAN, M.L. 1983. Isotopic evidence for U.K. Upper Permian mineralization by bacterial reduction of evaporites. *Nature,* **301,** 597–9.

—— & SMITH, F.W. (*this volume*). Mineralization in Upper Permian Carbonates at outcrop in Eastern England.

HSÜ, K.J. & SIEGENTHALER, C. 1969. Preliminary experiments on hydrodynamic movements indicated by evaporation and their bearing on the dolomite problem. *Sedimentology,* **12,** 11–25.

JAMES, A.N., COOPER, A.H. & HOLLIDAY, D.W. 1981. Solution of the gypsum cliff (Permian, Middle Marl) by the River Ure at Ripon Parks, North Yorkshire. *Proc. Yorks. geol. Soc.* **43,** 433–50.

JONES, K. 1969. *Mineralogy and geochemistry of the Lower and Middle Magnesian Limestones of County Durham*. Unpublished PhD thesis, University of Durham.

KALDI, J.G. 1980. *Aspects of the sedimentology and diagenesis of the Lower Magnesian Limestone (Permian) of Yorkshire, England*. Unpublished PhD thesis, University of Cambridge.

—— (*this volume*). Diagenetic environment of the Sprotbrough Member of the Cadeby Formation (Upper Permian) of England.

LONGMAN, M.W. 1980. Carbonate diagenetic textures from near surface diagenetic environments. *Bull Am. Ass. Petrol. Geol.* **64,** 461–87.

MCKENZIE, J.A., HSÜ, K.J. & SCHNEIDER, J.E. 1980 Movement of subsurface waters under the sabkha Abu Dhabi, U.A.E., and its relation to evaporite and dolomite genesis. In: ZENGER, D.H., DUNHAM J.B. & ETHINGTON, R.L. (eds) *Concepts and Models of Dolomitization. Spec. Publ. Soc. econ. Paleont Mineral., Tulsa*, **28,** 11–30.

RAYMOND, L.R. 1953. Some geological results from the exploration for potash in north-east Yorkshire. *Q Jl geol. Soc. London*, **108,** 283–306.

SEDGWICK, A. 1829. On the geological relations and internal structure of the Magnesian Limestone and the Lower portions of the New Red Sandstone series in their range through Nottinghamshire, Derbyshire, Yorkshire and Durham to the southern extremity of Northumberland. *Trans. geol. Soc London*, Series 2, 37–124.

SHINN, E.A., GINSBURG, R.M. & LLOYD, R.M. 1965 Recent supratidal dolomites from Andros Island, Bahamas. In: PRAY, L.C. & MURRAY, R.C. (eds) *Dolomitization and Limestone Diagenesis—a Symposium. Spec. Publ. Soc. econ. Paleont. Mineral., Tulsa*, **13,** 113–23.

SMITH, D.B. 1968. The Hampole Beds—a significant marker in the Lower Magnesian Limestone of Yorkshire, Derbyshire and Nottinghamshire. *Proc. Yorks. geol. Soc.* **36,** 463–77.

—— 1970. The palaeogeography of the English Zechstein. In: RAU, J.L. & DELLWIG, L.F. (eds) *Third Symposium on Salt*, **1,** 20–3. Northern Ohio Geological Society.

—— 1974a. Permian. In: RAYNER, D.H. & HEMINGWAY, J.E. (eds) *The Geology and Mineral Resources of Yorkshire*, 115–44. Yorkshire Geological Society, Leeds.

—— 1974b. The stratigraphy and sedimentology of Permian Rocks at outcrop in North Yorkshire. *J. Earth Sci. Leeds*, **8,** 365–86.

—— 1976. The Upper Permian sabkha sequence at Quarry Moor, Ripon, Yorkshire. *Proc. Yorks. geol. Soc.* **40,** 639–52.

—— 1981. Bryozoan-algal patch-reefs in the Upper Permian Lower Magnesian Limestone of Yorkshire, northeast England. In: TOOMEY, D.F. (ed.) *European Fossil Reef Models. Spec. Publ. Soc. econ. Paleont. Mineral., Tulsa*, **30,** 187–202.

TUTE, J.S. 1870. On certain natural pits in the neighbourhood of Ripon. *Proc. Yorks. geol. polytech. Soc.* **5,** 2–7.

G.M. HARWOOD, Department of Geology, The University of Newcastle upon Tyne, Newcastle upon Tyne NE1 7RU.

Discussion contribution by D.B. Smith

Although some Mg^{2+} was presumably already present in the Cadeby Formation, enormous additional quantities of Mg^{2+} had to be introduced and Ca^{2+} removed in order to effect full dolomitization. Several authors have speculated that such large-scale dolomitization elsewhere may have required high Mg/Ca ratios in the refluxing brines and that such ratios might have resulted from precipitation of gypsum near the landward margin of the carbonate shelf prism. Has Dr Harwood found evidence of the former presence of such gypsum or does the source of the Mg^{2+} remain a mystery?

Reply

Evidence of former primary evaporites (gypsum, anhydrite and possibly halite) has been found within the Wetherby Member of the Cadeby Formation (Harwood 1980). Former primary evaporites (dominantly gypsum) are also present in the Sprotbrough Member in North Yorkshire (Harwood 1981). The volume of these evaporites is, however, small and would not be sufficient for pervasive dolomitization of the formation, although it is adequate for facies-restricted penecontemporaneous dolomitization. There is also considerable evidence for replacive, secondary, evaporites (mostly anhydrite) throughout the formation, although these formed after the pervasive phase of dolomitization (Harwood 1981 and *this volume*).

The source of the magnesium for the pervasive dolomitization phase was, I believe, the hypersaline water within the Zechstein Sea which covered the carbonate platform and in which the westernwards extension of the Hayton (Anhydrite) Formation was being deposited. The continuation of similar fluid migration pathways, with persistent hypersalinity, after dolomitization led to patchy replacement of the dolomite by sulphates, particularly along porous zones within the formation. Sulphur isotope values indicate that this replacive sulphate is either remobilized primary Zechstein evaporite sulphate or was formed from migrating pore fluids derived directly from the hypersaline Zechstein (Harwood & Coleman 1983).

Diagenesis of nearshore carbonate rocks in the Sprotbrough Member of the Cadeby (Magnesian Limestone) Formation (Upper Permian) of eastern England

J. Kaldi

SUMMARY: The petrographic textures in the upper member of the Cadeby (Magnesian Limestone) Formation demonstrate the combined effects of original depositional setting and later diagenetic alteration of a carbonate rock.

Dolomitization was the most ubiquitous diagenetic process to affect these rocks. Dolomites formed (1) penecontemporaneously, (2) perhaps as primary precipitates, (3) as cements during early diagenesis, and (4) as locally controlled late diagenetic replacement and cement. These dolomite textures were subsequently altered by the advent of new diagenetic environments. A drop in Zechstein sea-level exposed much of the upper (Sprotbrough) member to the influence of meteoric waters. The mixing of fresh and marine waters produced solutions capable of dissolving unstable carbonate matter, dolomitizing more resistant grains and replacing earlier dolomite textures with new ones. Generally, later dolomitization events produced coarser crystal textures than the earlier.

Evaporite minerals were formed as early nodules within supratidal sediments or as late diagenetic pore-filling cements in the sub-surface. Their dissolution led to the development of considerable vuggy porosity and the addition of Ca^{2+} to ground waters. Ca^{2+}-enriched fluids calcitized surrounding dolomites to produce dedolomite.

Porosity was controlled by the changes from one diagenetic environment to another. Primary pores were modified by cementation, recrystallization and pressure-solution. Secondary porosity was created by the dissolution of minerals that were unstable in pore fluids of the later diagenetic episodes.

The carbonate rocks of the Cadeby (Magnesian Limestone) Formation of eastern England have been extensively altered by diagenesis and several stages of textural development can be recognized; these include penecontemporaneous, early diagenetic and late diagenetic episodes. Commonly, these events can be related to the original depositional setting and many of the rocks contain diagenetic fabrics peculiar to one environment However, some diagenetic environments developed independently of depositional facies and resulted in varied textures and porosity styles within otherwise similar rocks.

Geological setting

The rocks of the upper (Sprotbrough) member of the Cadeby Formation (formerly the Lower Magnesian Limestone) were deposited along the western shoreline of the Upper Permian Zechstein Sea during Cycle 1 of the English Zechstein sequence (Fig. 1). Four depositional facies (Fig. 2) have been identified from outcrops and boreholes in eastern England: (I) sabkha (arid supratidal flat), (II) lagoon (shallow-subtidal semi-restricted basin), (III) oolite shoal/barrier complex, and (IV) open marine shelf.

Rocks of the sabkha facies have been described from subsurface and outcrop by Smith (1974a,b, 1976) and Kaldi (1980a) and from the subsurface by Fuzesy (1970). These deposits are characterized by inclusion-rich microcrystalline dolomites, stromatolites, intraclasts, desiccation cracks and tepee structures.

Lagoonal deposits in the Sprotbrough Member (Smith 1974a; Kaldi 1980a) consist of horizontally laminated pellet-wackestones or small-scale cross-bedded oolite grainstones. Thin clay layers are common between horizontal bedding planes, and oscillation ripple-marks, bioturbation and centimetre-size subspherical vugs occur locally.

The oolite shoal complex is by far the most distinctive and thickest facies unit in the Sprotbrough Member. It consists of several sub-environments, including large-scale (12 m sets) cross-stratified ooid sandwaves, spillover lobes and quiet-water intershoal areas (Kaldi, *this volume*). The 12 m cross-strata set this figure as a minimum water depth (Smith 1974a).

Open marine shelf carbonates of the Cadeby Formation were discussed by Fuzesy (1970), Goossens (1973), Smith (1974a,b) and Kaldi (1980b). Rocks of this facies were formed in normal marine conditions under perhaps 50–100 m of water; they consist of alternating clay-rich and lime-rich beds that were subsequently partially dolomitized.

From Harwood, G.M. & Smith, D.B. (eds), 1986, *The English Zechstein and Related Topics*, Geological Society Special Publication No. 22, pp. 87–102.

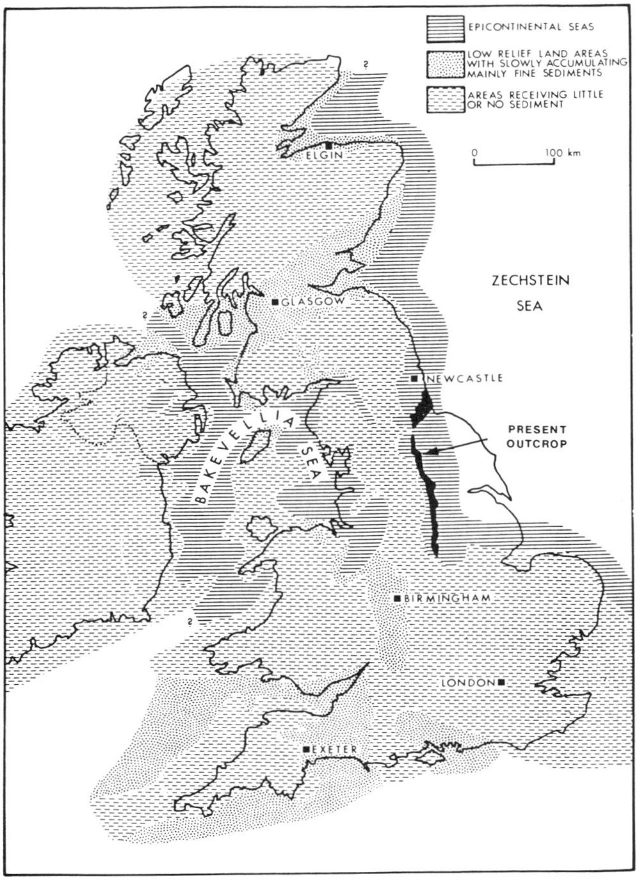

FIG. 1. Generalized Upper Permian palaeogeography in Britain, showing the present outcrop of the Cadeby Magnesian Limestone Formation (after Smith, in Pattison *et al.* 1973).

The following descriptions and discussions are grouped according to the depositional facies summarized above. Facies are identified by both their primary sedimentary features and petrographic characteristics.

Sabkha (Facies I)

Rocks of the sabkha (and supratidal flat) facies are found at only a few exposures in the study area (see Smith 1976; Kaldi 1980a, p. 20). Borehole equivalents have been described by Fuzesy (1970, p. 57; Smith 1974a,b) and were also examined in this study.

The predominant rock type of this facies is a microcrystalline groundmass of dolomite having anhedral mosaic texture: finely crystalline to mesocrystalline subhedral to euhedral dolomite rhombs are scattered throughout. Commonly, the coarser grains contain inclusions of the finer material, making the surface slightly rough (Fig. 3e). Many rhombs have corroded edges or hollow centres (Fig. 3a,b). Skeletal grains are generally absent from this unit. Dolomitized ooids, pellets and intraclasts occur rarely and exhibit a turbid microcrystalline texture. Small (<75 μm) grains of quartz and vague discontinuous laminae of unidentified brown algal-organic material are scattered sparingly throughout. In the subsurface this facies contains irregular centimetre-scale nodules of gypsum with anhydrite inclusions (Fig. 4c). Where organic-rich laminae are present, they are deformed by the gypsum nodules. Subhorizontal fractures, containing fibrous (satin-spar) gypsum emanate from many nodules. At

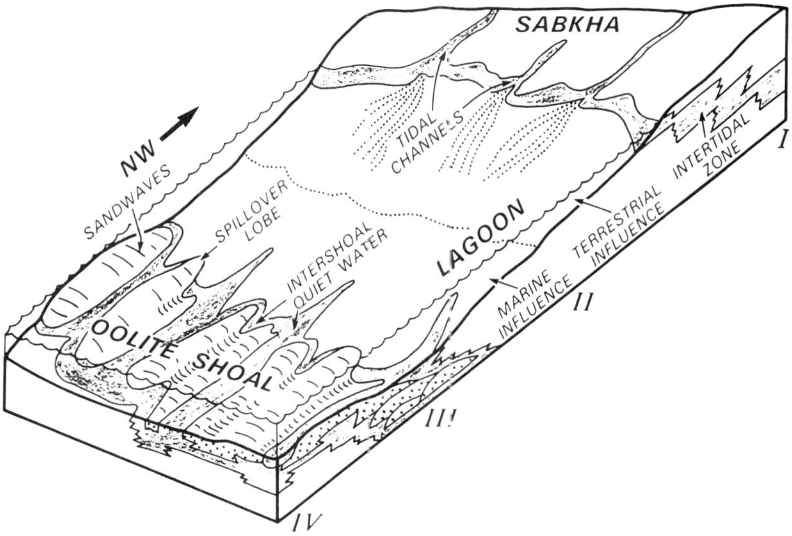

FIG. 2. The four facies of nearshore carbonate environments of the Cadeby Magnesian Limestone Formation (schematic); Facies IV is of open marine shelf sediments shown here undergoing progradational burial by shoal sediments of Facies III.

several exposures the rock contains irregular vugs, some lined with calcite, that correspond in size to the gypsum nodules (Fig. 4d); these vugs too were probably filled with gypsum and can be correlated with the samples from the subsurface.

Dolomitization

Almost all recent dolomites are being formed by the penecontemporaneous replacement of calcite or aragonite in evaporite-associated supratidal zones (sabkhas) and lagoons (see Deffeyes *et al.* 1964; Shinn 1964; Illing *et al.* 1965; Shinn *et al.* 1965; Friedman & Sanders 1967; Kinsman 1969; Friedman 1980; McKenzie *et al.* 1980; Patterson & Kinsman 1981). The mechanism for penecontemporaneous dolomitization necessitates hypersaline brines in which the Mg/Ca ratio is higher than in normal seawater. Two models have been proposed for the formation of the dolomitizing brines: (1) seepage reflux (Adams & Rhodes 1960; Deffeyes *et al.* 1964; Berner 1965; Murray 1969; Lucia 1972; Zenger 1972); and (2) evaporative pumping/capillary movement (Illing *et al.* 1965; Shinn *et al.* 1965; Friedman & Sanders 1967; Hsü & Siegenthaler 1969; Kinsman 1969; Friedman 1980).

Penecontemporaneous supratidal dolomites are commonly microcrystalline, aphanitic, imperfectly ordered and chemically impure. This is caused by (1) the abundance of sites of nucleation, (2) rapid crystallization of the dolomite preventing the 1 to 1 cation ordering of Ca and Mg from developing, and (3) the presence of 'foreign' ions such as Na, S, Cl and K in solution (Folk & Land 1975). Petrographically it is almost impossible to identify penecontemporaneous dolomites with certainty, and it is even more difficult where extensive sequences have been completely dolomitized and several episodes of dolomitization are suspected (Chilingar *et al.* 1979, p. 525); geochemical data are required and a strong correlation between dolomite content and a supratidal depositional environment needs to be established before attributing a penecontemporaneous sabkha origin to a dolomite rock. Textoris (1969) summarized some of the conditions diagnostic of supratidal dolomitization and included the occurrence of dolomite closer to terrigenous sources and limestone farther away, the presence of algal stromatolites, the presence of evaporite moulds and the presence of dolomicrite replaced to various degrees by euhedral dolomite.

A penecontemporaneous origin for the dolomite of Facies I rocks of the Sprotbrough Member is suggested by textures typical of supratidal dolomites. These include (1) the generally small crystal size of the groundmass, (2) the virtual absence of skeletal grains, (3) the presence of probable algal matter and (4) the association with displacive evaporites.

The presence of coarser-grained dolomite within the microcrystalline groundmass suggests subsequent episodes of dolomitization. The origi-

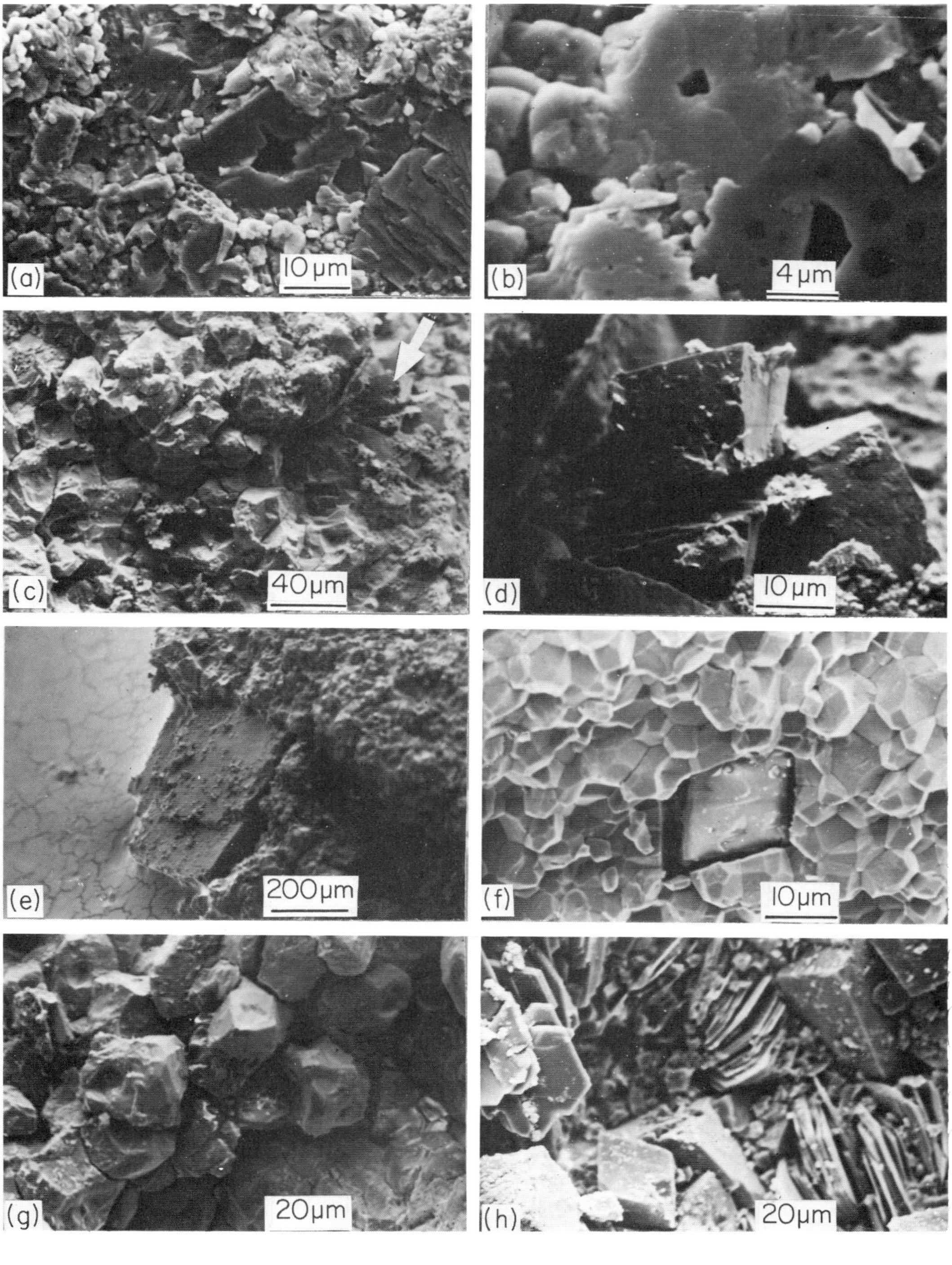
(a)
10 μm
(b)
4 μm
(c)
40μm
(d)
10μm
(e)
200μm
(f)
10μm
(g)
20μm
(h)
20μm

nal sediments were formed as dolomicrite in waters with high Mg/Ca ratios, but because of rapid crystallization and competition with other ions, these dolomicrites were fairly unstable. A later dolomitization event probably involved fluids with considerably lower Mg/Ca ratios and resulted in new larger crystals and clearer textures. McKenzie's (1981) stable isotope data from the sabkha at Abu Dhabi demonstrates that ordering within the dolomite crystal lattice increases with age. This suggests a molecular layer-by-layer increase in ordering within individual crystals as they grow outward from the nucleus, with each successive layer being slightly closer to stoichiometry. Overgrowths of stable dolomite then formed on the less stable dolomicrite cores. With prolonged exposure to meteoric water, the less stable (i.e. soluble) nucleus was dissolved, leaving hollow rhombs (Fig. 3a,b). A similar explanation was proposed for hollow-cored crystals in dolomites in the Late Palaeozoic strata of Bear Island (Folk & Siedlecka 1975) and the Cretaceous Edwards Formation of Texas (Longman & Mench 1978).

Evaporite diagenesis

The morphology, structure and petrography of the evaporites of Facies I of the Sprotbrough Member are similar to those of recent sulphates forming in the peritidal zone of the Arabian Gulf, where gypsum commonly replaces anhydrite over large areas of the inland parts of the sabkhas (Illing *et al.* 1965; Shearman 1978). That the gypsum is secondary is indicated by the presence of enclosed anhydrite relics. A volume increase associated with the replacement of the anhydrite by gypsum, similar to that described by Blatt *et al.* (1972, p. 505), caused the surrounding dolomitic sediment to be pushed aside and subhorizontal fractures to open; primary gypsum (satin spar) was then precipitated within these fractures as they opened. The timing and sequence of replacement may have proceeded in one of two ways: (1) intertidal gypsum may have been replaced by anhydrite at depth and the anhydrite, in turn, was replaced by gypsum on uplift, or (2) supratidal nodular anhydrite may have been replaced by gypsum near the surface. The second mechanism is preferred because of (1) the association with a supratidal environment, and (2) the greater likelihood of creating subhorizontal fractures in the shallow subsurface rather than at depth where the rock would be under considerable overburden pressure.

Porosity and dedolomite

Surface weathering of rocks of the sabkha facies of the Sprotbrough Member caused the dissolution of a considerable volume of evaporites. This resulted in the formation of solution residues (Smith 1976) and the creation of vuggy 'evaporo-mouldic' porosity (Fig. 4d). The dissolution of calcium sulphate minerals caused Facies I dolomites to be flooded by Ca^{2+}-rich fluids which, in turn, led to the creation of dedolomite (discussed in greater detail later, and by Smith 1976).

The evolution of the sabkha facies is generally part of a regressive cycle. This is especially apparent at Quarry Moor (SE 308691), where Smith (1976) recognized ooids within the lower part of the section and considered that '. . . most—and perhaps all the dolomites and limestones at Quarry Moor were formerly oolitic'; he also presented several lines of evidence for the existence of a sabkha at this locality.

Modern sabkhas and other exposures and boreholes through supratidal strata of the Sprotbrough Member do not have such a multitude of ooids, suggesting that the sequence at Quarry Moor may be atypical. Smith's description of the ooid textures and the appearance of the ooids in his plate 52 are similar to the textures within Facies III (discussed later). This suggests that the original depositional environment at Quarry Moor was a topographically high ooid sand body

FIG. 3. Scanning electron micrographs of dolomite textures. (a,b) Finely crystalline dolomite within dolomicrite. Note that centres of dolomite rhombs have been dissolved; dolomite that originally occupied the leached cores probably formed penecontemporaneously in the peritidal environment. The more stable rims were deposited in schizohaline, or lower salinity, waters. (c) Dolomitization in the semi-restricted lagoon facies: the inclusion-rich texture of the anhedral, finely crystalline dolomite and the presence of interboundary sheet pores suggest that this is a replacive fabric after a finer-grained precursor. This texture probably formed while salinities and Mg/Ca ratios of interstitial fluids were still relatively high. Euhedral dolomite rhombs (arrow and cavity) were formed at a later time in lower salinity waters. (d) Detail of dolomite rhombs pointed out in (c). (e) Mesocrystalline dolomite rhomb which has grown in dolomicrite. The finer matter has been incorporated into the large crystal, making the surface slightly irregular. (f) Preferential dedolomitization of microcrystalline dolomite. Note the partially leached and etched dolomite rhomb amidst calcite polyhedra. This texture formed when sediments, with textures such as seen in (c), (d) and (e), were bathed in freshwater solutions with very high Ca/Mg ratios. (g) Interboundary sheet pores. Thin, uncemented, planar pores between adjacent crystals are not fractures, but rather a neomorphic crystal growth phenomenon. (h) Authigenic kaolinite and euhedral dolomite rhombs within microstylolite. Material surrounding this solution seam is mesocrystalline anhedral dolomite.

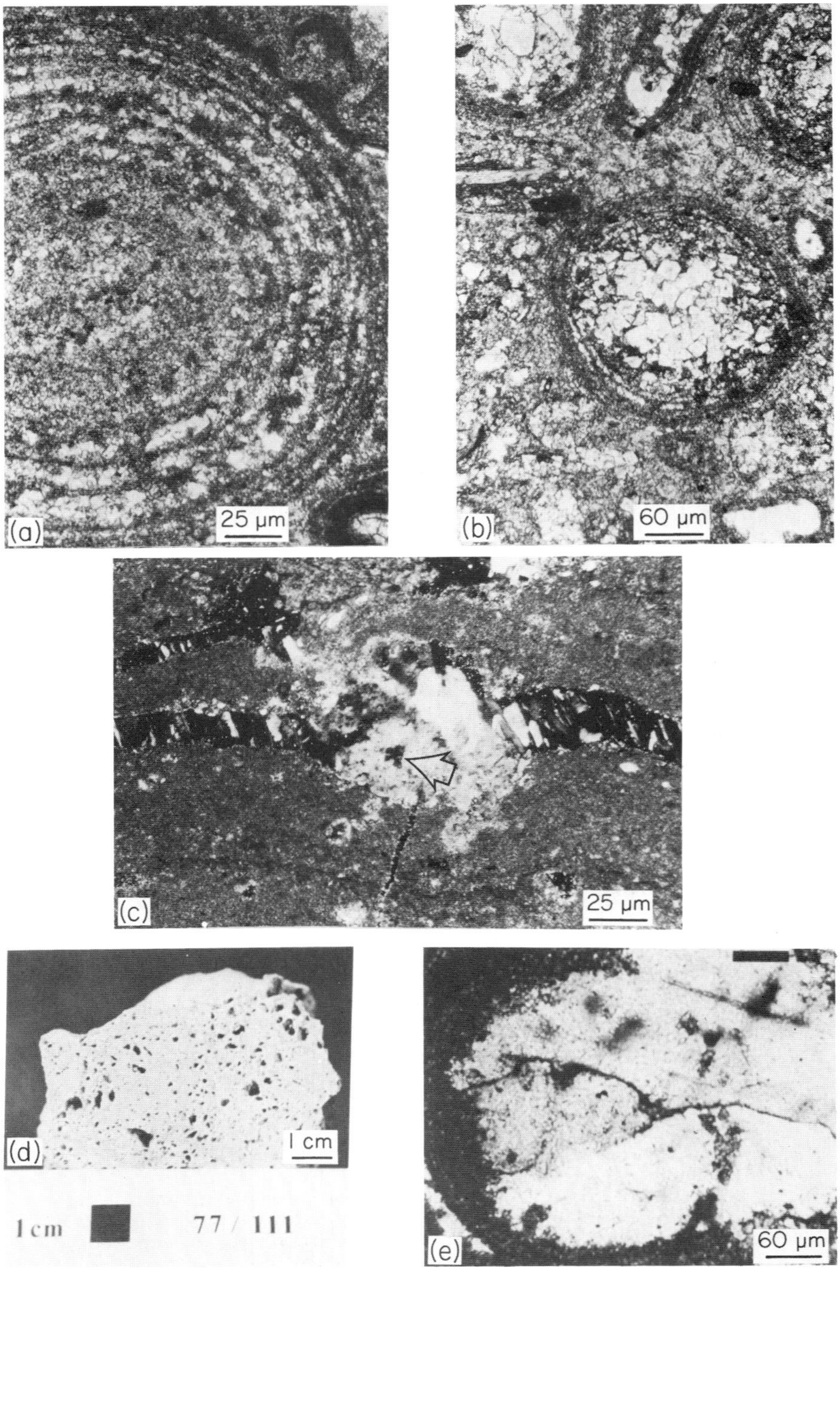
(a)
25 μm
(b)
60 μm
(c)
25 μm
(d)
1 cm
1 cm
77 / 111
(e)
60 μm

that became emergent as the Zechstein Sea shrank and marginal facies belts migrated eastwards.

Lagoon (Facies II)

The rocks of this facies are characterized by finely crystalline anhedral mosaic dolomites that commonly contain inclusions of finer grained carbonate matter. Finely crystalline rhombic dolomite and rhomb-shaped cavities that probably once were occupied by dolomite, are scattered throughout the groundmass (Fig. 3c,d). Clay layers and stylolites are abundant and many of the stylolites contain authigenic kaolinite and euhedral finely crystalline to mesocrystalline clear dolomite rhombs (Fig. 3h). Completely dolomitized skeletal remains, burrows, pellets, ooids and pisoids are locally present. Silt- to fine-sand size quartz grains with edges corroded and embayed by dolomite are scattered through the groundmass. Scanning electron microscope examination of rocks of this facies reveals tiny ($< 1\ \mu m$) cracks (not visible by transmitted light microscopy) between dolomite anhedra (Fig. 3c,g).

Dolomitization

The inclusion-rich character of the anhedral finely crystalline groundmass suggests that it replaced a finer grained precursor; probably it formed while salinites and Mg/Ca ratios were relatively high. The euhedral dolomite rhombs (Fig. 3d) probably were formed later, when salinites and Mg/Ca ratios of the surrounding fluids were lower.

The mechanism of the first episode of dolomitization of this facies is problematical. The absence of supratidal features and the presence of oscillation ripples suggest that the sediment had a subtidal origin. One possible model is seepage reflux in an evaporitic lagoon (Adams & Rhodes 1960) where dense hypersaline brines sank and flowed seaward, dolomitizing the aragonite and calcite sediments through which they passed. This process might result in the accumulation of subtidal evaporites (mostly gypsum; Schreiber 1978), but no bedded gypsum deposits are known from outcrop or borehole in the Sprotbrough Member.

Another possible origin for the dolomite in this facies is by primary precipitation in the manner proposed by Behrens & Land (1972) for Baffin Bay, Texas. However, primary dolomite is rare, if it occurs at all, in recent sediments, and proving that it occurred in ancient rocks has proved to be singularly difficult. Nevertheless, several similarities between the rocks of Facies II and the primary precipitated dolomites of Behrens & Land (1972) may justify analogies with their model: (1) both environments are subtidal semi-restricted lagoons showing no signs of supratidal exposure; (2) original crystals from both are microcrystalline and anhedral; (3) dolomites at Baffin Bay contain higher amounts of Sr^{2+} than ancient dolomites; the relatively high Sr^{2+} content of a sample analysed from this facies (Kaldi 1980a) may indicate a similar origin; (4) primary dolomites are associated with high pH environments; the replacement of quartz grains by dolomite may be indicative of such an environment for quartz dissolves under alkaline conditions and carbonate precipitates in its place (Oppenheimer & Master 1965).

It is difficult to determine which of the two models best explain the origin of the dolomite rocks of Facies II and it is likely that both processes (and perhaps others also) may have had an influence.

Stylolites

Stylolites (solution seams) characterize many exposures of this facies and are less common in other facies; they may be traceable for up to 20 m

FIG. 4. Miscellaneous dolomite and evaporite textures. (a) Alternating laminae of dolomite and calcite within an ooid. Dolomitization was probably late diagenetic and localized within organic layers where decomposition of chlorophyll released Mg^{2+} ions. (b) Centripetal increase in crystal size suggests a cement (void fill) origin for the dolomite within this allochem (an ostracod?). Note the similar texture within fenestrae of bryozoan fragment at top of photograph. Allochems are surrounded by oolitic coats having some dolomitized laminae. (c) Gypsum nodule with inclusions of relic anhydrite and satin spar (fibrous gypsum). Probable diagenetic history: nodular anhydrite grew displacively in peritidal dolomicrite and gypsum later replaced the anhydrite; the associated volume changes opened sub-horizontal fractures which were filled as they opened by fibrous gypsum. There are no anhydrite relics within the satin spar. Crossed nicols. (d) Probable outcrop equivalent of (c). Note the many irregular vugs, which at one time were filled with gypsum or anhydrite. (e) Replacive gypsum. Note relic dolomite ooid in poikilotopic gypsum. Gypsification was probably late diagenetic, associated with the dissolution of pre-existing evaporites. The subsequent flooding of underlying sediments with $CaSO_4$-saturated waters caused this unusual texture. Cross nicols.

or be only a few centimetres long. Stylolites have also been formed along planes of bedding so that where units are cross-stratified, solution seams are parallel with the primary laminae.

The stylolites of this facies resemble the non-sutured seam solutions of Wanless (1979). These occur as swarms of fine clay seams, discontinuous seams or scattered microstylolites (thin undulate surfaces with relief of 20–40 μm). Clay particles, platy silt and dolomite are present along many of these surfaces.

Oldershaw & Scoffin (1967) noted the lateral passage of pressure-solution seams into primary clay deposits. Because solution seams and microstylolites tend to form along pre-existing primary argillaceous layers, it is likely that primary clay deposits also initiated the formation of the solution seams of Facies II. Occasional flooding of the upper Permian landmass could have brought the fine-grained clastic sediments into the lagoon.

Stylolites owe their origin to pressure-solution. Clays or chemical impurities within carbonate minerals undergoing pressure-solution migrate towards solution surfaces and remain there as concentrates.

The occurrence of kaolinite solution concentrates probably evidences the migration of ions from silica and aluminium-rich impurities in the rock; these impurities were probably derived from the occasional input of clastics into the lagoon. The euhedral crystal shapes and relatively random distribution of the kaolinite books suggest that most of the kaolinite is authigenic. The presence of authigenic kaolinite in a carbonate succession is unusual and has been suggested as evidence of fluctuating fluid chemistry in a particular environment (Bartow & Sims 1975).

Dolomite associated with stylolites of Facies II commonly is mesocrystalline, euhedral and limpid, and cannot be explained as a simple residue from solution of the surrounding rock. The rhombs are clearly late diagenetic, developing after or during the formation of authigenic clay minerals (kaolinite 'books' are displaced by dolomite; Fig. 3h). This dolomite may be attributed to the presence of clay minerals other than kaolinite within the stylolite. Wanless (1979) considered the origin of dolomite within stylolites to be 'intimately associated' with microenvironments existing within the solution seam, and Kahle (1965) suggested that clays promote the formation of dolomites, acting as catalysts by providing a source of magnesium ions. Clay minerals, especially smectite on conversion to illite, may release appreciable quantities of Mg^{2+}. The Mg^{2+} originates either from inter-lattice sites or during ion exchange (McHarge & Price 1982) in the course of diagenesis.

Porosity

Porosity and permeability in this facies is very limited because of the aphanitic dolomite textures, the presence of clays and the effects of pressure-solution; despite this limitation, there seems to have been sufficient fluid migration through the system to permit several stages of diagenesis. This may be explained by the tiny cracks between dolomite crystals (Fig. 3c,g). Similar cracks have been described by Wardlaw (1976) and Longman & Mench (1978) who suggested the term 'interboundary sheet pores' (ISPs). ISPs are narrow uncemented planar pores between adjacent crystals that are observed only with the scanning electron microscope and are too small to be seen in thin-sections; they form because crystal growth is inhibited in confined intergranular spaces. Wardlaw (1976) suggested that the displacive growth of some crystals indicated the presence of a water film capable of supporting a shear stress and that this water film was never replaced by crystalline matter when two crystals grew together. Longman & Mench (1978) suggested that interboundary sheet pores are probably widespread in sedimentary rocks; if this is so, they may furnish a partial explanation for the puzzling problem of how fluids can move through these and other apparently tightly cemented rocks.

Dedolomite

Dedolomite in Facies II occurs as a mosaic of calcite polyhedra containing sparsely disseminated partially leached and corroded dolomite rhombs (Fig. 3f). This texture is the result of the preferential calcitization of relatively unstable microcrystalline dolomite that formed in highly saline waters. Coarser dolomite rhombs which precipitated later, probably in mixed meteoric-marine waters, were more in equilibrium with the dedolomitizing fluids and were preserved. The dedolomitizing waters probably had very high Ca/Mg ratios (Al-Hashimi & Hemingway 1973) resulting from the dissolution of evaporites either after uplift (Smith 1976) or at depth (Land & Prezbindowski 1981).

A model for the development of recrystallization textures in dolomites and dedolomites as a function of progressive freshwater diagenesis and Mg/Ca ratio is presented in Fig. 5.

Oolite shoal/barrier complex (Facies III)

At most exposures this facies consists of large cross-stratified ooid sandwaves and the generally flat-bedded intershoal areas between them. The sandwaves are made up of well sorted 100–150 μm hollow-centred ooids, and the intershoal areas consist of mixtures of ooids, pisoids, pellets, grapestones and skeletal fragments. Grains from both environments have isopachous fringes of finely crystalline limpid dolomite. Kaldi & Gidman (1982) demonstrate that this dolomite grew as an early diagenetic cement, and formed during a fall in Zechstein Sea level which exposed the sandwaves to a mixture of meteoric and marine fluids.

Dolomites containing 'ghost' ooids are common in the southernmost parts of the study area. Ghost ooids are recognizable in thin-section as concentric brown laminae which commonly surround quartz grains; the quartz clearly served as nuclei (Fig. 6b). Skeletal debris and other primary carbonate grains are conspicuously absent in these southern areas.

The obliteration of primary textures in these rocks is probably a function of repeated episodes of dolomitization, and the abundance of clastic grains suggests a proximity to the shoreline. The influence of the Permian Mansfield River (Sherlock 1911) in this region may have been quite substantial; seasonal flooding of the river perhaps on to an extensive floodplain and into the

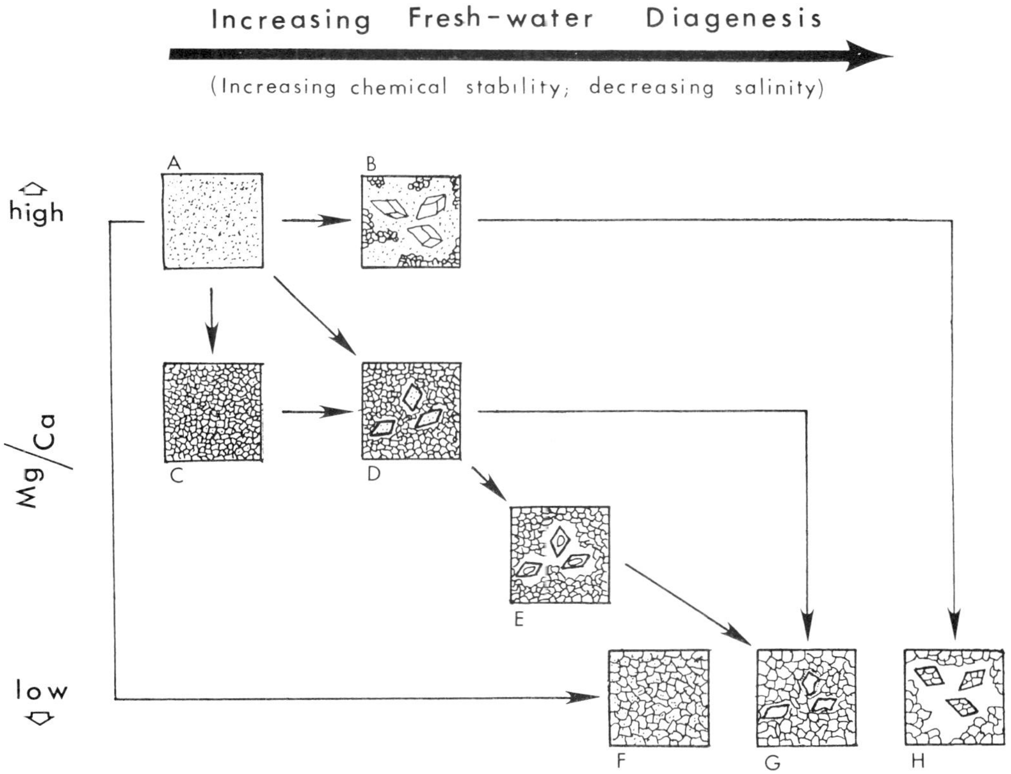

FIG. 5. Development of recrystallization textures in dolomite and dedolomite. (a) Hypersaline penecontemporaneous dolomicrite (microcrystalline dolomite) formed under high Mg/Ca ratios. Dolomicrite is not well ordered because of rapid crystallization and competition with other ions. (b) Limpid dolomite rhombs; formed when Mg/Ca ratios remained high, but salinity decreased during dilution. (c) Dolomicrospar (finely crystalline dolomite); probably formed as early diagenetic 'reflux' dolomite. Example: Fig. 3(c). (d) Inclusion-rich dolomite rhombs in matrix of dolomicrospar; formed when fresh waters diluted sediment-type A or C. Example: Fig. 3(e). (e) Dolomite rhombs with leached cores. Centres of rhombs formed in sediment-type A or C; more stable rims formed in D. Under increasing freshwater diagenesis, less stable cores dissolved. Example: Fig. 3(b). (f–h) Dedolomite textures, formed when earlier dolomite was flushed by fluids with very low Mg/Ca ratios. These low ratios were probably caused by Ca^{2+} in solution from dissolved gypsum or anhydrite. (f) Dolomite inclusion-rich microspar (finely crystalline calcite); (g) Etched relic dolomite rhombs, with irregular boundaries, in microspar. Example: Fig. 3. (h) Rhomb shapes of precursor limpid dolomite pseudomorphed by microspar.

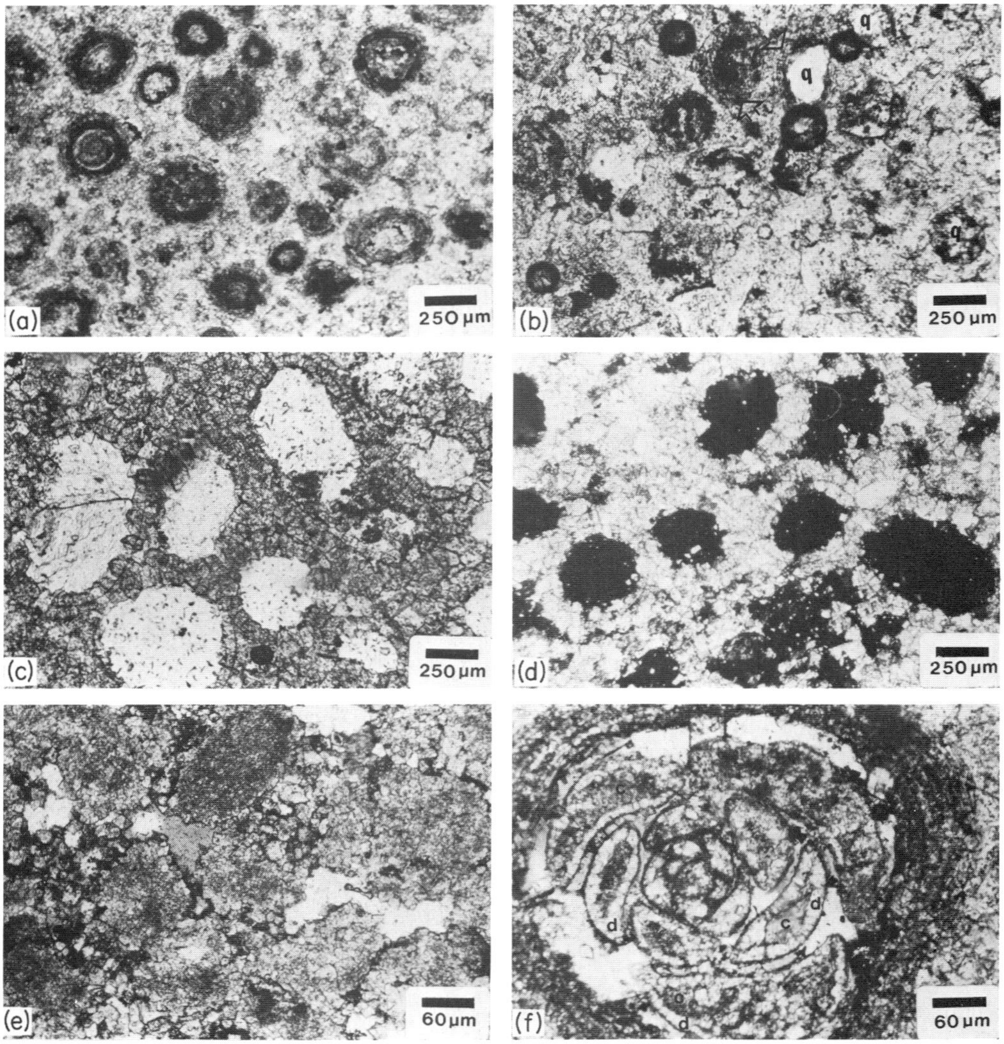

FIG. 6. Ooid textures and associated porosity types. (a) Calcite ooids within dolomicrospar matrix. Matrix preferentially dolomitized because of its original greater porosity (which allowed the passage of dolomitizing fluids) and higher surface area to pore fluid ratio. (b) 'Ghost' ooids in mesocrystalline anhedral mosaic dolomite. This material was probably deposited near the shoreline and underwent several episodes of dolomitization which destroyed most primary textures. Ooids are only recognizable by concentric organic laminae. Note abundant quartz-grain nuclei of ooids. (c) Oomouldic porosity in dolomicrospar. This texture is probably the result of freshwater flushing of sediment similar to (a). The more soluble calcite was preferentially dissolved, leaving the dolomite matrix. Intact oomoulds suggest that sediment was lithified at time of calcite dissolution. (d) Inter- and intra-particle oomouldic porosity. Framework of dolomite cement-reinforced micrite envelopes formed from mixing of marine and meteoric waters in ooid sand bodies. Crossed nicols. (e) Ooids that have undergone at least two stages of dolomitization (see also Fig. 7c). Mixed marine and meteoric waters probably precipitated coarser euhedral dolomite rhombs and also dissolved any calcite matrix, leaving the substantial interparticle porosity. (f) Intra-particle porosity and preferential dolomitization in *Glomospira sp.* Note well-preserved internal detail. Chambers of this foraminifer served as conduits for interstitial dolomitizing fluids; parts of the chamber walls adjacent to fluid pathways are now dolomite (d), whereas wall interiors are still calcite (c). Also note oolitic coating around foraminifer.

lagoons, followed by a return to normal marine conditions, could have caused repeated schizohaline dolomitization in the manner suggested by Folk & Land (1975).

Another diagenetic texture found in some samples from the dolomitized ooid sand bodies is shown in Figs 6(e) and 7(c). In this rock the gross morphology of the ooids has been preserved but their internal details have been obliterated; grains consist of tightly interlocking anhedral microcrystalline interiors and loosely packed euhedral mesocrystalline exteriors. The microcrystalline dolomite may have developed when sabkha-like hypersaline conditions dominated local areas during the emergence of the Sprotbrough Member. As meteoric conditions became established, fresh waters mixed with saline pore fluids to produce the coarser rhombs near the grain margins. The lack of cement around these grains may be related to ooid mobility. Grains stabilized by organic activity such as algal binding are the best sites for marine cementation (Bathurst 1975; Longman 1980). Most of the mobile ooid shoals of the Bahamas lack cements, presumably because the constant movement of the grains restricts the growth of organic sheaths. Similarly, these uncemented grains from the Sprotbrough Member were probably formed in the more active parts of the ooid sand bodies.

Dedolomite

Calcified dolomite within the oolite shoal/barrier complex facies occurs as patchy stringers of

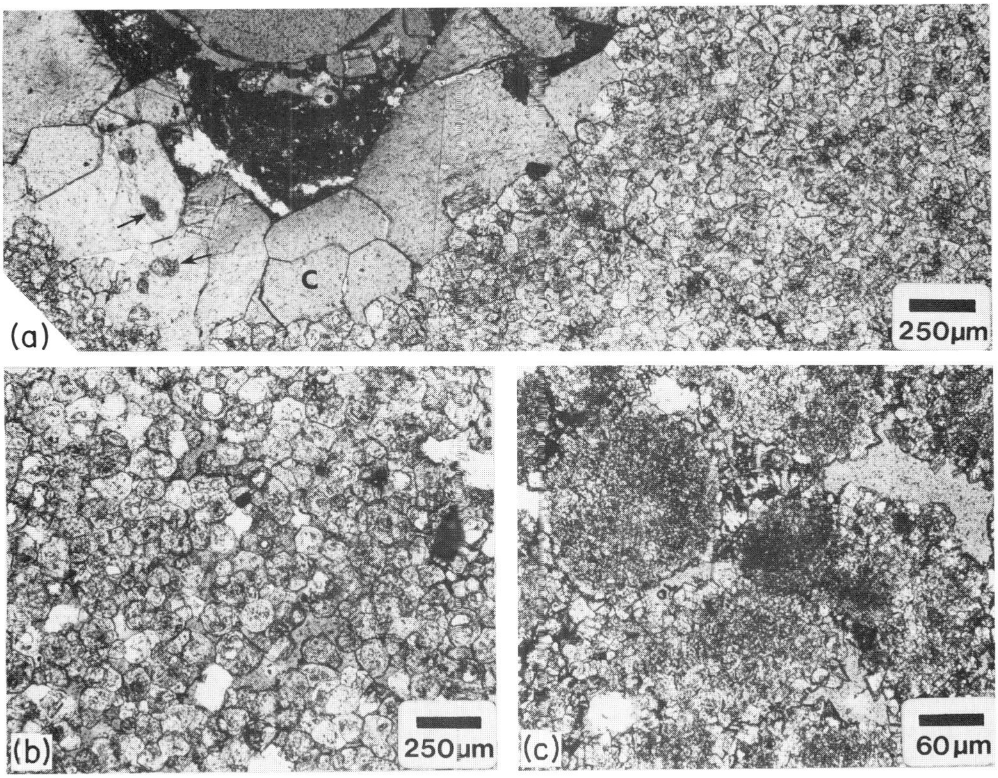

FIG. 7. Textures of dolomitized ooids. (a) Mesocrystalline anhedral dolomite (right) surrounds a burrow (left) Organic matter (dark) is ringed by calcite (C) which has grown both replacively (arrows point to inclusions of relic dolomite) and displacively (large calcite crystal deforms laminae of organic material). The tightly crystalline dolomite obliterates primary textures (see b) for about 1 cm around the burrow. (b) 1 cm to the right of (a). Dolomitization has taken place without destroying the gross oolitic texture. Several superficial (single coat) ooids, some with quartz nuclei, are clearly visible. Note the relatively high interparticle porosity. (c) Two stages of dolomitization in ooids. Internal textural details of ooids have been obliterated by interlocking microcrystalline anhedral dolomite which probably formed soon after deposition. Mesocrystalline euhedral rhombs near grain margins formed later, probably when mixed marine and meteoric waters passed through the sediment. Though lacking internal structure, the grains can be identified as ooids by their size, shape and sorting.

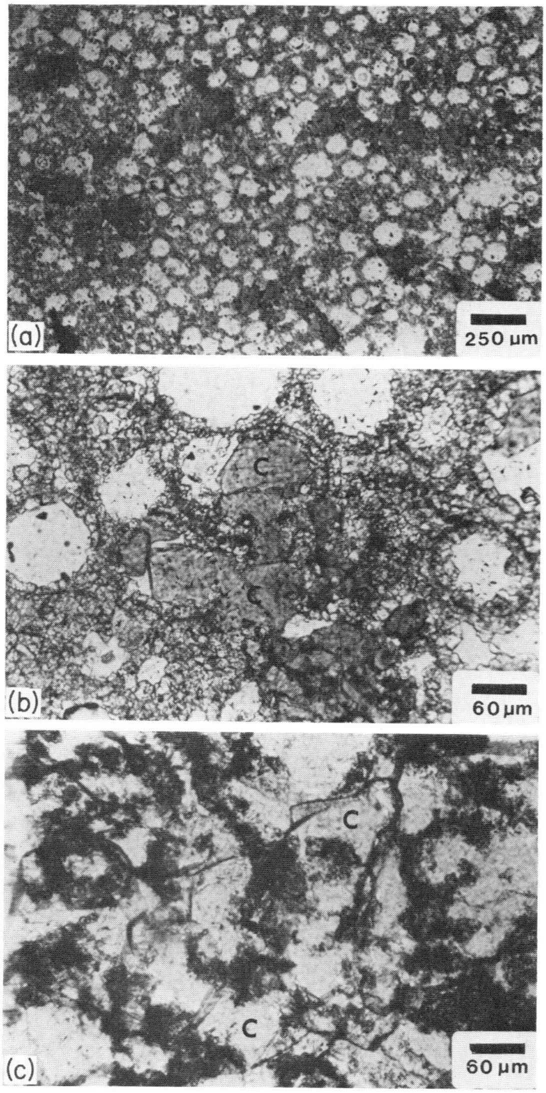

FIG. 8. Dedolomite textures. (a) Patchy distribution of dedolomite within hollow ooids. (b) Detail of patchy dedolomite with hollow ooids showing dolomite cement (surrounding micrite envelopes) replaced by calcite (c). (c) Nearly total dedolomitization of oolitic rock; calcite (c) has replaced dolomitized ooids. Relic dolomite, with vaguely discernible primary grain shapes, forms inclusions within the calcite.

calcite within otherwise dolomitized oolitic sediments (Fig. 8a), or as calcite-replaced isopachous dolomite cements surrounding hollow ooids (Fig. 8b). Locally, oolitic sediments have undergone nearly total dedolomitization and coarse equant calcite has replaced dolomite. Relic dolomite ooids with vaguely discernible grain shapes occur as inclusions in the calcite (Fig. 8c).

Porosity

The creation and modification of porosity in sediments of the oolite shoal facies was controlled by primary depositional setting, secondary post-depositional diagenesis (early and late) and tertiary surface-related erosional processes. These zones correspond with Choquette & Pray's (1970) 'eogenetic', 'mesogenetic' and 'telogenetic' zones of porosity development.

Sediment from inactive ooid areas consists of a framework of dolomitized micrite envelopes (Fig. 6d). Inter- and intraparticle mouldic porosity was created when allochem interiors were dissolved by meteoric waters; late diagenetic mesocrystalline dolomite partially filled some moulds but left considerable intercrystalline porosity; telogenetic calcite cement filled some of these residual voids (Kaldi & Gidman 1982). Dedolomitization was locally patchy, only slightly reducing porosity (Fig. 8a), or nearly total, occluding most primary and secondary pores (Fig. 8c).

Grains that formed in active ooid environments did not develop an early diagenetic fringing cement. The early diagenetic dolomitization of grain interiors prevented the formation of mouldic voids but served to retain substantial interparticle porosity. Later diagenetic dolomitization reduced this porosity somewhat, but the coarse euhedral texture of the late diagenetic crystals led to the preservation of some intercrystalline pores (Fig. 6e).

Porosity occlusion is most pronounced in rocks from the southern parts of the study area, where the original sediment was probably a fairly porous grainstone. Mixed-water fluids passed through the sediments repeatedly during early diagenesis, altering them and creating their present tightly interlocking crystal textures (Fig. 6b). The only fluid migration paths through this rock after the early events were late fractures, solution seams or interboundary sheet pores.

Burrows locally led to modification of porosity and served as fluid channels. This resulted in preferential late diagenetic alteration and porosity occlusion within sediments immediately adjacent to the burrows (Fig. 7a).

Open marine shelf (Facies IV)

The open marine shelf facies is known chiefly from the subsurface. On the basis of distinctive lithology and inferred depositional environments, this facies is divided into mutually gradational lower and upper units.

The characteristics of the lower unit are detailed by Fuzesy (1970), Smith (1974a) and Kaldi (1980b); the upper unit was examined in detail in this study. It consists of calcitic allochems in a groundmass of finely crystalline dolomite (Fig. 6a). Ooids, pisoids and oolitically coated fossil fragments (Fig. 4b) typify the allochems, and coated lumps composed of algal-bound and encrusted aggregates of ooids, pisoids or fossil fragments are common; coatings around ooids, pisoids or lumps commonly are made up of alternating laminae of dolomite and calcite (Fig. 4a). Fossils include tests of foraminifera, ostracods and fragments of bryozoans. Many skeletons are partially dolomitized; the dolomitized areas are everywhere adjacent to intraparticle pores (Fig. 6f). Several fossils are filled with 'baroque' or 'saddle' dolomite (i.e. dolomite having undulose extinction) (Fig. 4b). Gypsum occurs as subspherical and irregularly shaped masses, as fibrous veins within the sediment or as intercrystalline pore-fill between dolomite grains. Gypsum blebs commonly contain inclusions of relic anhydrite or, less commonly, vague traces of former dolomite allochems (Fig. 4e).

Dolomitization

Dolomitization within the lower unit of this facies has been shown by Kaldi (1980b) to have been late diagenetic and to have taken place in the deep subsurface. Dolomitizing fluids were authigenic connate brines and the patchy dolomitization was restricted to areas within the immediate influence of these fluids.

Dolomitization of carbonate within the upper unit was incomplete. The character of the dolomite matrix and limestone allochems (Fig. 6a) suggest preferential dolomitization of an originally more porous matrix. The preservation of the limestone ooids indicates that early digenetic meteoric waters (which would have either dissolved or mixed with marine fluids and dolomitized the calcite) never affected this sediment. In outcrop equivalents of this rock, post-exhumation freshwater flushing has dissolved the more soluble calcite allochems and left the dolomite matrix (Fig. 6c). The matrix was retained as a sturdy framework, suggesting that it was at least partly lithified at the time of calcite dissolution.

The alternation of laminae of calcite and dolomite within algal-coated grains invites explanation by reference to Gebelein & Hoffman's (1973) cryptalgal dolomitization model; this mechanism involves alternating algal-rich and sediment-rich layers wherein the algal layers were dolomitized as a result of the localized release of Mg^{2+} (concentrated in chlorophyll) whereas sediment-rich layers remained as limestone.

The coarse 'saddle' dolomite filling allochem interiors is interpreted as a late diagenetic cement. The potential of this dolomite texture as a geothermometer (indicative of temperatures of 60°–150°C) was noted by Radke & Mathis (1980), and its common association with deep sulphate-bearing carbonates and hydrocarbons was noted by Beales (1971) and Choquette (1971).

Dolomitization of the open marine shelf facies

is attributed to late diagenetic processes; its patchiness suggests that there was a limited volume of dolomitizing fluids available. These fluids only altered rock immediately adjacent to solution pathways or within locally dolomitizing microenvironments. The sediments of this facies were unaltered limestones during the deposition of the overlying EZ1 evaporites. The evaporites created overburden pressure that compacted and lithified the underlying sediment and also prevented the upwards escape of formation fluids. As a result connate brines passed laterally or downwards through the rock, but were not recharged by surface solutions. Jodry (1969) described similar late diagenetic dolomitization textures from a Silurian reef carbonate-evaporite complex in Michigan; in these rocks, too, connate brines were prevented from escaping upwards by overlying impermeable evaporites. In both examples, fluids migrated only laterally or downwards, taking advantage of fractures, solution channels or patches of more porous carbonate material; these pathways were preferentially dolomitized.

Porosity

Porosity is very poorly preserved in rocks of this facies. Late diagenetic dolomite cemented many fractures and solution channels. The lack of meteoric flushing left many potential moulds filled with calcite. Considerable intercrystalline porosity may have remained after dolomitization. However, the slow seepage of sulphate-rich solutions resulted in the precipitation of gypsum in many of these residual pores.

The rocks of this facies owe their low porosity and permeability to the precipitation in fluid pathways of dolomite and sulphates by evaporite-rich fluids: these fluids were generated at basin margins and travelled down-dip towards the basin centre. The open marine facies of the Sprotbrough Member extends for a considerable distance eastward into the English Sector of the North Sea Basin (Taylor & Colter 1975). There may be a lateral limit to the area affected by porosity occlusion. Beyond this belt it is likely that fractures and secondary pores have been preserved within rocks of this facies, making it a potential hydrocarbon reservoir.

Conclusions

The Cadeby Formation has undergone a complex history of diagenesis. Dolomite was formed in various ways in the different facies of the Sprotbrough Member: (1) penecontemporaneously within supratidal sediments, (2) as a primary precipitate or by reflux in the semi-restricted lagoon, (3) by replacement of micrite and as early and late cements in the oolite shoal facies, and (4) as locally controlled subsurface replacement of limestone within the open shelf deposits. The dolomite textures formed in these settings were subsequently locally altered by the superimposition of new diagenetic environments. New environments that were associated with early diagenesis commonly involved increasing freshwater solutions, decreasing salinity and lower Mg/Ca ratios. This resulted in aggrading neomorphism, limpid rhombs and rhombs with dissolved cores within sabkha and lagoon sediments.

At least two generations of evaporites were formed within these rocks. Penecontemporaneous anhydrite and gypsum grew interstitially within supratidal sediments and have a displacive nodular habit. Late diagenetic evaporites resulted from the solution and remobilization of earlier deposits into the subsurface; these later sulphates exhibit varied void-filling and replacement textures.

Extensive flushing of dolomite by meteoric waters enriched in Ca^{2+} from dissolved evaporites and limestone caused dedolomitization; this was either patchy or pervasive, depending on local hydrological conditions.

Porosity in the Sprotbrough Member was controlled by the nature of the primary sediments and the diagenetic environments to which these sediments were subjected. Supratidal and lagoonal sediments were fine-grained and had low primary porosities. The dissolution of evaporites from within sabkha sediments resulted in poorly interconnected vuggy pores. Recrystallization of dolomicrites from the lagoon facies formed a very tight rock having only interboundary sheet pores for fluid migration.

Pressure solution caused considerable quantities of carbonate to be mobilized and reprecipitated. Stylolites developed preferentially along primary clay layers in the lagoonal deposits. Authigenic clays and dolomite which grew within solution concentrate seams contributed to a substantial decrease in porosity.

The development of a sturdy framework and the subsequent meteoric dissolution of unstable grains created considerable inter- and intraparticle porosity in the oolite shoal facies. Late diagenetic dolomite cements, calcite cements and dedolomite partially occluded these pores.

In the limestones of the open shelf facies, remobilization of overlying evaporites filled many late diagenetic intercrystalline pores and fractures with gypsum. Some fracture and secondary porosity may be preserved toward the basin centre within this facies.

The detailed study of the relationships between original depositional settings and subsequent diagenetic environments in the Sprotbrough Member of the Cadeby Formation demonstrates the profound effects of each of these factors on the resulting rock. The importance of these variables is increased when it is realized that all the textures and structures described from these rocks occur in a very limited stratigraphical setting and most were formed in a geologically short period. Many aspects of diagenesis remain to be learned. As a better understanding of deposition and diagenesis in marine carbonates is gained, the models and ideas presented here will undoubtedly have to be modified.

ACKNOWLEDGMENTS: I thank R.G.C. Bathurst, J.A. Dickson, I.J. Fairchild, P.F. Friend and G.M. Harwood for much valuable advice and encouragement. I am especially indebted to D.B. Smith for his years of guidance and thoughtful and constructive criticism. This work represents part of a PhD thesis carried out at the University of Cambridge. Financial assistance provided by Steetley Minerals (UK) Ltd, Sigma Xi, Clare Hall, Cambridge and the Cambridge Philosophical Society are gratefully acknowledged.

References

ADAMS, J.E. & RHODES, M.L. 1960. Dolomitization by seepage refluxion. *Bull. Am. Ass. Petrol. Geol.* **44,** 1912–20.

AL-HASHIMI, W.S. & HEMINGWAY, J.E. 1973. Recent dedolomitization and the origin of the rusty crusts of Northumberland. *J. sedim. Petrol.* **43,** 82–91.

BARTOW, J.A. & SIMS, J.D. 1975. Authigenic kaolinite associated with Tertiary oil-bearing sandstones, California coast ranges, U.S.A. *Proc. IX int. Congr. Sedimentology, Nice,* **7,** 9–15.

BATHURST, R.G.C. 1975. *Carbonate Sediments and their Diagenesis.* Elsevier, Amsterdam.

BEALES, F.W. 1971. Cementation by white sparry dolomite. In: BRICKER, O.P. (ed.) *Carbonate Cements,* 330–8. Johns Hopkins Press, Baltimore.

BEHRENS, E.W. & LAND, L.S. 1972. Subtidal Holocene dolomite, Baffin Bay, Texas. *J. sedim. Petrol.* **42,** 155–61.

BERNER, R.A. 1965. Dolomitization of the Mid-Pacific Atolls. *Science,* **147,** 1297–9.

BLATT, H., MIDDLETON, G. & MURRAY, R. 1972. *Origin of Sedimentary Rocks.* Prentice-Hall, New Jersey.

CHILINGAR, G.V., ZENGER, D.H., BISELL, H.J. & WOLF, K.H. 1979. Dolomites and dolomitization. In: LARSEN, G. & CHILINGAR, G.V. (eds) *Diagenesis in Sediments and Sedimentary Rocks,* 425–537. Developments in Sedimentology, 25a. Elsevier, Amsterdam.

CHOQUETTE, P.W. 1971. Late ferroan dolomite cement, Mississippian carbonates, Illinois Basin, U.S.A. In: BRICKER, O.P. (ed.) *Carbonate Cements,* 339–46. Johns Hopkins Press, Baltimore.

—— & PRAY, L.C. 1970. Nomenclature and classification of porosity in sedimentary carbonates. *Bull. Am. Ass. Petrol. Geol.* **54,** 207–50.

DEFFEYES, K.S., LUCIA, F.J. & WEYL, P.K. 1964. Dolomitization – observations on the island of Bonaire, Netherlands Antilles. *Science,* **143,** 678–9.

FOLK, R.L. & LAND, L.S. 1975. Mg/Ca ratio and salinity: two controls over crystallization of dolomite. *Bull. Am. Ass. Petrol. Geol.* **59,** 60–8.

FOLK, R.L. & SIEDLECKA, A. 1974. The 'schizohaline' environment: its sedimentary and diagenetic fabrics as exemplified by Late Paleozoic rocks of Bear Island, Svalbard. *Sediment. Geol.* **11,** 1–15.

FRIEDMAN, G.M. 1980. Dolomite is an evaporite mineral: evidence from the rock record and from sea-marginal ponds of the Red Sea. In: ZENGER, D. H., DUNHAM, J.B. & ETHINGTON, R.L. (eds) *Concepts and Models of Dolomitization. Spec. Publ. Soc. econ. Paleont. Mineral.* **28,** 69–80.

—— & SANDERS, J.E. 1967. Origin and occurrence of dolostones. In: CHILINGAR, G.V., BISSELL, H.J. & FAIRBRIDGE, R.W. (eds) *Carbonate Rocks, Origin, Occurrence and Classification,* 267–348. Elsevier, Amsterdam.

FUZESY, L.M. 1970. *Petrology of the Lower Magnesian Limestone in the neighbourhood of Selby, Yorkshire.* Unpublished PhD thesis, University of Cambridge.

GEBELEIN, C.D. & HOFFMAN, P. 1973. Algal origin of dolomite laminations in stromatolite limestones. *J. sedim. Petrol.* **43,** 603–13.

GOOSSENS, R.F. 1973. Coal reserves in the Selby area. *Min. Eng. London,* **132,** 237–46.

HSU, K.J. & SIEGENTHALER, C. 1969. Preliminary experiments on hydrodynamic movement induced by evaporation and their bearing on the dolomite problem. *Sedimentology,* **12,** 11–25.

ILLING, L.V., WELLS, A.J. & TAYLOR, J.C.M. 1965. Penecontemporaneous dolomite in the Persian Gulf. In: PRAY, L.C. & MURRAY, R.C. (eds) *Dolomitization and Limestone Diagenesis. Spec. Publ. Soc. econ. Paleont. Mineral.* **13,** 89–111.

JODRY, R.L. 1969. Growth and dolomitization of Silurian reefs, St. Clair County, Michigan. *Bull. Am. Ass. Petrol. Geol.* **53,** 957–81.

KAHLE, C.F. 1965. Possible roles of clay minerals in the formation of dolomite. *J. sedim. Petrol.* **35,** 448–53.

KALDI, J. 1980a. *Aspects of the sedimentology of the Lower Magnesian Limestone (Permian) of eastern England.* Unpublished PhD thesis, University of Cambridge.

—— 1980b. The origin of nodular structures in the Lower Magnesian Limestone (Permian) of Yorkshire, England. In: FUCHTBAUER, H. & PERYT, T.M. (eds) *The Zechstein Basin with Emphasis on Carbo-*

nate Sequences. Contr. Sedim. **9,** 45–60.

—— 1982. Reservoir properties, depositional environments and diagenesis of the Mississippian Midale Beds, Midale Field, South-eastern Saskatchewan. In: CHRISTOPHER, J. & KALDI, J. (eds) *Fourth Int. Williston Basin Symp. Spec. Publ. Sask. geol. Soc.* **6,** 211–16.

—— (*this volume*) Sedimentology of sandwaves in an oolite shoal complex in the Cadeby (Magnesian Limestone) Formation (Upper Permian) of Eastern England.

—— & GIDMAN, J. 1982. Early diagenetic dolomite cements: examples from the Permian Lower Magnesian Limestone of England and the Pleistocene carbonates of the Bahamas. *J. sedim. Petrol.* **52,** 1073–82.

KINSMAN, D.J.J. 1969. Modes of formation, sedimentary associations, and diagnostic features of shallow-water and supratidal evaporites. *Bull. Am. Ass. Petrol. Geol.* **53,** 830–40.

LAND, L.S. & PREZBINDOWSKI, D. 1981. Origin and evolution of saline formation waters, Lower Cretaceous carbonates, south-central Texas (Abstract). *Bull. Am. Ass. Petrol. Geol.* **65,** 947. Ann. Meet., San Francisco.

LONGMAN, M.W. 1980. Carbonate diagenetic textures from near surface diagenetic environments. *Bull. Am. Ass. Petrol. Geol.* **64,** 461–87.

LONGMAN, M.W. & MENCH, P.A. 1978. Diagenesis of Cretaceous limestones in the Edwards aquifer system of south-central Texas: a scanning electron microscope study. *Sediment. Geol.* **21,** 241–76.

LUCIA, F.J. 1972. Recognition of evaporite–carbonate shoreline sedimentation. In: RIGBY, J.K. & HAMBLIN, W.K. (eds) *Recognition of Ancient Sedimentary Environments. Spec. Publ. Soc. econ. Paleont. Mineral.* **16,** 160–91.

MCHARGUE, T.R. & PRICE, R.C. 1982. Dolomite from clay in argillaceous or shale-associated marine carbonate. *J. sedim. Petrol.* **52,** 873–86.

MCKENZIE, J.A. 1981. Holocene dolomitization of calcium carbonate sediments from the coastal sabkhas of Abu Dhabi, U.A.E., a stable isotope study. *J. Geol.* **89,** 185–98.

——, HSÜ, K.J. & SCHNEIDER, J.F. 1980. Movement of subsurface waters under the sabkha, Abu Dhabi U.A.E., and its relation to evaporative dolomite genesis. In: ZENGER, D.H., DUNHAM, J.B. & ETHINGTON, R.L. (eds) *Concepts and Models of Dolomitization. Spec. Publ. Soc. econ. Paleont. Mineral.* **28,** 11–30.

MURRAY, R.C. 1969. Hydrology of south Bonaire, N.A., a rock selective dolomitization model. *J. sedim. Petrol.* **39,** 1009–13.

OLDERSHAW, A.E. & SCOFFIN, T.P. 1967. The source of ferroan and non-ferroan calcite cements in the Halkin and Wenlock Limestones. *Geol. J.* **5,** 309–20.

OPPENHEIMER, C.H. & MASTER, M. 1965. On the solution of quartz and precipitation of dolomite in sea water during photosynthesis and respiration. *Allg. Mikrobio.* **5,** 48–51.

PATTERSON, R.J. & KINSMAN, D.J.J. 1981. Hydrological framework of a sabkha along Arabian Gulf. *Bull. Am. Ass. Petrol. Geol.* **65,** 1457–75.

PATTISON, J., SMITH, D.B. & WARRINGTON, G. 1973. A review of late Permian and early Triassic biostratigrapy in the British Isles. In: LOGAN, A. & HILLS, L.V. (eds) *The Permian and Triassic Systems and their Mutual Boundary. Mem. Can. Soc. Petrol. Geol.* **2,** 220–60.

RADKE, B.M. & MATHIS, R.L. 1980. On the formation and occurrence of saddle dolomite. *J. sedim. Petrol.* **50,** 1149–68.

SCHREIBER, B.C. 1978. Introduction—classification of marine evaporites. In: DEAN, W.E. & SCHREIBER, B.C. (eds) *Marine Evaporites. Soc. econ. Paleont. Mineral. Short Course No. 4, Oklahoma,* 1–5.

SHEARMAN, D.J. 1978. Evaporites of coastal sabkhas. In: DEAN, W.E. & SCHREIBER, B.C. (eds) *Marine Evaporites. Soc. econ. Paleont. Mineral. Short Course No. 4, Oklahoma,* 6–42.

SHERLOCK, R.L. 1911. The relationship of the Permian to the Trias in Nottinghamshire. *Q. J. geol. Soc. Lond.* **67,** 75–117.

SHINN, E.A. 1964. Recent dolomite, Sugarloaf Key, Florida. In: GINSBURG, R.N. (ed.) *South Florida Sediments,* 26–33. Geol. Soc. Am. Guidebook Field Trip 1 Ann. Mtg., Florida.

SHINN, E.A., GINSBURG, R.N. & LLOYD, R.M. 1965. Recent supratidal dolomite from Andros Island, Bahamas. In: PRAY, L.C. & MURRAY, R.C. (eds). *Dolomitization and Limestone Diagenesis. Spec. Publ. Soc. econ. Paleont. Mineral. Tulsa.,* **13,** 112–23.

Smith, D.B. 1974a. Permian. In: RAYNER, D.H. & HEMINGWAY, J.E. (eds) *The Geology and Mineral Resources of Yorkshire,* 115–44. Yorkshire Geological Society, Leeds.

—— 1974b. The stratigraphy and sedimentology of Permian rocks at outcrop in north Yorkshire. *J. Earth Sci. Leeds geol. Ass.* **8,** 365–86.

——1976. The Upper Permian sabkha sequence at Quarry Moor, Ripon, Yorkshire. *Proc. Yorks. geol. Soc.* **40,** 639–52.

TAYLOR, J.C.M. & COLTER, V.S. 1975. Zechstein of the English Sector of the southern North Sea Basin. In: WOODLAND, A.W. (ed.) *Petroleum and the Continental Shelf of North-West Europe, Vol. 1 Geology,* 249–63. Applied Science, London.

TEXTORIS, D.A. 1969. Supratidal origin of Appalachian Basin dolostone (Abstract). *Spec. Pap. geol. Soc. Am.* **121,** 470–1.

WANLESS, H.R. 1979. Limestone response to stress; pressure solution and dolomitization. *J. sedim. Petrol.* **49,** 437–62.

WARDLAW, N.C. 1976. Pore geometry of carbonate rocks as revealed by pore casts and capillary pressure. *Bull. Am. Ass. Petrol. Geol.* **60,** 245–57.

ZENGER, D.H. 1972. Significance of supratidal dolomitization in the geological record. *Bull. geol. Soc. Am.* **83,** 1–11.

J. KALDI, Shell Canada Ltd, Calgary Research Centre, PO Box 2506, Calgary, Alberta T2P 3S6, Canada.

Mineralization in Upper Permian carbonates at outcrop in eastern England

G.M. Harwood & F.W. Smith

SUMMARY: Upper Permian carbonates in eastern England at outcrop feature sub-economic syngenetic and epigenetic Pb-Zn-Cu-Fe-Ba-F mineralization. Pb-Zn-Fe sulphides occur in vugs and breccias of sedimentary origin whereas much barite replaces the host carbonate. Vein mineralization is rare. The majority of occurrences are within the Cadeby Formation (Lower Magnesian Limestone; EZ1 Ca)

More than 70 occurrences of mineralization have been recorded, clustered on centres at Farnham, Bramham, Edlington and Mansfield. Host carbonates at the first three of these centres overlie major cross-cutting structures in pre-Permian strata; these structures were possible flow channels for mineralizing solutions. Mineralization in the Mansfield area can be explained by the upwards leakage, on a broad scale, of formation fluids from Carboniferous strata that overlie the South Pennine basement high.

Two Zechstein carbonate units, the Cadeby Formation (former Lower Magnesian Limestone; EZ1 Ca) and the Brotherton Formation (former Upper Magnesian Limestone; EZ3 Ca) crop out in the eastern England counties of Yorkshire, Derbyshire and Nottinghamshire. The mineralization discussed in this paper is restricted to the Cadeby Formation; only one, no longer exposed, mineralized locality, has been recorded from the Brotherton Formation (Marshall 1856; Hunt 1884). The Marl Slate, the English equivalent of the Kupferschiefer, does not crop out in the counties covered in this paper although it is present in the subsurface further east and crops out further north in County Durham. No metalliferous minerals are currently exploited from the carbonates, although there are records of former mining in the Whitwell area (Pilkington 1789) and at Farnham (Marshall 1856; Hunt 1884).

In this paper we attempt to demonstrate that various episodes of mineralization took place in the Cadeby Formation and that the mineralization resulted from fluids of diverse origin. Where mineralized occurrences cluster within a geographical area, we discuss controls on location and possible genesis.

The Cadeby Formation comprises a lower, Wetherby Member, and an upper, Sprotbrough Member. Both consist of nearshore carbonates deposited on a shallow water platform; facies are described by Harwood (1981b) and summarized in Table 1. In the north of the area the Cadeby Formation thins on to a Permian topographic high (Smith 1974) whereas in the south it thins out against the gently eastwards and northwards sloping early Permian land surface. The Cadeby Formation averages 30–40 m thick at outcrop and dips eastwards at about 2°. The formation is dolomitized throughout the area discussed in this paper in all but two exposures. Exposure is generally poor and detailed correlation is commonly difficult. Many quarries and railway cuttings formerly described in the literature have now been filled in. Several mineralized localities documented in this paper were from temporary exposures.

Mineralization in the Cadeby Formation was both penecontemporaneous (syngenetic) and epigenetic. Penecontemporaneous mineralization is facies-dependent and the minerals may be present in trace amounts over a wide area. Epigenetic mineralization, in general, shows less relationship to facies, and the minerals present differ between individual occurrences or between the individual geographical areas considered here. Both penecontemporaneous and epigenetic mineralization may have occurred in the same area and, rarely, in the same exposure. All mineralized localities known to the authors, by personal field work and literature searches, are shown on Fig. 1.

Penecontemporaneous mineralization

Calcitized anhydrite nodules

Lagoonal wackestones of the Wetherby Member at Wormald Green (SE 300650) contain calcitized displacive anhydrite nodules with sphalerite, marcasite, barite, and galena. Harwood (1980) presents petrographic evidence for calcitization and concurrent mineralization during penecontemporaneous diagenesis. This interpretation is further supported by stable (carbon, oxygen and sulphur) isotope determinations (Harwood & Coleman 1983). Similar replaced displacive anhydrite nodules containing sphalerite are found at Mansfield Woodhouse (SK 532648) (Table 1).

From HARWOOD, G.M. & SMITH, D.B. (eds), 1986, *The English Zechstein and Related Topics*, Geological Society Special Publication No. 22, pp. 103–111.

TABLE 1. Summary of relationships between mineral assemblages and facies within the Cadeby Formation and with structural elements in the underlying strata

Mineral assemblage	Location(s)	Cadeby Formation Member and host facies	Timing	Relation to structure
Sphalerite ± marcasite ± calcite ± galena ± barite	Wormald Green (Fig. 2) Mansfield Woodhouse (Fig. 5)	Wetherby Member; lagoon wackestones with displacive evaporite nodules	Penecontemporaneous	None apparent
Stratabound galena	Nottinghamshire (Fig. 5)	Wetherby Member; boundary of lower clay-rich lagoon sequence with lagoon wackestones	Penecontemporaneous	None apparent
Barite in hydraulic breccia	Well, N. Yorkshire	Wetherby Member; lagoon wackestones	Penecontemporaneous	None apparent
Copper mineralization	Farnham (Fig. 2)	Wetherby Member; ?lagoon wackestones	Epigenetic	Associated with intersection of major basement structures and post-Permian folding and fracturing
Barite ± fluorite ± haematite ± copper minerals	Bramham (Fig. 3)	Wetherby and Sprotbrough Members; little or no facies dependence	Epigenetic; post-dolomitization and after some fracturing	Overlies pre-Permian faults and folds
Galena + cerussite ± calcite	Edlington (Fig. 4)	Sprotbrough Member; breccia facies	Epigenetic; post-dolomitization and after some fracturing	Overlies pre-Permian Don Monocline and associated fracture belt
Barite with rare galena	Whitwell (Fig. 5)	Mostly in Sprotbrough Member ooid grainstone shoals	Epigenetic; post-dolomitization and after some faulting	Overlies pre-Permian Whitwell Anticline
Barite (dispersed) ± galena ± marcasite ± Fe oxides	Nottinghamshire (Fig. 5)	Wetherby Member; lagoon wackestones with evaporites and siliciclastic-rich carbonates	Epigenetic; post-sucrose dolomite formation and after evaporite dissolution	No apparent direct relationship
Galena ± wulfenite ± asphaltite ± rare sulphides ± rare barite	Nottinghamshire (Fig. 5)	Sprotbrough Member; ooid grainstone shoals and siliciclastic-rich carbonites	Epigenetic post-dolomitization	No apparent direct relationship

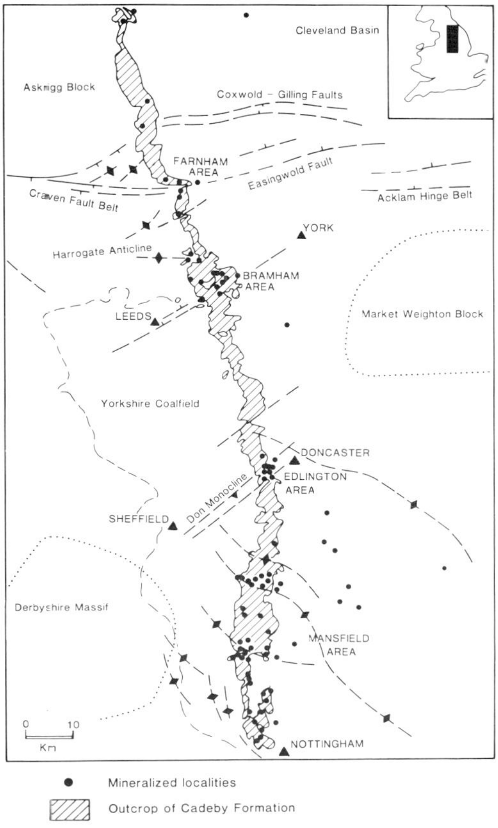

FIG. 1. Crop of the Cadeby Formation showing the occurrence of mineralization and structural trends in the Permian and underlying strata.

Stratabound galena

Galena is present as replacive cubes in bivalve mounds, on irregular fracture surfaces and disseminated through the upper 30 cm of a partially dolomitized grey clay wackestone/mudstone lagoon facies. Galena is the only mineral present. This facies, previously known as the 'Lower Marls', is present near the base of the Wetherby Member in north Nottinghamshire, Derbyshire and southern Yorkshire both at outcrop and in the subsurface (Edwards 1967; Smith *et al.* 1973). Above the level of the galena occurrences, the Cadeby Formation consists of more oxic, buff-coloured completely dolomitized lagoon mudstones and wackestones with little clay. Sulphur isotope analysis (Harwood & Coleman 1983) suggests that the sulphur originated from bacterial reduction of seawater sulphate. This, together with occurrences of galena over a wide area at the same facies boundary (Fig. 5), suggests penecontemporaneous galena mineralization (Table 1). Epigenetic barite-calcite-galena mineralization, present in some parts of the same area (Ineson *et al.* 1972), is found higher in the Cadeby Formation.

Hydraulic brecciation and barite cementation

A unique, barite-cemented breccia pipe is found directly above a penetrative flame structure in lagoonal wackestones and packstones of the Wetherby Member at Well, North Yorkshire (SE 266813). The breccia is clast-supported at the margins but matrix-supported in the centre, suggesting instantaneous cementation. The small crystal size ($< 50\ \mu m$) of the barite also suggests multiple site nucleation in a very short time. Dewatering and vertical fluid expulsion, causing hydraulic fracturing of the overlying lithified dolomites, is considered to have taken place during penecontemporaneous diagenesis (Harwood 1981a). Harwood further suggests that the barium was formerly present in the lattice structure of gypsum crystals; evidence of dewatering was found only in areas where penecontemporaneous evaporite dissolution had taken place.

Epigenetic mineralization

When the epigenetic mineral localities within the Cadeby Formation are plotted on a map of the outcrop of the Cadeby Formation, four geographical groupings can be discerned (Fig. 1). These are centred on Farnham, Bramham and Edlington with a broader spread around Mansfield. Between the groups there are areas where there is little or no epigenetic mineralization. The degree of exposure of the Cadeby Formation is sufficiently constant throughout the area to assure the authors that this grouping is real and not a factor of exposure density. Further east, in the subsurface, mineral records are more suspect. Mineralization in borehole cores through the Cadeby Formation may have been overlooked; altogether cores are few. Records of mineralization east of the outcrop cannot, therefore, be taken to be an unbiased indication of mineral distribution.

Underlying structural control has been demonstrated to be an important factor in localizing Ba-F-Pb-Zn mineralization in carbonate rocks of equivalent age to the Cadeby Formation in County Durham (Hirst & Smith 1974). We would similarly relate the four centres of mineralization described here spatially to structures in the pre-Permian strata and to basement blocks.

The Farnham area

Mineralization

Several records exist of copper minerals and mining in this area (Sedgwick 1829; Marshall 1856; Hunt 1884). Marshall (1856) described the mined ore as occurring 'in masses, seldom exceeding the size of a large potato'. Chalcopyrite, bornite, covellite and malachite are known to have been present in nodules and on joint coatings. The 'nodules' could have been void fills after dissolution of displacive evaporites. Surface exposures in the area are of Sprotbrough Member ooid and pellet grainstones; shaft depths of up to 15 m (16 yards) documented by Marshall (1856) would have reached Wetherby Member carbonates at about the same stratigraphic level as the displacive anhydrite nodules at Wormald Green a few kilometres further north (Fig. 2). D.B. Smith (pers. comm., 1983) states the top of one of these shafts started just below the base of the Sprotbrough Member. Soil sample traverses by H.V. Warren (pers. comm., 1980) over drift-free slopes of the Cadeby Formation show copper values of up to 0.1% (total extractable Cu). Total copper values decreased where the Cadeby Formation is covered by several metres of drift. The Spellow Hill Borehole core log records 0.63 m of hard limestone with pyrite and galena in the Wetherby Member, indicating that mineralization continues eastwards in the subsurface.

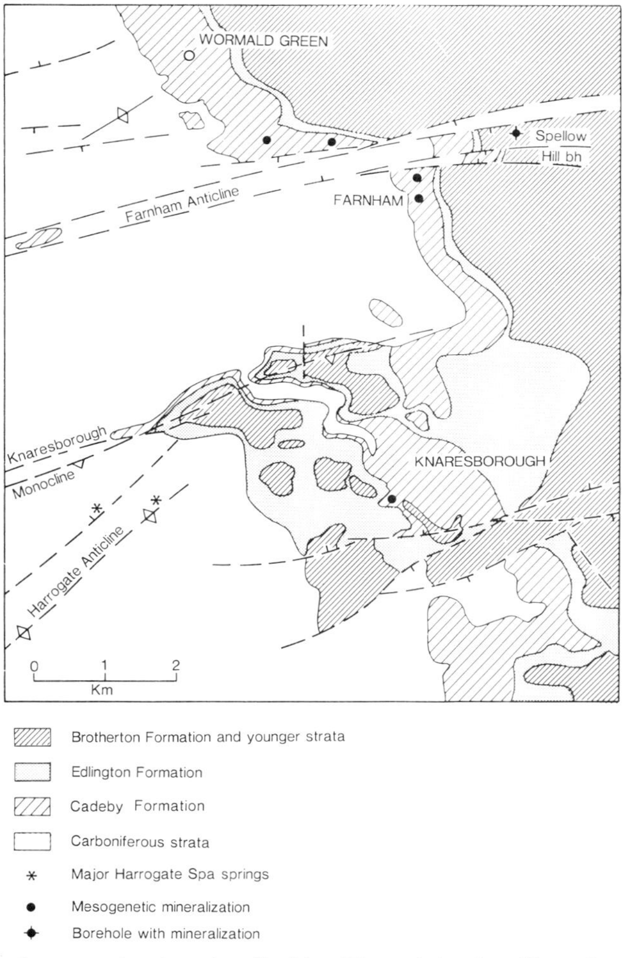

FIG. 2. Map of the Farnham area showing mineralized localities and structure. Formation boundaries are from Dundas (1980), Fox-Strangeways (1908), published BGS maps and personal interpretation.

Structural control

The Farnham area contains the largest fold structures (the Knaresborough Monocline and the Farnham Anticline) known at outcrop in the Cadeby Formation. These folds overlie the northeastwards plunging Harrogate Anticline, a late Carboniferous structure. Warm, saline springs are present above faults on the flanks of the Harrogate anticline in Harrogate itself (Fox-Strangeways 1908). The axis of the Farnham anticline also follows the eastern continuation of the Craven Fault Zone (Fig. 2). This zone is a major Hercynian fracture lineation which forms the southern margin of the Askrigg Block (Fig. 1) and is associated with mineralization in Carboniferous strata to the west (e.g. the Greenhow and Grassington mining districts; Dunham & Stubblefield 1945; Dunham 1959). East of the Farnham area, the Acklam Fault Belt (Kent 1974) and its associated Coxwold–Gilling Faults (Fig. 1) forms the northern margin of the Market Weighton Block (Bott *et al.* 1978).

The copper and lead mineralization is localized in the faulted Farnham Anticline (Fig. 2). This occurrence of base metals in spatial association with conjectured major, and complex, basement faulting, together with the possibility of a corner to corner juxtaposition of two basement blocks, suggests the likelihood of a deep-rooted source for mineralization in the Farnham area (Table 1).

The Bramham area

Mineralization

Barite mineralization in the Bramham area is widespread in the Cadeby Formation and is also present in underlying Namurian sandstones (Versey 1925; Barratt 1975). In the Cadeby Formation barite occurs: (i) with calcite and marcasite/goethite replacing original displacive anhydrite nodules in supratidal carbonates; (ii) in supratidal and intertidal carbonates, patch reefs and shallow water peloid grainstones as vug linings, joint coatings and irregular veinlets and (iii) as large scale replacive flats (up to 200 m long by 1 m thick) in both cryptalgal laminites and peloid grainstone shoals. There is little or no facies control of the larger barite occurrences although in places replacement flats are confined between thin clay beds. Within the flats barite replaces the dolomite host rock but is also found in cross-cutting pink-white colloform veins and intergrown with fluorite and hematite/goethite. The crystal form of the replacive barite is somewhat dependent on the crystal size of the dolomite precursor; spherulitic barite replaced dolomicrites whereas bladed barite formed where there are dolomicrospars or more coarsely crystalline dolomites. The barite spherules form 'pseudopisoids' which are rarely related to former carbonate pisoids. J. Barratt (pers. comm. 1980) has also found traces of chalcopyrite, azurite and malachite in the area, perhaps of similar origin to the malachite reported from Newton Kyme (Marshall 1856), an occurrence thought to have been in the Brotherton Formation. No base metals other than copper have been noted in the Bramham area (Table 1).

A source of sulphate from Permian seawater or evaporites is suggested both from the presence of barite in former anhydrite nodules and from sulphur isotope analysis of barite which replaced cryptalgal laminites (Harwood & Coleman 1983).

Structural control

On a local scale, the degree of exposure is too poor to reveal any intimate relationship between mineralization and faults which cut the Permian strata. On a broader scale, however, the barite occurrences overlie both the extension of the Wharfe Anticline and the NE–SW trending Roundhay and Farsley-Scarcroft fault zones (Fig. 3) (Edwards *et al.* 1950). Together these

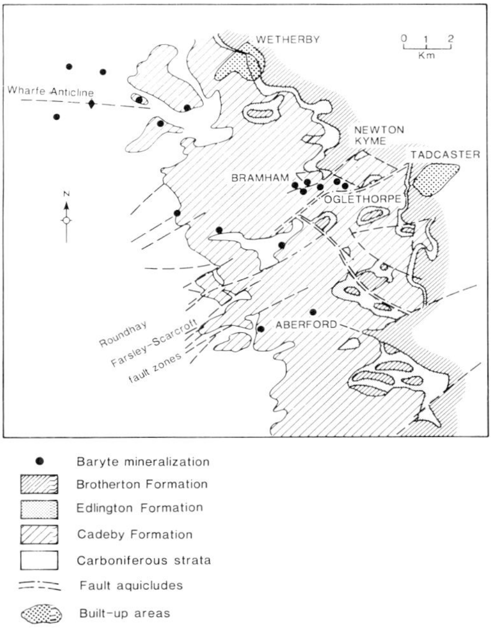

FIG. 3. Barite mineralization in the Bramham area. Localities compiled from Barratt (1975) and the authors' research. Geological boundaries based on Aldrick (1978) and published BGS maps.

structures, mainly of Hercynian age, mark the northern boundary of the exposed Yorkshire–Nottinghamshire coalfield (Fig. 1). The Bramham district lies over, and just north of, this boundary. Present-day groundwater flow in the area is fault-controlled (Aldrick 1978). Known fault aquicludes in the Permian (Aldrick 1978) isolate at least one fault-bounded (south of Oglethorpe) that contains no recorded mineralization. This suggests that movement of the mineralizing fluids may have been subject to similar constraints to the present-day groundwater flow. Saline springs, with a different water chemistry from the present-day groundwater (Edwards *et al.* 1950; Aldrick 1978), are present in the area. For example, at Boston Spa, 3 km north of Bramham, spring water has been reported with 410 g l^{-1} $BaCl_2$ and to be fetid with hydrogen sulphide (Edwards *et al.* 1950). The barium may originate from Carboniferous formation waters (cf. Fox Strangeways 1908; Anderson 1945) channelled through structures along and near the coalfield boundary.

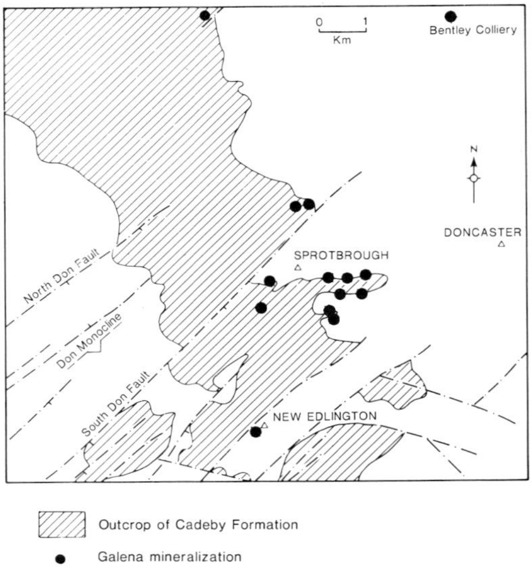

FIG. 4. Galena mineralization and structure in the Edlington area. Formation boundaries from published BGS maps.

The Edlington area

Mineralization

In this area ooid grainstone shoals in the upper part of the Sprotbrough Member formed rounded domes, some of which are now exhumed as hillocks (Harwood 1981b); similar domes have been recorded further south at Gildingwells (Edwards 1967). A discontinuous breccia, comprising clasts of ooid grainstones together with some clasts of continental-derived green marls, is found between shoal crests. The breccia ranges in thickness from 1 cm to nearly 1 m and is overlain by the continental clastic sediments of the Edlington Formation. Sedimentological evidence shows that the breccia was deposited subaqueously and later cemented by isopachous fringing cements (Harwood 1981b). Galena, now partially altered to cerussite, and later calcite occur throughout the breccia in interclast voids. Galena is only present in the underlying grainstones at one locality (New Edlington Brick Works Quarry (SK 532986), Fig. 4) where it occurs on the upper surface of the grainstones and in burrow voids. Galena mineralization reaches its maximum recorded intensity at this locality with an overall grade of 0.2% Pb throughout 0.8 m. Individual samples here were found to contain up to 3% Pb. Deans (1961) also noted stringers of slightly uraniferous asphaltite in the breccia. In the subsurface, galena was also recorded from the top of the Cadeby Formation at Bentley Colliery Shaft (Mitchell *et al.* 1947; Fig. 4).

Structural control

Although the main galena localities in the area are strata-controlled, the discovery of galena on NE–SW trending fractures in grainstones and packstones of the Wetherby Member in a quarry at Sprotbrough (SE 533015) is an indication of possible structural control. The mineralized area overlies the Don Monocline (Figs 1 and 4; Mitchell *et al.* 1947). This structure trends NE–SW and has a south-easterly downthrow of at least 600 m in the underlying Coal Measures. Associated faults were reactivated after lithification of the Cadeby Formation (Fig. 4) and provide possible flowpaths for lead-bearing fluids (Table 1).

The Mansfield area

Mineralization

Four different mineral assemblages are found in an area around Mansfield (Fig. 5), together with a number of uncertain, or poorly recorded, occurrences. The four assemblages comprise (1) penecontemporaneous stratabound galena; (2) vein, colloform and replacement barite with trace galena (Ineson *et al.* 1972); (3) dispersed barite crystals with few sulphides (e.g. Aldred 1969) and

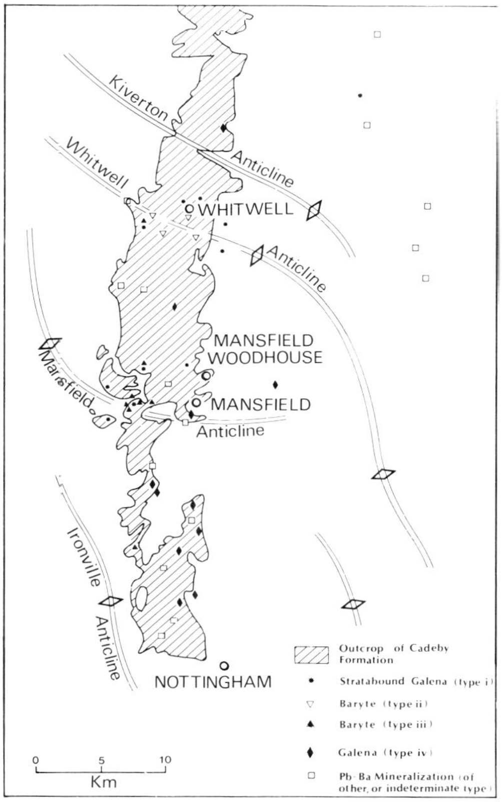

FIG. 5. Mineralization assemblages in the Mansfield area. Fold trends in the underlying Carboniferous strata are also delineated (after Edwards 1951).

(4) stratiform galena-wulfenite mineralization (Deans 1961). Although assemblage (1) is probably penecontemporaneous, the other three assemblages are epigenetic.

The vein, colloform and replacement barite mineralization (assemblage 2) occurs mainly in ooid grainstones of the Whitwell area (Ineson *et al.* 1972). Galena is not common, despite records of lead mining (Pilkington 1789). The mineralization is clearly epigenetic since it post-dates both dolomitization and some fracturing. Robinson & Ineson (1979) have shown the barite sulphur isotopes to be unrelated to Permian seawater or evaporites. Ineson *et al.* (1972) suggested that sub-Permian barium-rich fluids were focussd upwards by the Hercynian age Whitwell Anticline (Fig. 5) and flowed into the Permian strata along fractures.

The dispersed barite mineralization with sparse sulphides (assemblage 3) occurs as vug infills, intercrystalline cement and as minor replacement of dolomite by barite and includes some localities previously documented by Aldred (1969). Sulphides found are galena and marcasite; iron oxides are also present. The minerals occur in lagoonal wackestones of the Wetherby Member, commonly partially filling vugs formed by evaporite dissolution; rarely fossil moulds are also mineralized. Mineralization post-dates sucrose dolomite formation since it corrodes the dolomite crystal margins. Barite sulphur isotopes are not directly related to Permian seawater sulphate (Harwood & Coleman 1983).

The fourth assemblage, described by Deans (1961), comprises very low grade disseminated galena, cerussite, wulfenite and asphaltite with traces of barite, chalcopyrite, malachite, pyrite and sphalerite. The mineralization is found in the topmost 1–2 cm of the Cadeby Formation. Host facies are ooid grainstones and quartz-rich grainstones of the Sprotbrough Member. The immediate control of this mineralization was trapping of upward-migrating fluids by overlying impermeable clays and marls of the Edlington Formation.

Controls of mineralization in the Mansfield area

Mineralization in the form of these three epigenetic assemblages is found over a broad area (Fig. 5), in different facies and overlying diverse Hercynian structures (Figs 1 and 5). In places, as at Whitwell, it has been suggested that known sub-Permian structures have focused and channelled mineralizing solutions upwards (e.g. Ineson *et al.* 1972) One unifying feature, however, is that the whole mineralized area overlies the eastwards extension of the Derbyshire basement massif, over which significant eastwards movement of Carboniferous formation waters has been proposed (Downing 1967; Downing & Howitt 1969). Research into present-day groundwater hydrology and chemistry in the Carboniferous strata by those authors has identified zones of different salinity and chemistry. Downing & Howitt (1969, p. 24[illegible]) interpreted these zones as the probable results of dilution of formation fluids by meteoric water 'moving eastwards and down-dip through aquifers from the outcrop areas and discharging into Permo–Triassic rocks, where these overlie Carboniferous aquifers at depth'. Certain Upper Carboniferous formation brines are known to be enriched in barium in many British coalfields, whereas others are sulphate-rich. Most analyses (Anderson 1945; Edmunds 1975) are from the north-east coalfield, where published values for chloride-rich brines average 625 g l^{-1} Ba. Barium-enriched brines have also been recorded in the Derbyshire and Nottinghamshire coalfield (Anderson 1945; Edmunds 1975). Edmunds discovered that coalfield formation water chemistry differed markedly between beds, suggesting little or no passage of fluids between beds. In the east Midlands Downing & Howitt (1969) recorded that, in general, Coal Measures groundwaters are chloride-rich whereas those of the deeper Lower Carboniferous strata have a higher sulphate content. Where previously separated flows from these different groundwater sources discharge into the Cadeby Formation (Downing & Howitt 1969), it would be reasonable to predict fluid interaction and consequent mineral deposition. In this way, formation fluid migration could result in the low grade mineralization found in the Mansfield area.

If this is the correct interpretation of the source of the mineralizing fluids, there is little prospect for substantial base metal ore deposits in the Mansfield area. The base metal content of analysed coalfield brines is low compared with their average barium content (Edmunds 1975), and burial temperatures of the Lower Carboniferous formation fluids are unlikely to have been sufficiently high to cause substantial leaching of base metals from aquifer sediments.

The maximum dilution, or leakage, of Carboniferous formation fluids has occurred in the area south from Gainsborough to Grantham (Downing & Howitt 1969). This corresponds to the broad area of mineralization in the Cadeby Formation centred on Mansfield. In contrast, north of Gainsborough, towards Doncaster, there has been virtually no flushing of the Carboniferous aquifers. Pore fluids there contain up to 24.6% total dissolved solids (TDS) and the overlying Cadeby Formation in that area is correspondingly unmineralized (Fig. 1) except in the fault/monocline-associated occurrences at Edlington.

Conclusions

We have detailed no single episode of mineralization in the Cadeby Formation; rather we infer several episodes centred on discrete areas. Penecontemporaneous mineralization is facies-related; epigenetic mineralization is facies-related only where voids, formed during carbonate diagenesis, provide sites of mineral precipitation. The timing of the various episodes of mineralization can be related to the diagenetic history of the Cadeby Formation; both facies-relationships and timing are summarized in Table 1.

Penecontemporaneous mineralization was not spatially related to the structure of the pre-Permian basement. In contrast, epigenetic mineralization has been demonstrated to be structure-related. In at least two of the mineralized centres, Farnham and Edlington, deep-rooted structures may have tapped uprising hydrothermal fluids from the sub-Carboniferous basement. We propose that the mineralization centred on the Mansfield area is related to a broad zone of discharge of Carboniferous formation fluids; these fluids are potentially a source of barium rather than base metals.

The variety of both mineral assemblages and styles of mineralization both within and between areas renders correlation of mineralization episodes impracticable. The Cadeby Formation of the Yorkshire Province does, however, overlie a region of high heat flow (Brown *et al.* 1980) and, with several high porosity facies, may be considered a target for future mineral exploration.

ACKNOWLEDGMENTS: This paper in part results from the PhD dissertation of G.M. Harwood at Leeds and the Open Universities. We thank Mike Nutting for aid in mineralization location in Nottinghamshire and D.B. Smith and D.M. Hirst for help with the older mineralization records. We also thank Elizabeth Walton for her accurate typing of the final draft.

References

ALDRED, K. 1969. Occurrences of galena and other minerals in the area west of Mansfield, Nottinghamshire. *Mercian Geol.* **3,** 35–8.

ALDRICK, R.J. 1978. The hydrogeology of the Magnesian Limestones in Yorkshire between the River Wharfe and the River Aire. *Q. Jl eng. Geol.* **11,** 193–201.

ANDERSON, W. 1945. On the chloride waters of Great Britain. *Geol. Mag.* **82,** 267–73.

BARRATT, J. 1975. *Geology of the Ballynoe Baryte deposit, Ireland.* Unpublished PhD thesis. University of London, Imperial College.

BOTT, M.H.P., ROBINSON, J. & KOHNSTAMM, M.A. 1978. Granite beneath Market Weighton, east Yorkshire. *J. geol. Soc. London,* **135,** 535–43.

BROWN, G.C., CASSIDY, J., OXBURGH, E.R., PLANT, J., SABINE, P.A. & WATSON, J.V. 1980. Basement heat flow and metalliferous mineralisation in England and Wales. *Nature,* **288,** 657–9.

DEANS, T. 1961. A galena-wulfenite-uraniferous asphaltite horizon in the Magnesian Limestone of Nottinghamshire. *Min. Mag.* **32,** 705–15.

DOWNING, R.A. 1967. The geochemistry of groundwaters in the Carboniferous Limestone in Derbyshire and the East Midlands. *Bull. geol. Surv. G.B.* **27,** 289–307.

—— & HOWITT, F. 1969. Saline groundwaters in the Carboniferous rocks of the English East Midlands in relation to the geology. *Q. Jl eng. Geol.* **1,** 24–69.

DUNDAS, D.L. 1980. The sand and gravel resources of

the country east of Harrogate, North Yorkshire *Min. Assess. Rep. Inst. Geol. Sci. No.* **70,** 57 pp.

DUNHAM, K.C. 1959. Epigenetic mineralization in Yorkshire. *Proc. Yorks. geol. Soc.* **32,** 1–30.

—— & STUBBLEFIELD, C.J. 1945. The stratigraphy, structure and mineralization of the Greenhow mining area, Yorkshire. *J. geol. Soc. London*, **100,** 209–68.

EDMUNDS, W.M. 1975. Geochemistry of brines in the Coal Measures of north east England. *Trans. Inst. Min. Metall.* **84,** B39–52.

EDWARDS, W. 1951. The Concealed Coalfield of Yorkshire and Nottinghamshire, 3rd edn. *Mem. geol. Surv. G.B.* HMSO, London.

—— 1967. Geology of the country around Ollerton. *Mem. geol. Surv. G.B.* HMSO, London.

——, MITCHELL, G.H. & WHITEHEAD, T.H. 1950. Geology of the District north and east of Leeds. *Mem. Geol. Surv. G.B.* HMSO, London.

FOX-STRANGWAYS, L. 1908. Geology of the country north and east of Harrogate. *Mem. geol. Surv. G.B.* HMSO, London.

HARWOOD, G.M. 1980. Calcitised anhydrite and associated sulphides in the English Zechstein First Cycle Carbonate (EZ1 Ca). In: FUCHTBAUER, H. & PERYT, T.M. (eds) *The Zechstein Basin with Emphasis on Carbonate Sequences, Contr. sedim.* **9,** 61–72.

—— 1981a. Baryte mineralization related to hydraulic fracturing in English Permian Z1 carbonates. In: *Proc. Int. Symp. Central European Permian, Warsaw 1978*, 379–87.

—— 1981b. *Controls of mineralization in the Cadeby Formation (Lower Magnesian Limestone).* Unpublished PhD Thesis, Open University.

—— & COLEMAN, M.L. 1983. Isotopic evidence for U.K. Upper Permian mineralization by bacterial reduction of evaporites. *Nature*, **301,** 597–9.

HIRST, D.M. & SMITH, F.W. 1974. Controls of baryte mineralisation in the Lower Magnesian Limestone of the Ferryhill area, Co. Durham. *Trans. Inst. Min. Metall.* **83,** B49–55.

HUNT, R. 1884. *British Mining.* Crosby Lockwood & Co., London.

INESON, P.R., RICHARDSON, R.T. & WOOD, G.H. 1972. A baryte-galena vein in the Magnesian Limestone at Whitwell, Derbyshire. *Proc. Yorks. geol. Soc.* **39,** 139–50.

KENT P.E. 1974. Structural history. In: RAYNER, D.H. & HEMINGWAY, J.E. (eds) *The Geology and Mineral Resources of Yorkshire*, Yorkshire Geological Society, Leeds. 13–28.

MARSHALL, W. 1856. Notice of carbonate of copper occurring in the Magnesian Limestone at Newton Kyme, near Tadcaster. *Trans. geol. Soc. London*, **2,** 140.

MITCHELL, G.H., STEPHENS, J.V., BROMEHEAD, C.E.N. & WRAY, D.A. 1947. Geology of the country around Barnsley, Yorks. *Mem. geol. Surv. G.B.* HMSO, London.

PILKINGTON, J. 1789. *A View of the Present State of Derbyshire, with an Account of its most Remarkable Antiquities*, 2nd edn. Marriot, Derby.

ROBINSON, B.W. & INESON, P.R. 1979. Sulphur, oxygen and carbon isotope investigations of lead-zinc-baryte-fluorite-calcite mineralisation, Derbyshire, England. *Trans. Inst. Min. Metall.* **88,** B107–17.

SEDGWICK, A. 1829. On the geological relations and internal structure of the Magnesian Limestone and the lower portions of the New Red Sandstone series in their range through Nottinghamshire, Derbyshire, Yorkshire and Durham to the southern extremity of Northumberland. *Trans. geol. Soc. London*, Series 2, **3,** 37–124.

SMITH, D.B. 1974. The stratigraphy and sedimentology of Permian rocks at outcrop in North Yorkshire. *J. Earth Sci. Leeds*, **8,** 365–86.

SMITH, E.G., RHYS, G.H. & GOOSSENS, R.F. 1973. Geology of the country around East Retford, Workshop and Gainsborough. *Mem. geol. Surv. G.B.* HMSO, London.

VERSEY, H.C. 1925. The beds underlying the Magnesian Limestones in Yorkshire. *Proc. Yorks. geol. Soc.* **20,** 200–14.

G.M. HARWOOD, Department of Geology, The University, Newcastle upon Tyne NE1 7RU.
F.W. SMITH, Hallgarth Grange, Kirk Merrington, Spennymoor, Co. Durham DH6 7HZ.

The Trow Point Bed—a deposit of Upper Permian marine oncoids, peloids and columnar stromatolites in the Zechstein of NE England

D.B. Smith

SUMMARY: An extensive thin bed of oncoids, peloids and columnar stromatolites occurs at the top of the Cycle 1 carbonate of the English Zechstein sequence in NE England; here termed the Trow Point Bed, it lies on the lower part of the basin-margin slope, basinward of a linear shelf-edge reef in the Ford Formation. The bed is generally 0.05–0.20 m thick but locally ranges to 0.60 m; it rests on an uneven surface that has a local relief of up to 15 m and slopes exceeding 35° in places. It contains a sparse marine fauna of foraminifera, ostracods and bivalves; there are only a few traces of current action and sorting, and terrigenous sediment nowhere comprises more than 2% of the bed.

The Trow Point Bed is generally similar to deposits described at a comparable stratigraphical position in Germany and Poland and although environmental indications are partly mutually contradictory, it seems likely that the bed accumulated in oxic conditions on the basin floor under water perhaps 25–100 m deep.

Stratigraphy and distribution

The Trow Point Bed is an extraordinarily extensive thin dolomitized bed of oncoids, peloids and subsidiary columnar stromatolites at the top of the Cycle 1 carbonate of the Zechstein sequence in NE England (Fig. 1); it ranges from 3 m to more than 20 m above the base of the EZ 1 carbonate according to the stratigraphic level of an underlying slide plane (see Smith 1970b). The bed lies on the lower part of the basin margin slope, entirely basinward of a linear shelf-edge reef in the Ford Formation; it has not been recognized within 3 km of the foot of the reef wall.

Oncoids at this stratigraphical position in the English Zechstein sequence have been recognized (Smith & Francis 1967, plate 13A; Magraw 1975, 1978; unpublished National Coal Board records) in the cores of several National Coal Board offshore boreholes (see Fig. 2 for distribution); elsewhere oncoids at a similar stratigraphical position have been reported from cored boreholes in the southern North Sea (Taylor & Colter 1975), northern Germany (Fuchtbauer 1968; Richter-Bernburg 1982) and Poland (Peryt & Peryt 1975 and other authors).

On land in NE England the oncoid bed is exposed for about 1 km in coastal cliffs between Trow Point (NZ 386 667) and Frenchman's Bay, South Shields, and less continuously for a further 0.3 km to the SE (Smith 1970a, p. 76); farther north it has been reported in Marden Quarry near Cullercoats (Land 1974, p. 110). Trow Point has been chosen as the type locality because the bed is readily accessible there and displays most of its typical range of variation; it differs at Trow Point from most of its offshore provings, however, in that it has been partly or wholly dedolomitized during or after dissolution of the overlying Hartlepool Anhydrite.

Thickness and field relationships

The Trow Point Bed is unusually varied in thickness. At outcrop it is most commonly 0.05–0.20 m, but ranges locally to 0.60 m; it is absent from or not readily recognizable over perhaps 5–10% of its outcrop. This variation in thickness is closely related to the hummocky character of the top of the underlying slide sequence (Smith 1970b, fig. 8), the bed being thickest and most uniform in the hollows and thinning sharply against and over the steepest and tallest slide blocks and other eminences (Fig. 3); in places it clings to primary slopes of 40° or more. Variation in thickness appears to be less sharp offshore, presumably because there the bed rests on more distal (less hummocky) parts of the slide sequence. Some variations in the thickness partly result from stylolitization.

In the South Shields area the base of the Trow Point Bed has a medium-scale primary relief of at least 15 m, partly because of drape over slide blocks but mainly because of the irregularity of the underlying slide plane. Its base is also uneven in detail and in offshore boreholes B2 and B8 it features cracks up to 8 cm deep and 1 cm across that are partly filled with sediment of the bed itself. in a surface exposure (3941 6602) 1.1 km SE of Trow Point the surface bears a 3–8 cm pattern of grooves resembling desiccation cracks. Cracks

From HARWOOD, G.M. & SMITH, D.B. (eds), 1986, *The English Zechstein and Related Topics*, Geological Society Special Publication No. 22, pp. 113–125.

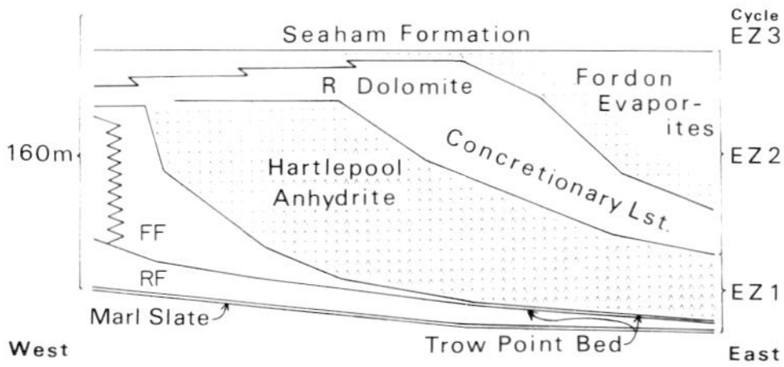

FIG. 1. Stratigraphy of English Zechstein Cycles 1 and 2 in NE England, showing the position of the Trow Point Bed. RF = Raisby Formation; FF = Ford Formation, R Dolomite = Roker Dolomite Formation. The section is about 15 km long.

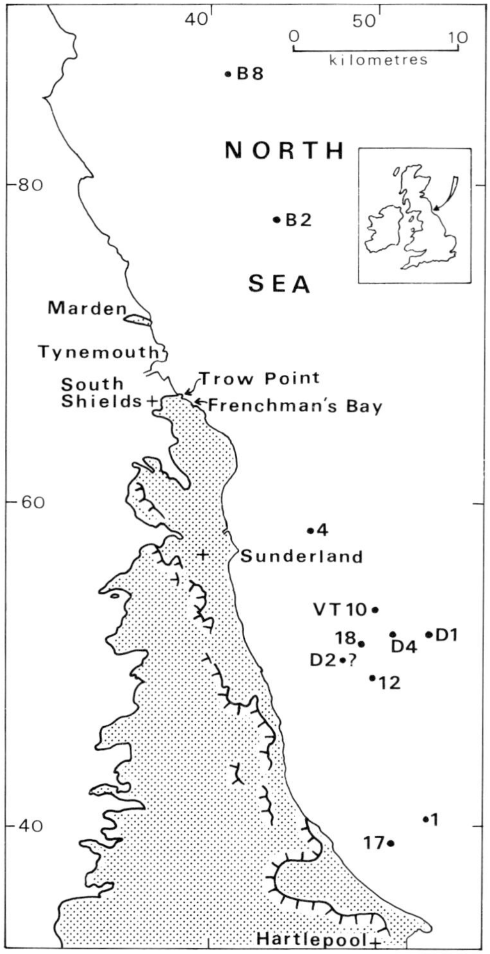

FIG. 2. Part of NE England showing places mentioned in the text and boreholes proving the Trow Point bed. Outcrop of Zechstein strata stippled; the discontinuous toothed line shows the approximate basinward margin of the shelf-edge reef in the Ford Formation.

in the uppermost surface of the Cycle 1 carbonate have also been recorded by Taylor & Colter (1975, plate 2F) in North Yorkshire and by Taylor (1980, p. 101 and fig. 8a) in the southern North Sea.

General description

The Trow Point Bed is mainly a poorly sorted packstone containing coated grains (Fig. 4), most of which are probably algal in origin, but it is diversified by patches of laminar, colleniform and columnar stromatolitic boundstone. The coated grains are prevalent and most varied in low-relief parts of the deposit and where it fills hollows, whereas the colleniform and columnar stromatolites tend to occur preferentially on upstanding areas. In places parts or all of the bed are composed of microcrystalline carbonate but there is a common tendency for an upwards increase in grain size and for columnar structures and other laminated rock to be concentrated in upper parts of the bed. No undoubted algal mats were found in the Trow Point Member during this study, although possible algal mats were noted in it in the cores of Offshore Borehole 18. Cross-lamination is uncommon and desiccation cracks, fenestral fabric, flat-pebble conglomerates and channels appear to be absent.

The top of the Trow Point Bed is generally sharp and slightly uneven. It is most irregular where secondary displacive veins and wedges of fibrous gypsum are abundant and in places the whole bed has been brecciated by such veins. Compaction varies widely, ranging from almost nil to marked (Fig. 4). Stylolites and microstylolites are similarly variable and are absent in many places; commonly microstylolites encircle the grains without much evidence of accompanying loss of volume, but elsewhere much thinning and interpenetration of grains is apparent (Fig. 4d); in places many or most of the grains appear to have been eliminated by pressure-solution and convergence of the intervening residues has given rise to black shaly argillaceous carbonaceous dolomite that superficially resembles algal mats.

Petrography

The fine detail of most of the Trow Point Bed has been obscured by complex diagenesis, including early replacement of secondary dolomite by anhydrite and subsequent widespread dedolomitization and calcitization of anhydrite. Enough detail generally remains, however, to suggest that data revealed by microscope study of the least

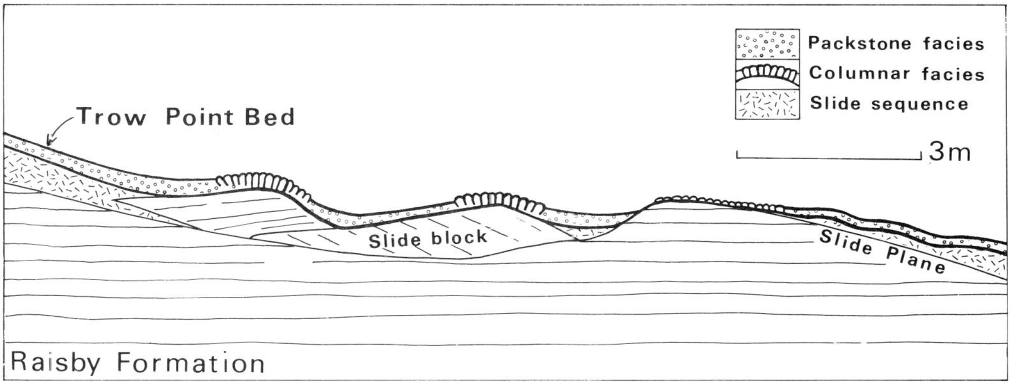

FIG. 3. Mutual relationships of facies of the Trow Point Bed and underlying strata (schematic).

altered specimens are representative of the bed as a whole, though the relative proportions of the several grain types vary greatly. Oncoids and peloids comprise most of the larger grains and they are accompanied by small proportions of spherulites, compound grains, uncoated bioclasts and derived rock fragments; within the micritic matrix there are many small oncoids, peloids and bioclasts and up to 2% of silt-grade quartz. Replacive fluorite, baryte and pyrite occur in the bed in several boreholes and in at least one surface exposure.

Components of the packstone facies

Oncoids

Subspherical to oblate irregularly concentrically laminated microcrystalline simple oncoids (Fig. 5) are a major component of much of the Trow Point Member. They are commonly 1.0–1.3 mm across but range to more than 3 cm, and contain up to 60 laminae (although 10–20 is more usual); in most thin-sections oncoids with widely different numbers of laminae are present. The laminae each comprise a couplet of relatively clear microcrystalline carbonate inside a thinner more turbid finer-grained layer and range from barely perceptible to strongly marked; the former presumably indicate relatively regular accretion and the latter more discontinuous growth. Many laminae are incomplete. Neither voids nor clasts (including fossils) occur within or between the laminae of any of the simple oncoids seen.

No nucleus was seen in many of the oncoids but this may be a function of the plane of sectioning and most of them contain nucleii that comprise more than half the grain (making these superficial oncoids); nucleii include subangular to subrounded lithoclasts and bioclasts but most are of slightly worn rimmed peloids or spherulites (see below). Nucleii composed of fragments of contemporaneously fractured oncoids have not been observed. Inner laminae are generally uneven, and incomplete laminae, eccentric laminae and minor nonsequences are relatively common here.

Most of the oncoids in the Trow Point Bed lack radial fabric but a small proportion are weakly radial in one or more laminae (Fig. 5b); this weak radial fabric is most apparent where laminae are thick and distinct, and appears to stem mainly from variations in turbidity rather than from differences of crystal size or orientation. Only rarely are radial elements continuous across the intervening cryptocrystalline lamina. No unambiguous evidence of a former radial mineral fabric has been noted in the cortices of any of the oncoids seen and the radial structure may have been inherited from coatings of algal filaments or, more probably, from infilling of centripetal endolithic borings.

Throughout the Trow Point Bed, oncoids up to a few millimetres in diameter tend to be more regular and symmetrical than their larger counterparts, and the largest oncoids are generally oblate and are botryoidal in their outer layers; basal flattening or asymmetrically low positioning of the nucleii is extremely uncommon even in the largest oncoids.

Peloids

These are the dominant grain-type in much of the Trow Point Bed and comprise apparently structureless near-spherical aggregates of turbid microcrystalline carbonate; most are 0.7–1.0 mm across, with an observed maximum of 1.2 mm. The peloids generally lack a well-defined rim (Fig. 6a) but some less regular peloids have an unlaminated rim up to 0.1 mm thick of finer-grained more turbid carbonate from which slightly tor-

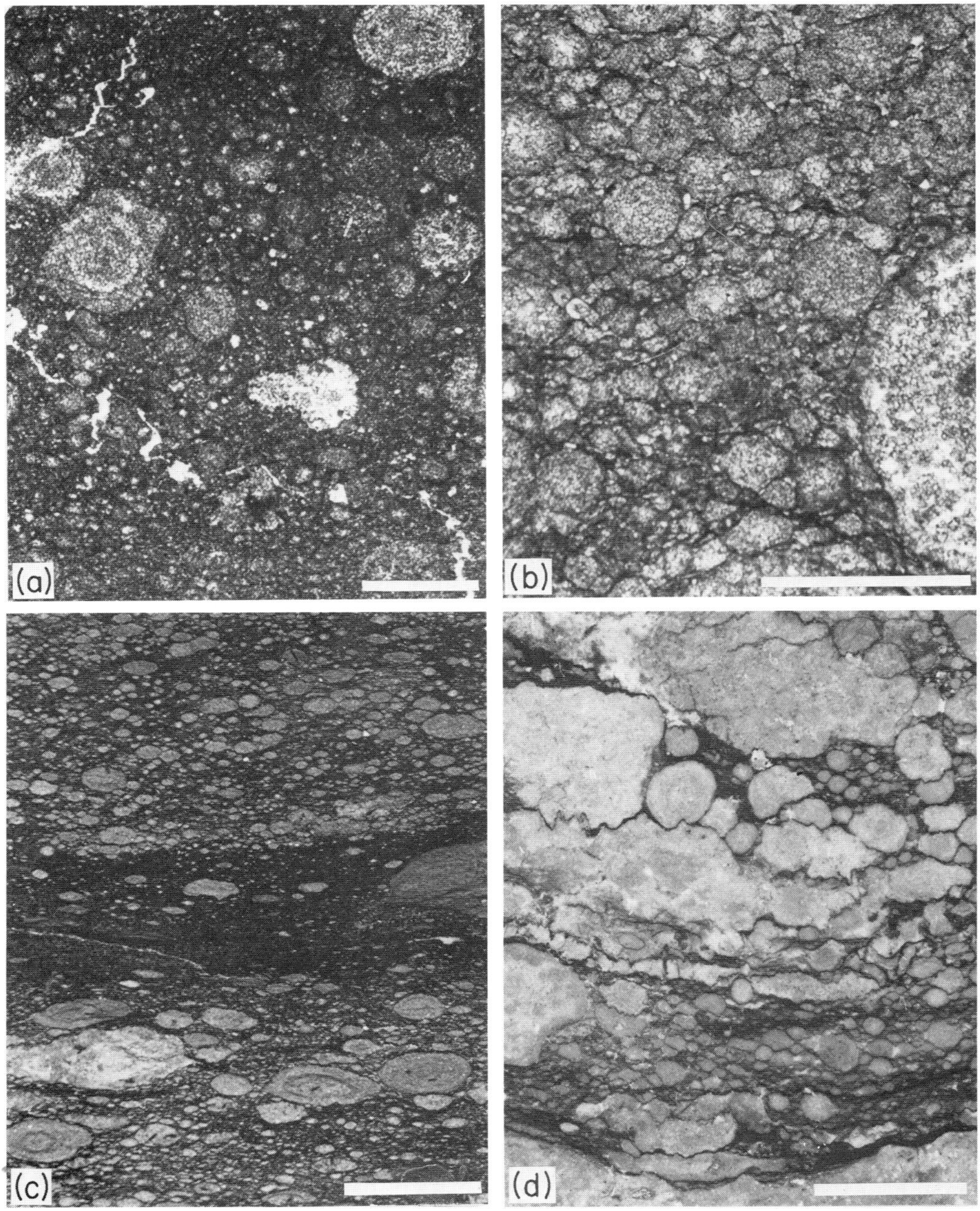

FIG. 4. Typical packstone facies of the Trow Point Bed, showing a range of grain types. (a) Lower part of bed, depth 263.60 m, borehole D4 (NGR NZ 5000 5201). British Geological Survey thin-section E 54516. Scale 2 mm. (b) Middle of bed, mid-north side of Frenchman's Bay, South Shields (NGR NZ 3889 6617); E 54517. Scale 1 mm. (c) Compressed oncoids and lithoclasts near top of Trow Point Bed. Scale 0.5 cm. Offshore Borehole 12, depth 269.75 m (NGR NZ 4993 4896). Sunderland Museum specimen TWCMS D408.4. Photograph by T. Pettigrew. (d) Stylolitic contacts between grains in Trow Point Bed. Scale 1 cm. Offshore Borehole 1, depth 311.05 m (NGR NZ 5333 4043).

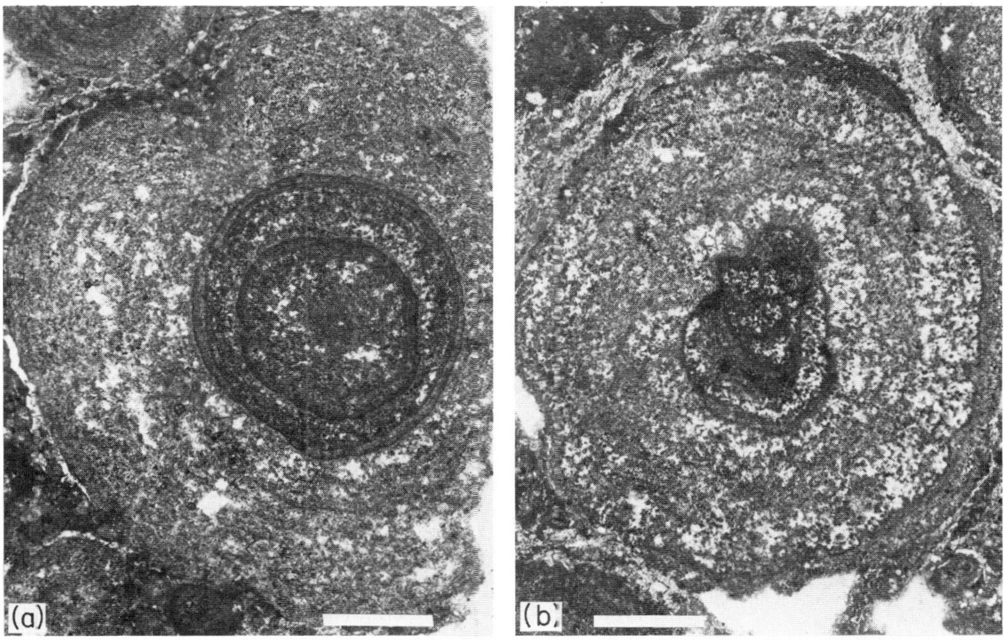

FIG. 5. Typical oncoids from the Trow Point Bed, both from one thin-section (E 54517, mid-north side of Frenchman's Bay, South Shields; NGR NZ 3889 6617). Scale 1 mm. Note the weak radial structure in parts of (b).

tuous but roughly radial processes extend inwards (Fig. 6b,c); micritization by boring algal or other organisms is a possible cause of these rims. Many peloids of both types have a faint concentric structure and it is possible that these were originally oncoids from which the structure was almost obliterated during diagenesis; where they lie in a rock that also contains strongly concentric oncoids some basic difference between the peloids and oncoids seems likely. Where all the grains are unlaminated or only faintly laminated, however, a common origin by homogenization of original oncoids seems possible.

A small proportion of grains in some samples have subspherical to irregular bodies comprising an angular to subrounded nucleus of cryptocrystalline carbonate—probably a derived rock fragment—with a microcrystalline coating similar to that of the peloids. These coatings generally lack clear concentric layers but faint concentricities are apparent in some and there is a complete gradation between these weakly layered grains and oncoids with nucleii. It is possible that some or all of the unlaminated cored peloids may once have been concentrically laminated.

Spherulites

Spherulites up to 0.8 mm across comprise a small proportion of grains with a radial fabric but without strong concentric elements; it is clear from their incorporation as nuclei into some oncoids that the spherulites are primary grains. As in the partly radial oncoids, the radial appearance results from alternating rays of clear and turbid microspar and there is no unambiguous evidence in the thin-sections seen of a former radial mineral fabric that might indicate precursor aragonite; such a precursor cannot be excluded, however. The spherulites range from very pale to very dark in thin-section depending on the proportion of turbid crystals present (Fig. 7). They are commonly irregular in outline and most have a turbid micritic rim resulting, perhaps, from the boring activity of endolithic algae or other organisms.

Compound oncoids

Concentrically-coated botryoidal to oblate oncoids up to 5 cm across are scattered unevenly amongst the generally smaller grains of the Trow Point Bed and in places they form much of the bed (Fig. 8a). Many of these bodies comprise groups of oncoids linked together by confluent outer laminae, but some have a core containing aggregates of several types of grain (Fig. 8b), implying contemporary cementation, fracturing

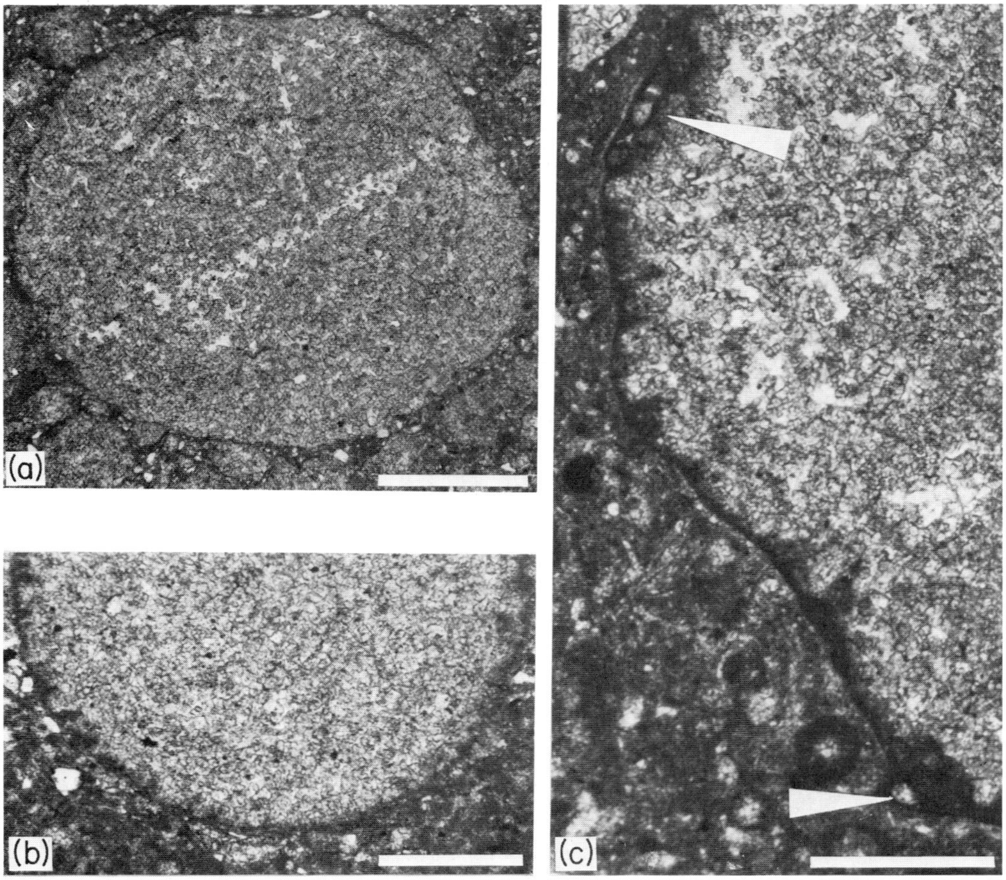

FIG. 6. (a) Unrimmed peloid of microcrystalline carbonate with faint concentric internal lamina and calcite (white) after acicular anhydrite. Scale 0.3 mm. Locality as for Fig. 4(a); E 54515A. (b) Peloid with marked micritic rim possibly caused by boring endoliths. Scale 0.3 mm. Locality as for Fig. 4(a); E 54515B. (c) Peloid with micritized (? endolithically-bored) rim and probable sessile tubular foraminifera (top left and bottom right). Scale 0.2 mm. Locality as for Fig. 4(a); E 54515B.

and reworking. The botryoidal exterior of some of the larger compound oncoids stems from blister-like secondary protuberances in the lamination, and of other results from the incorporation of adjoining smaller grains.

Skeletal grains and indigenous fossils

A marine fauna dominated by complexly coiled foraminifera and well-preserved ostracods is abundant in the Trow Point Bed, especially in its upper parts; almost all the fossils are uncoated and uncrushed even where the oncoids are strongly compressed. Small probable bivalves occur sparingly but remains of vascular plants, crinoids, bryozoans, brachiopods, gastropods, and fish and faecal pellets have not been noted; no evidence of invertebrate burrowing has been recognized. Cyst-like bodies up to 0.3 mm across are abundant in the matrix of most specimens (see discussion under 'matrix').

The almost ubiquitous presence of fossils and the incorporation of some shell fragments into oncoids shows that the fauna was contemporaneous and suggests that the benthonic skeletal remains probably lie close to their life site. The remains of probable sessile tubular foraminifera (Fig. 7c) on a few grains in some samples are almost certainly indigenous.

Matrix

The matrix in the packstone facies, and the body of the rock where it is wackestone, generally comprises a heterogeneous, ill-sorted, unlayered to strongly layered mass of small examples of

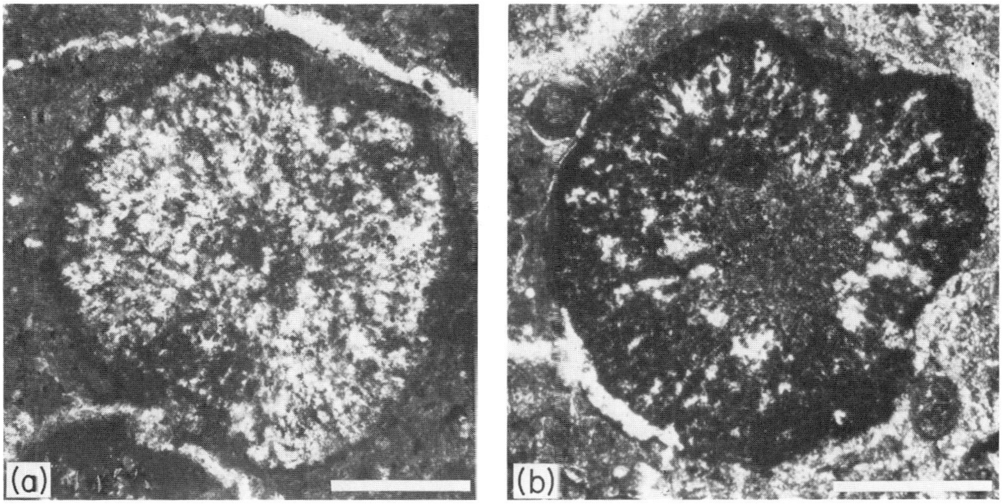

FIG. 7. Pale and dark-coloured spherulites with micritic rims possibly caused by boring endoliths. Note the irregularity of rim of (b), possibly resulting from bioerosion. Scale 0.3 mm. Locality as for Fig. 4(b); E 54517.

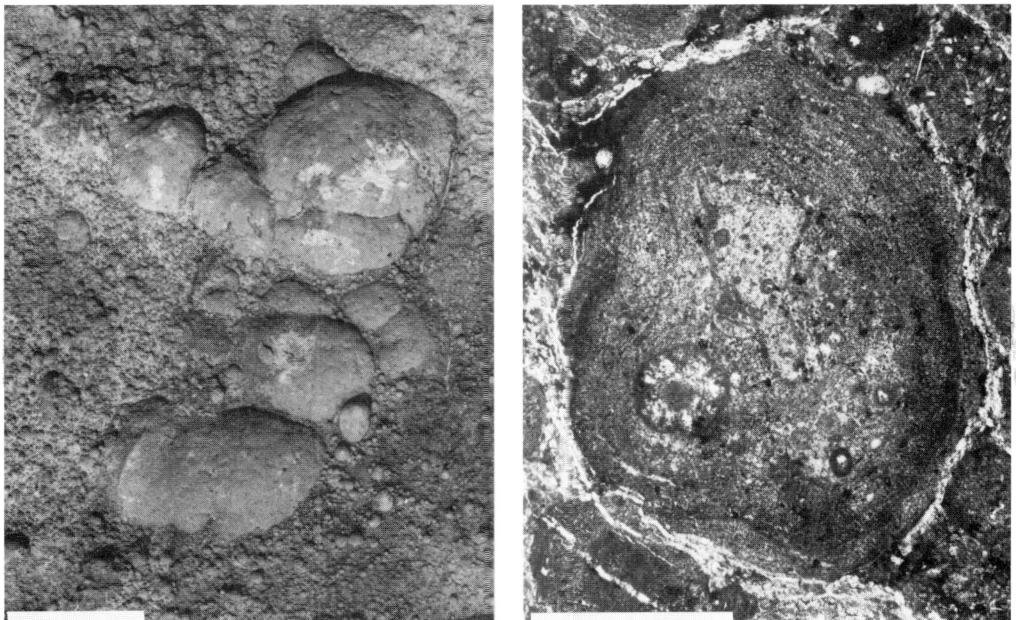

FIG. 8. Plan view of group of large compound oncoids at the top of Trow Point Bed. Scale 2 cm. Sunderland Museum specimen TWCMS:D 420 from rock platform 230 m SE of Trow Point (NGR NZ 3859 6650). Photograph by T. Pettigrew. (b) Compound oncoid with core containing an angular micrite clast (top), a spherulite and a small bioclast (centre). Scale 1.5 mm. Locality as for Fig. 4(b); E 54517.

some or all of the main grain types together with interstitial turbid micrite and microspar. The small cyst-like bodies described noted above are commonly the most abundant constituent, up to 40% of the matrix in some samples, but brown or black generally concordant carbonaceous laminae are a major component where the rock has been strongly compressed. The cyst-like bodies comprise a relatively thick cortex of crytocrystalline turbid carbonate and a core of crudely radial or centripetal drusy microspar; they may be of organic origin but could also have resulted from diagenetic alteration (dissolution and infilling or sparitization?) of the unstable nucleii of primary peloids. Voids are uncommon in the matrix except where sulphate has been dissolved.

Subangular to subrounded quartz grains generally less than 0.1 mm across but exceptionally up to 0.2 mm are present almost throughout the matrix of the Trow Point Bed and small detrital micas occur in some samples. The quartz is most abundant in fine-grained parts of the deposit, where it approaches 2% of its volume, but it is also abundant where the rock has been stylolitically condensed. No siliciclastic grains in the bed are coated and very few were found within or between the laminae of any of the carbonate grains except as components of reworked lithic clasts. An aeolian origin appears possible for the quartz.

Components of the boundstone facies

Laterally-linked hemispherical stromatolites

Upper parts of the Trow Point Bed at several exposures feature extensive sheets of laterally-linked hemispherical laminar stromatolites. Hemispheres range from a few centimetres across near the base of the laminar rock to up to 60 cm in domes produced by the upwards coalescence of stacked hemispheroids (Fig. 9). The hemispheres and domes have a relief of up to about 10 cm although most are 2–5 cm high, and they contain few entrapped grains of any type.

Columnar stromatolites

Perhaps 10–15% of the Trow Point Bed at outcrop is formed of columnar stromatolites

FIG. 9. Plan view of laterally-linked hemispherical stromatolites passing upwards into broadly confluent bun-shaped low domes at top of bed. Scale 20 cm. NE side of Trow Point.

either in closely linked groups or as isolated columns in the packstone facies.

The grouped columns comprise parallel-sided, inverted-conical or slightly club-shaped forms individually generally 2–3 cm across and 5–8 cm high (Fig. 10); observed maxima are 5 cm across and 15 cm high; none of those seen is branched and only a few doubtful examples of confluence were noted. The columns have sharp steep margins against rocks of the enclosing packstone facies and most are separated by relatively narrow pockets of fine-grained detritus similar to the matrix of the packstones. The columns are formed of dense aggregates of micrite and microspar, almost lacking in clasts and only weakly laminated; the lamination is fine, non-crenulate and generally slightly convex upwards, with only a slight downturn at the column margins. Although many columns are vertical or nearly so, there is a marked tendency for groups of 10–50 columns to form slightly radiating arrays and for these groups to form part of larger radial groups cloaking eminences on the substrate; most such eminences are formed by blocks in the underlying slide sequence and the slopes cloaked by the columnar stromatolites are locally quite steep. In some places the grouped columnar stromatolites lie in the upper part of the bed, overlying and flanked by packstones; elsewhere they form the whole bed, commonly in two distinct layers.

The lamination of the grouped columnar stromatolites is different from that of any of the oncoids in the Trow Point Bed, and presumably was created by different organisms in an appreciably different microenvironment; the columns clearly projected above the surrounding seafloor, but it is not clear to what extent they projected at any one time. No fallen columns were found during this study, suggesting either early cementation or low contemporaneous relief; the lack of laminae on the sides of columns implies the latter.

Isolated columnar stromatolites in the packstone facies are more variable in shape than the columns in groups. In plan they are circular or oval and up to 5 cm across but in profile they are barrel-shaped, pear-shaped, biscuit-shaped or compound and up to 12 cm high. The columns originate from any level in the bed, although there is a tendency at any one place for them to rise from one or two preferred levels; at least some are nucleated on to large oncoids. The lamination in the columns (and presumably also the constructing organisms and the inferred microenvironment), is similar to that in the oncoids and field relationships suggest that during growth these columns projected only slightly above the surrounding sediment surface.

Other algal stromatolites

In addition to hemispherical and columnar stromatolites, thin beds and lenses of smooth finely

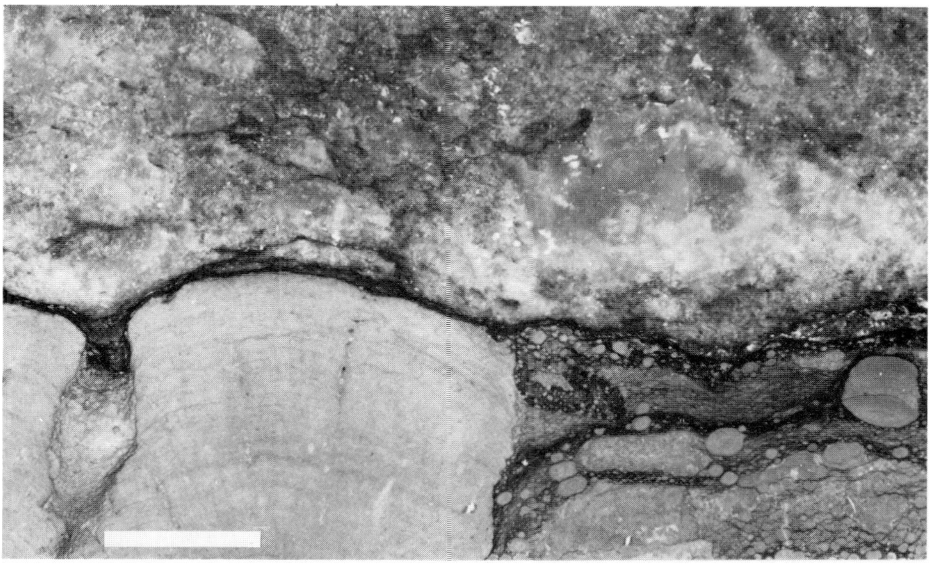

FIG. 10. Contact of Trow Point Bed with gypsified Hartlepool Anhydrite, showing columnar stromatolite boundstone abutting against stylolitized oncoidal packstone with carbonaceous layers resembling algal mats and with small isolated columnar stromatolite (on right). Scale bar 2 cm. Locality as for Fig. 7(b).

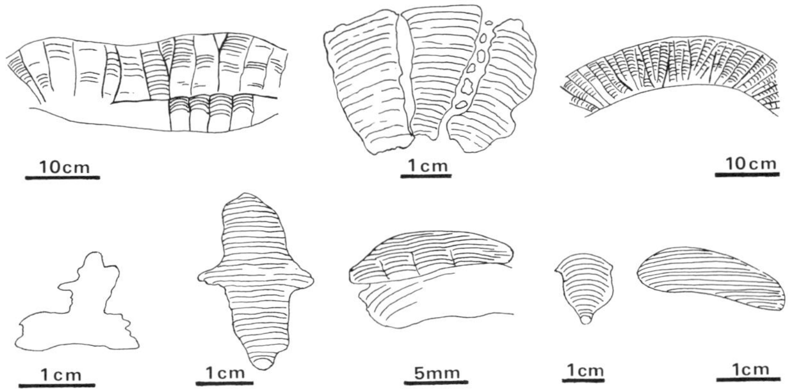

FIG. 11. Sketches of grouped (upper row) and isolated columnar stromatolites from surface exposures and borehole cores of the Trow Point Bed. The irregular isolated column (bottom left) appears to have no lamination but contains much secondary calcite after replacive anhydrite.

laminated carbonate are thinly scattered in parts of the packstone facies. They lack identifiable algal structures but their lithology and associations favour an algal origin.

Comparison with other ancient oncoid-bearing deposits

Although the literature contains many mentions of ancient oncoids and peloids, few accounts fully describe individual grain types, their mutual relationships and their depositional environment. Exceptions include the accounts of oncoids in the Zechstein by Fuchtbauer (1968), Taylor & Colter (1975) and Peryt and his co-workers (Peryt & Peryt 1975; Peryt 1977, 1978a,b, 1981; Peryt & Piatkowski 1976, 1977a,b; Peryt & Wazny 1980; Wagner *et al.* 1978; Oszczepalski 1980).

Oncoids recorded in northern Germany by Fuchtbauer (1968) and Richter-Bernburg (1982) are at the top of the Z1 Zechstein Limestone in the same stratigraphical and palaeogeographical position as the Trow Point Bed; Fuchtbauer recognized that they were unlike oncoids found in shallow-water parts of the Z2 and Z3 carbonates and were associated with foraminifera and ostracods; he also recorded echinoderms, not yet found in the Trow Point Bed, and noted that the dolomite of the oncoids had a composition consistent with a salinity slightly above that of normal seawater. Taylor & Colter (1975) and Taylor (1980) also recorded oncoids of apparently algal origin near the top of the basinal Cycle 1 carbonate in several southern North Sea boreholes; they too found foraminifers and crinoids amongst the oncoids and noted, as did Fuchtbauer, no evidence of reworking. Taylor & Colter reached a similar conclusion to Fuchtbauer that the oncoids were deposited in quiet water perhaps 30–150 m deep.

The oncoids described by Peryt and his associates are also in the upper part of the Zechstein Limestone, and are concentrated in a belt at the foot of the basin slope and on basin-floor eminences. From descriptions and many illustrations it is clear that most or all of the varieties of oncoids and peloids found in the Trow Point Bed are also present in Poland in up to four thin beds of sandy argillaceous dolomite that also contain scattered foraminifers (including encrusting forms), brachiopods, gastropods, crinoids and nautiloids; in places columnar stromatolites (some with encrusting foraminifers) and algal mats containing peloids, bryozoans and gastropods also occur. Arguing from data from closely-spaced boreholes, Peryt & Wazny (1980) suggest that the basinal Zechstein Limestone in western Poland accumulated in no more than 70 m of water and probably generally less than 40 m, with phases of shallowing and local emergence leading to the superimposition of vadose effects near the end of Ca1 deposition.

The common juxtaposition of grains and structures with differing and apparently contradictory indications of depositional environment and with complex histories of growth, encrustation, erosion, calcification and perhaps transport was clearly recognized in the Polish Zechstein 1 carbonates, leading Peryt & Wazny (1980, p. 304) to infer a number of sea-level oscillations in the closing phases of Ca1 deposition; they postulate enhanced salinity during the later stages of deposition in the restricted environment on the basin floor plain. Oszczepalski (1980, pp. 314, 321) considered that oncoids and algal mats were formed on the marginal and basin-floor highs only after the Zechstein 1 Sea became shallower and more saline than in its early stages.

The similarities of the Trow Point Bed and the oncoidal rocks described from the southern North Sea, northern Germany and western and northern Poland is such that these varied occurrences must be regarded as probable correlatives. Unusual widespread environmental conditions must be postulated for the Zechstein basin at this time, but it is not clear whether these conditions were established in Poland earlier than elsewhere or whether the oncoidal rocks in England are a condensed equivalent of the whole of the multiple sequence in Poland.

Comparison with modern oncoid-bearing deposits

In papers dealing with oncoids and other stromatolitic growth forms it is customary to seek modern analogues and to use these analogues as a basis for drawing conclusions on the depositional environment of the supposed fossil equivalent; sooner or later the example of the superbly documented algal deposits of Shark Bay, Western Australia, is generally introduced. But for Palaeozoic stromatolites this well-worn approach has serious drawbacks for, as Monty (1973) pointed out, blue-green cyanophytic algae since the late Cretaceous have been outcompeted in marine environments by red algae and have been increasingly restricted to special situations such as lakes, rivers, lagoons and tidal flats. Unless the Palaeozoic rocks happen to have been formed in just such a special situation, then the supposed analogue can be at best partial and at worst misleading; for example, it is customary to infer that columnar stromatolites in ancient rocks were formed in a hypersaline high subtidal to intertidal environment similar to that of Shark Bay, ignoring the evidence of columnar stromatolites believed from their geological context to have formed under at least some metres of water of normal or only slightly enhanced salinity on the plunging slopes of Devonian reefs in the Canning Basin of Australia (Playford & Cockbain 1969), on the basin floor of the Proterozoic Athapuscow aulacogen of Canada (Hoffman 1974, p. 862), and at the margins of the Zechstein Westerstein reef of Germany (Paul 1980, fig. 6). Similarly, whilst allowing for the lower incidence of submarine sampling, it appears that almost all modern oncoids are formed in peritidal or high subtidal environments, leading to a widespread assumption that ancient oncoids are probably also peritidal or high subtidal. I conclude here that the apparent lack of a full modern analogue of the Trow Point Bed is probably an indication that no such analogue exists.

Depositional environment of the Trow Point Bed

Facts and inferences relating to interpretation of the depositional environment of the Trow Point Bed may be summarized as follows:

(1) It lies basinward of a shelf-edge reef which has a steep seaward slope demonstrably more than 60 m high and probably 100 m (Smith 1981). Assuming no sea-level change and a seaward bottom slope away from the reef, a minimum water depth exceeding 100 m is probable at Trow Point; isostatic settlement of the reef might have reduced this figure.

(2) It lies on top of a major submarine slide sequence that has moved eastwards down a basin-margin slope at least 7 km long; if this slope were at an angle of 1°, a post-slide water depth of at least 120 m at Trow Point would be indicated.

(3) It rests on a depositional surface that has a primary relief exceeding 15 m in coastal exposures near South Shields; a minimum water depth exceeding 15 m is therefore indicated for low-lying parts of the deposit unless it is diachronous. In places the underlying surface bears a pattern of grooves and fissures resembling desiccation cracks.

(4) It is a heterogeneous deposit of ill-sorted mainly cyanophytic algal oncoids and peloids with sessile algal stromatolites concentrated on eminences on the substrate. Many of the grains show evidence of complex growth histories with probable standstill phases marked by micritized cortices. Bioerosion apparently greatly exceeded mechanical erosion.

(5) The oncoids are quite different in appearance from oncoids in Zechstein grainstones inferred from other evidence to have formed in agitated shallow water, and few are nucleated on to algal chips or fragments of earlier oncoids.

(6) The interpretation of turbid rims on many grains as the product of endolithic boring algae and other organisms may limit those grains (unless transported) to depths where endolithic algae can synthesize. Swinchatt (1965) noted that most such grains on the

Great Barrier reef are concentrated in water less than 40 m deep. Foraminiferal borings are not thus limited.

(7) Depth limits for the survival and growth of cyanophytes are poorly documented and estimates differ widely. Brock (1973) quotes a minimum light requirement of 2100–2800 lux which, according to Sverdrup *et al.* (1942, quoted by Brock) is reached today as deep as 100 m only in the clearest ocean water and more commonly is not reached below 50 m. During normal marine conditions the Zechstein Sea, with its low terrigenous input, was probably exceptionally clear and the 100 m figure may have been appropriate; lesser light penetration would have been achieved if salinity rose towards the end of Cycle 1. These figures would, of course, not be relevant if it could be shown that bacteria contributed to the growth of the concentric grains, but it is clear that growth would have been extremely slow, whatever the organisms, near the minimal light-penetration limits.

(8) It contains a marine fauna mainly of organisms tolerant of abnormal salinity, and typically stenohaline forms such as bryozoans, brachiopods and echinoids have not been found. The two main species of foraminifera, *Agathammina* and *Nodosaria*, typify low-energy environments (Peryt 1977, p. 61). The absence of gastropod and fish remains may explain why the cyanophytes apparently suffered little browsing.

(9) It has yielded no evidence of emersion or subaerial weathering and no indicators of peritidal sedimentation. Although the intimate mixing of many different types of grain probably indicates some transport, the presence of a mud-grade matrix, generally poor sorting and lack of cross-lamination in the algal grains all suggest relatively weak bottom currents.

(10) It is immediately overlain by the thick Hartlepool Anhydrite, thought (Taylor & Fong 1969; Taylor & Colter 1975; Taylor 1980) to show features consistent with displacive formation in a supratidal environment. Probable time-equivalent algal pisoids are also present on parts of the basin floor (J.C.M. Taylor, pers. comm.) where parts of the Cycle 1 anhydrite are laminated and may be deep water deposits.

The stratigraphical, palaeogeographical, faunal and sedimentological evidence, although conflicting, leaves little doubt that the Trow Point Bed accumulated (probably slowly) on the lower part of the basin-margin slope and the basin floor of the English Zechstein Sea in an area that had previously lain under more than 100 m of water. This does not prove that the bed itself accumulated under this depth of water, but its visible relief at outcrop when taken in conjunction with the apparent lack of wave action even on the eminences strongly suggests depths in the hollows of at least 25 m. Regardless of whether sea-level fell or not, it thus seems likely that the Trow Point Bed accumulated under moderate depths and perhaps down to or below the 100 m limiting depth suggested by Brock (1976) for the growth of cyanophytes; the apparent absence of the deposit from parts of the basin-floor plain may be a consequence of a depth there exceeding 100 m. The suggested depth bracket of 25–100 m is close to that of 40–70 m suggested from different evidence for the comparable Polish deposits by Peryt & Wazny (1980) and of more than 30 m by Fuchtbauer (1968) for the German equivalent. Salinity slightly above normal is indicated by the fauna, which, by its presence and character, also suggests that bottom waters were oxic even though anoxic conditions probably prevailed a short distance below the sediment surface.

The apparently contradictory depth evidence of the possible desiccation cracks in the underlying surface and of displacive fabric in the immediately succeeding Hartlepool Anhydrite, if correctly interpreted, can only be reconciled with that of the other indicators if one or more sea-level falls of 100 m or so is invoked. The case for such falls has been advocated by the writer (Smith 1970a, 1980) and is supported by the recognition of vadose modification of possible equivalents of the Trow Point Bed in parts of Poland (Peryt & Piatkowski 1976, 1977a,b). Doubts persist, however, especially over the interpretation of the depth of water under which the thick marginal anhydrite was formed, and until these doubts have been resolved the number, extent and amplitude of relative sea-level changes remains uncertain. Such sea-level changes clearly carry the possibility of shallower water accumulation of the Trow Point Bed than other indicators suggest, would allow two-way migration of the oncoid facies, and would imply a measure of diachroneity.

ACKNOWLEDGMENTS: The author is greatly indebted to Drs G.M. Harwood and T.M. Peryt for reading early drafts of the script and for their many helpful and constructive comments, and to Mr T.H. Pettigrew for providing some of the specimens and illustrations and much helpful discussion.

References

BROCK, T.D. 1973. Lower limit for the existence of blue-green algae: evolutionary and ecological implications. *Science*, **179**, 480–3.

FUCHTBAUER, H. 1968. Carbonate sedimentation and subsidence in the Zechstein Basin (Northern Germany). In: MULLER, G. & FRIEDMANN, G.M. (eds) *Recent Developments in Carbonate Sedimentology in Central Europe*, 196–204. Springer-Verlag, Heidelberg.

HOFFMAN, P. 1974. Shallow and deepwater stromatolites in Lower Proterozoic platform-to-basin facies change, Great Slave Lake, Canada. *Am. Ass. Petrol. Geol.* **58,** 856–67.

LAND, D.H. 1974. The geology of the country around Tynemouth. *Mem. geol. Surv. G.B.* London.

MAGRAW, D. 1975. Permian beds of the offshore and adjacent coastal area of Durham and south-eastern Northumberland. *J. geol. Soc. London*, **131,** 397–414.

—— 1978. New boreholes into Permian beds off Northumberland and Durham. *Proc. Yorks. geol Soc.* **42,** 157–83.

MONTY, CL.L.V. 1973. Precambrian background and Phanerozoic history of stromatolitic communities an overview. *Annls Soc. géol. Belg.* **96,** 585–624.

OSZCZEPALSKI, S. 1980. Palaeogeography, sedimentation and mineralization of the Z1 carbonate series (Zechstein) in the western part of the Fore-Sudetic Monocline (western Poland). In: FUCHTBAUER, H. & PERYT, T.M. (eds) *The Zechstein Basin with Emphasis on Carbonate Sequences*, 307–23. Schweizerbart'sche Verlagsbuchhandlung, Stuttgart.

PAUL, J. 1980. Upper Permian algal stromatolite reefs. Harz Mountains (F.R. Germany). In: FUCHTBAUER. H & PERYT, T.M. (eds) *The Zechstein Basin with Emphasis on Carbonate Sequences*, 253–68. Schweizerbart'sche Verlagsbuchhandlung, Stuttgart.

PERYT, T.M. 1977. Environmental significance of foraminiferal-algal oncolites. In: FLUGEL, E. (ed.) *Fossil Algae, Recent Research and Developments*, 61–5. Springer-Verlag, Berlin.

—— 1978a. Microfacies of the carbonate sediments of the Zechstein Werra and Stassfurt cyclothems in the Fore-Sudetic Monocline. *Stud. geol. Pol.* **54,** 1–88 (in Polish).

—— 1978b. Sedimentology and palaeoecology of the Zechstein Limestone (Upper Permian) in the Fore-Sudetic area (western Poland). *Sediment. Geol.* **20,** 217–43.

—— 1981. Former aragonitic submarine hemispheroids associated with vadose deposits, Zechstein Limestone (Upper Permian), Fore-Sudetic area, western Poland. *Neues. Jb. Geol. Paläeont. Mh.* **9,** 559–70.

—— & PERYT, D. 1975. Association of sessile tubular foraminifera and cyanophytic algae. *Geol. Mag.* **112,** 612–4.

PERYT, T.M. & PIATKOWSKI, T.S. 1976. Osady caliche w wapieniu cechsztynskim zachodniej czesci syneklizy perybaltychiej. *Kwart.geol.* **20,** 525–37.

—— 1977a. Stromatolites from the Zechstein Limestone (Upper Permian) of Poland. In: FLUGEL, E. (ed.) *Fossil Algae, Recent Research and Developments*, 124–35. Springer-Verlag, Berlin.

—— 1977b. Algal-vadose pisoliths in the Zechstein Limestone (Upper Permian) of northern Poland. *Sediment. Geol.* **19,** 275–86.

PERYT, T.M. & WAZNY, H. 1980. Microfacies and geochemical development of the basin facies of the Zechstein Limestone (Ca 1) in western Poland. In: FUCHTBAUER, H. & PERYT, T.M. (eds) *The Zechstein basin with Emphasis on Carbonate Sequences*, 279–306. Schweizerbart'sche Verlagsbuchhandlung, Stuttgart.

PLAYFORD, P.E. & COCKBAIN, A.E. 1969. Algal stromatolites: deepwater forms in the Devonian of Western Australia. *Science*, **165,** 1008–10.

RICHTER-BERNBURG, G. 1982. Stratogenese der Zechsteinkalkes am Westharz. *Z. dt. geol. Ges.* **133,** 381–401.

SMITH, D.B. 1970a. Permian and Trias. In: HICKLING, G.A. (ed.) *The Geology of Durham County. Trans. nat. Hist. Soc. Northumb.* **41,** 66–91.

—— 1970b. Submarine slumping and sliding in the Lower Magnesian Limestone of Northumberland and Durham. *Proc. Yorks. geol. Soc.* **38,** 1–36.

—— 1970c. The palaeogeography of the British Zechstein (Abstract). In: RAU, J.L. & DELLWIG, L.F. (eds) *Third Symposium on Salt*, 20–3. Northern Ohio Geological Society, Cleveland, Ohio.

—— 1972. Foundered strata, collapse-breccias and subsidence features of the English Zechstein. In: RICHTER-BERNBURG, G. (ed.) *Geology of Saline Deposits*, 255–69. *Proc. Hannover Symp. 1968.* Unesco, Paris.

—— 1980. The evolution of the English Zechstein Basin. In: FUCHTBAUER, H. & PERYT, T.M. (eds) *The Zechstein Basin with Emphasis on Carbonate Sequences*, 7–34. Schweizerbart'sche Verlagsbuchhandlung, Stuttgart.

—— 1981. The Magnesian Limestone (Upper Permian) reef complex of northeastern England. In: TOOMEY, D.F. (ed.) *European Fossil Reef Models. Spec. Publ. Soc. Econ. Palaeont Miner, Tulsa*, **39,** 161–86.

—— & FRANCIS, E.A. 1967. The geology of the country between Durham and West Hartlepool. *Mem. geol. Surv. G.B.* London.

SWINCHATT, J.P. 1965. Algal boring: a possible depth indicator in carbonate rocks and sediments. *Bull. geol. Soc. Am.* **80,** 1391–6.

SVERDRUP, J.C., JOHNSON, M.W. & FLEMING, R.H. 1942. *The Oceans, their Physics, Chemistry and General Biology*. Prentice-Hall, New York.

TAYLOR, J.C.M. 1980. Origin of the Werraanhydrit in the U.K. Southern North Sea—a reappraisal. In: FUCHTBAUER, H. & PERYT, T.M. (eds) *The Zechstein Basin with Emphasis on Carbonate Sequences*, 91–113. Schweizerbart'sche Verlagsbuchhandlung, Stuttgart.

TAYLOR, J.C.M. & FONG, G.I. 1969. Correlation of Upper Permian strata in east Yorkshire and Durham. *Nature*, **224,** 173–5.

—— & COLTER, V.S. 1975. Zechstein of the English sector of the Southern North Sea Basin. In: WOODLAND, A.W. (ed.) *Petroleum and the Continental Shelf of north-west Europe, vol. 1, Geology*, 249–63. Applied Science Publishers, Barking.

DENYS B. SMITH, 79 Kenton Road, Newcastle upon Tyne NE3 4NL and Department of Geology, University of Durham, South Road, Durham DH1 3LE.

Subsidence and foundering of strata caused by the dissolution of Permian gypsum in the Ripon and Bedale areas, North Yorkshire

A.H. Cooper

SUMMARY: Underground dissolution of thick gypsum beds in the Edlington Formation and Roxby Formation of the Zechstein sequence in North Yorkshire, England, has resulted in a 3 km-wide and 100 km-long belt of ground susceptible to foundering. Within this belt a large subsidence depression at Snape Mires, near Bedale, was largely filled with lacustrine deposits in the later part of the Late Devensian and during the Flandrian. South of Snape Mires the Nosterfield–Ripon–Bishop Monkton area has suffered about 40 episodes of subsidence in the past 150 years, and the presence of several hundred other subsidence hollows indicates considerable activity from the later part of the Devensian onwards. The linear and grid-like arrangement of these subsidence hollows indicates collapse at intersections in a joint-controlled cave system. Linear subsidence features at Snape Mires are also joint-controlled. The transition from anhydrite at depth to secondary gypsum near surface marks the down-dip limit of the subsidence-prone belt. Cavities are propagated upwards by roof collapse of caverns in the gypsum, leading to the formation of breccia pipes. Choking of the pipes can reduce the surface expression of the underground collapse, but the larger cavities are liable to produce pipes that reach the surface even at the eastern boundary of the 3 km-wide belt described. Further subsidence in the Ripon area is predicted and some suggestions for remedial measures are given.

Evidence of the presence of soluble rocks within the Permian sequence is provided from time to time by dramatic subsidence resulting from their subsurface dissolution. The most graphic record of the oldest event is in Mayhall's *Annals of Yorkshire* (1878, Vol. 1, p. 187) where it is recorded that in 1796: 'On October 16th about six o'clock in the morning, the inhabitants of Ripon were greatly alarmed by a violent earthquake, which shook almost every house in the town; a mile from which, near Littlethorpe, about three roods of ground sunk nineteen fathoms, and a large ash tree growing on the spot entirely disappeared. For some time the gulph continued to increase, and an immense body of water issued from it, which filled the inhabitants with fear, for as there were no coal pits in the neighbourhood, it was evidently a great natural convulsion.' Between 1796 and 1881 eight more subsidences were recorded by various authors (Table 1). Suggestions that the subsidence hollows were the result of gypsum dissolution (Tute 1868, 1870; Kendall & Wroot 1924) were supported by Smith (1972).

During the recent 1:10,000-scale geological resurvey of the area the author recorded several hundred subsidence hollows affecting Late Devensian and Flandrian deposits. Many of the local landowners and farmers remember holes appearing on their land and their recollections have allowed the record of subsidences to be updated (Table 2, Fig. 2).

Geology

The Upper Permian and Lower Triassic rocks of this part of Yorkshire have been documented by Smith (1974a,b) and typify the marginal Zechstein sequence in England (Fig. 1). They dip gently eastwards and are briefly described below with emphasis on the soluble strata.

The Cadeby (Magnesian Limestone) Formation forms a marked escarpment that extends southwards from Bedale to Knaresborough and beyond, passing to the west of Ripon; it is up to 65 m thick, rests unconformably on Carboniferous strata and is composed mainly of porous dolomite and dolomitic limestone, details of which are given by Smith (1974a,b). The rock is an excellent aquifer that commonly yields artesian water down dip, and is hydrogeologically uniform throughout its outcrop (Aldrick 1978). The formation is generally well jointed and water-widened fissures are numerous. Caves, however, are uncommon and only nine have been recorded in North Yorkshire. Most of these caves are small and the largest is of tension fissure origin (Lowe 1978). During the mapping few stream sinks and only a few springs were observed. The most notable example of an underground river is provided by the Skell, which at times sinks near Fountains Abbey and rises again at Hell Wath, 2 km downstream and 40 m lower (Lowe 1978).

The Edlington Formation is exposed in only a few scattered places and most of the details of it

From HARWOOD, G.M. & SMITH, D.B. (eds), 1986, *The English Zechstein and Related Topics*, Geological Society Special Publication No. 22, pp. 127–139.

TABLE 1. Localities and dates of historically recorded subsidence hollows.

Date	Locality	Source of information
16/10/1796	[SE 3245 6891]	Harrison, 1892
18/10/1796	"	Kendal & Wroot, 1924
c. 1800	"	Cameron, 1881
c. 1830	[SE 3375 6566]?	Tute, 1868
c. 1828 - 1838	"	Tute, 1870
c. 1830	[SE 3186 7260]	Tute, 1868
-/6/1836	"	Tute, 1870
c. 1830	"	Fox-Strangways, unpublished notebook
19/6/1834	"	Harrison, 1892
12/7/1835	"	Kendall & Wroot, 1924
c. 1844	[SE 3254 7193]	Tute, 1868
c. 1850	"	Tute, 1870
c. 1848	"	Fox-Strangways, unpublished notebook
1857	Nunwick?	Fox-Strangways, unpublished notebook
c. 1861 - 2	[SE 3122 7328]	Tute, 1868
Spring 1860	"	Tute, 1870
Spring 1860	"	Fox-Strangways, unpublished notebook
21/5/1860	"	Harrison, 1892
21/5/1860	"	Kendall & Wroot, 1924
1871	[SE 3207 7229]	Cameron, 1881
14/5/1871	"	Harrison, 1892
14/5/1871	"	Kendall & Wroot, 1924
1873	[SE 3176 7842]	Cameron, 1881
1865 onwards	[SE 3254 7326]	Cameron, 1879
1877	[SE 3244 7317]	Cameron, 1879
1979	[SE 3170 7192]	Acrill Newspapers

have been obtained from boreholes. At outcrop the formation is generally represented by 7–20 m of mudstone, brecciated in places, with traces of gypsum and solution residues after gypsum and halite. At depths of up to 100 m, in a belt extending to 3 km from the outcrop, gypsum with evidence of patchy dissolution is present in the sequence; here the lower part of the formation comprises up to 40 m of massive secondary gypsum after anhydrite, equivalent to the Hayton (Anhydrite Formation) (Smith 1974a). The upper part is composed of up to 30 m of mudstone with nodular and bedded gypsum; subordinate dolomitic limestone beds—possibly equivalent to the Kirkham Abbey Formation of Smith (1974a)—are also present. At greater depths, where the circulation of water is restricted, the gypsum passes into anhydrite. At Ripon Parks (SE 307 753) the gypsiferous beds are exposed beside the River Ure where they comprise 7.1 m of massive alabastrine and porphyroblastic gypsum overlain by 6.5 m of interbedded gypsum and mudstone (Forbes 1958; James *et al.* 1981); the gypsum shows many signs of dissolution by groundwater, with numerous water-widened fissures, small caves and pipes present in the rock face, and is undergoing rapid erosion and dissolution by the River Ure at present (James *et al.* 1981).

The Brotherton (Magnesian Limestone) Formation forms a low escarpment sub-parallel to that of the Cadeby Formation. The best exposures are near Burton Leonard (SE 3318 6345) and Ripon Parks (SE 3131 7392). It consists of pale grey to white dolomitic limestone with a few vughs and is thinly to very thinly bedded and normally well jointed, breaking into small angular fragments. In boreholes it is 8–12 m thick and in many places has yielded artesian water. Very

TABLE 2. Localities, dates and sizes of subsidence hollows; compiled from the recollections of farmers and landowners in the Ripon–Bedale area.

Date	Locality	Diameter m	Depth m
post 1870	[SE 3156 7261]	30	-
pre 1910	[SE 3122 7190]	-	slight
1910 - recent	[SE 3114 7232]	c. 20	c. 5
1910 - recent	[SE 3121 7239]	c. 20	c. 5
1910 - recent	[SE 3115 7238]	c. 25	c. 6
1910 - recent	[SE 3120 7234]	c. 30	7
c. 1910	[SE 3192 7253]	c. 30	-
1930 - 1935	[SE 3227 7615]	14 - 19	c. 5
c. 1935	[SE 3146 7488]	c. 14	c. 3
c. 1935	[SE 3146 7490]	c. 14	c. 3
c. 1935	[SE 3122 7406]	c. 15	-
c. 1935	[SE 3135 7403]	-	slight
c. 1937	c. [SE 3175 7217]	-	slight
1930 - 1940	c. [SE 2830 8100]	10 - 14	1
1939	[SE 3184 7473]	35 x 25	8 - 10
1940 - recent	[SE 3149 7250]	20	1
1940 - recent	[SE 3140 7244]	1.5	0.5
1940 - recent	[SE 3147 7242]	10	0.5
1940 - recent	[SE 3145 7247]	20	1
1946	[SE 3160 7132]	10	deep
1950 - 1953	[SE 3116 7614]	10	2 - 3
pre 1960	c. [SE 339 728]	-	slight
pre 1960	[SE 3345 7312]	-	1
1960 - 1965	[SE 3248 7336]	5	-
1960 - present	[SE 3272 7311]	2 - 4	-
1960 - present	[SE 3194 7317]	8	2+
1960 - present	[SE 3027 7928]	-	slight
1965 - present	[SE 3134 7228]	2	c. 1
1968 - 1969	[SE 2999 8383]	1	1
1968 & 1970	[SE 3153 7242]	4	3
c. 1970	[SE 3187 7198]	2	1
c. 1975	[SE 3163 7419]	2	0.75
c. 1975	c. [SE 2919 8643]	8 - 10	2 - 3
1978	[SE 3207 7778]	2	2
1978	[SE 3238 7185]	2	0.5
c. 1978	[SE 3255 7192]	-	slight
1979	[SE 3164 7107]	-	slight
1979 & 1980	[SE 3170 7192]	14	4
1980	[SE 3187 7043]	c. 4	c. 2.5
1980	[SE 3210 7783]	15	0.5
1981	[SE 3118 7317]	10	slight
1/2/1982	[SE 3237 7182]	12	9.7

few signs of dissolution of the rock were observed in the field, and no caves have been recorded locally.

The Roxby Formation is of very similar character to the Edlington Formation and is likewise poorly exposed. It ranges in thickness from about 15 m at crop to about 30 m at depth where gypsum and anhydrite are present. Eastwards, at depth, the formation consists of up to 10 m of massive anhydrite overlain by anhydritic siltstone and mudstone. The anhydrite is the correlative of the Billingham (Anhydrite) Formation (Smith 1974a) and towards the crop passes into a similar thickness of gypsum. The gypsum thins out westwards where the formation comprises mudstone with solution residues and collapse breccias.

The Sherwood Sandstone Group forms a transition from the Roxby Formation into the overlying Sherwood Sandstone, about 300 m thick. The group has basal beds of red-brown, fine- to medium-grained sandstones with lenses of mudflake conglomerate and beds of mudstone and siltstone. Above are red-brown, channelled and cross-bedded, fine- to medium-grained sandstones.

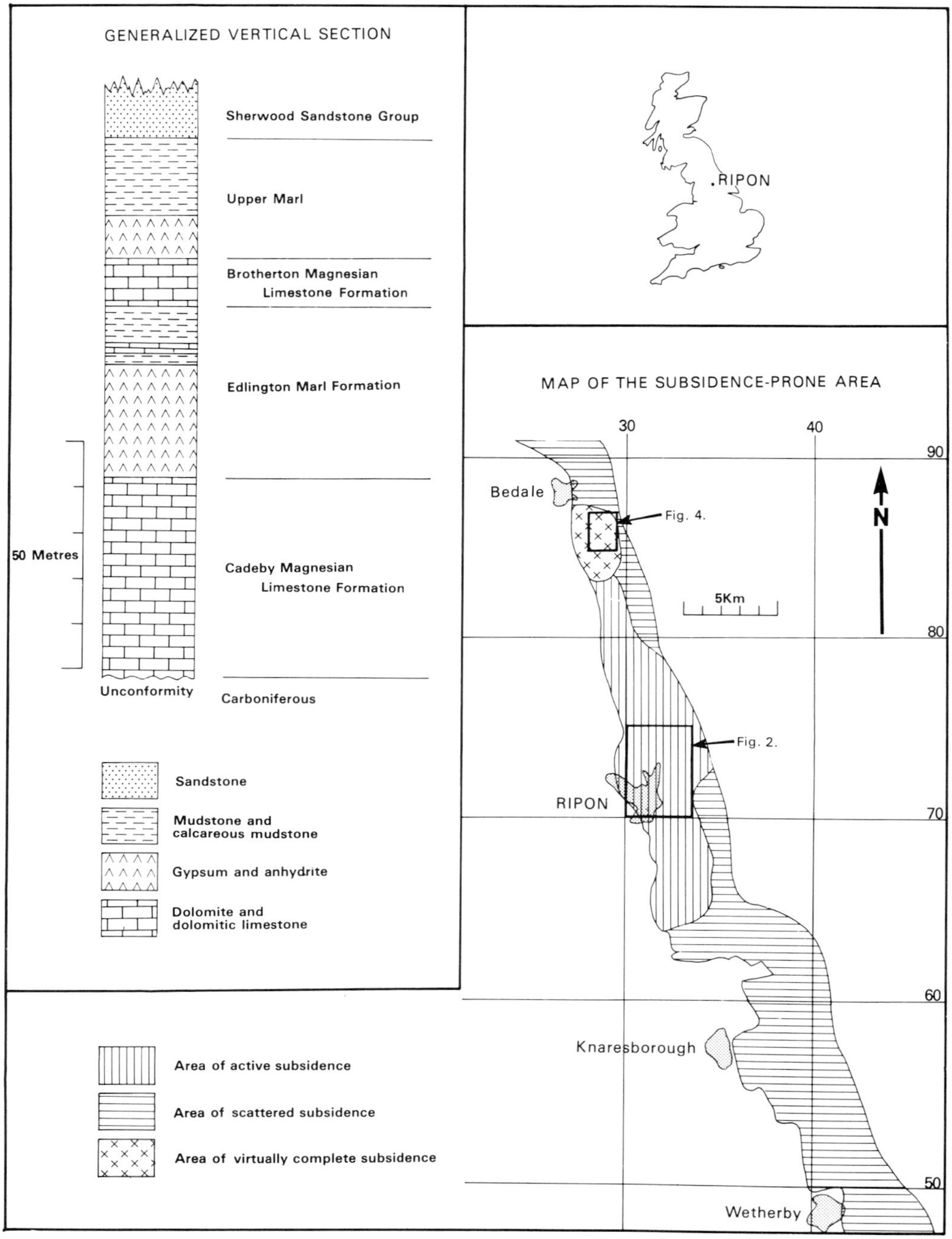

FIG. 1. Location map, generalized vertical section and map of the area prone to subsidence caused by gypsum dissolution.

Terminology

Subsidence features are found in many parts of the world where soluble rocks occur at or near surface. Extensive surface dissolution results in karst topography, but where substantial overburden is present subsidence hollows are the only visible effects of dissolution at depth. Terms used for these hollows include doline, sink, sinkhole, subsidence sinkhole, swallow, swallow hole, solution chimney, pipe cockpit, geological organ, polje and uvala. A review of the terminology is beyond the scope of this paper, and the terms 'subsidence hollow' and, for a large area, 'subsidence depression' are used here. The term 'breccia pipe' is used for the fill of the former cavity above a collapsed cavern.

The nature of the subsidence

The gypsum belt is about 3 km wide and extends from near Catterick to south of Doncaster (Smith 1972). Except for one record of subsidence at Burton Salmon, near Castleford (Edwards *et al.* 1940), most of the 40 recent subsidence hollows have formed in the area between Bishop Monkton, Ripon and Nosterfield. This active area also has hundreds of subsidence hollows that postdate the deposits of the Devensian glacial phase, suggesting a continuous history of subsidence during Flandrian time; most of these hollows are similar in size and shape to those formed recently, but in places they coalesce to form depressions up to several hundred metres across (Fig. 2). Near Dallamires Lane (SE 318 703), south of Ripon, a large area of peat 8 m below the adjacent fluvioglacial terrace occupies such a hollow. The origin of the large and small hollows is obvious where they have formed in flat fluvial or lacustrine deposits, but in areas mantled by glacial till it is difficult to distinguish old subsidence hollows from kettle holes. In areas of alluvium, only the recent hollows remain because of rapid filling by flood deposits. Within the subsidence belt there are also large drift-filled subsidence depressions, caused by the extensive dissolution of gypsum. These areas include Whitewater Common, south of Doncaster (Smith *et al.* 1973) and Snape Mires, near Bedale, which is discussed later. In a few places within the gypsum belt there is no surface evidence of subsidence and it must be assumed that any gypsum deposited there is still present or was dissolved before the last ice-age and the subsidence features obliterated.

The shape of a subsidence hollow depends upon whether there is solid competent rock or a substantial thickness of unconsolidated (usually glacial) deposits at the surface; the presence of competent rock results in the creation of steep-sided cylindrical shafts whereas conical depressions are formed in drift deposits.

A cylindrical hole about 14 m across and 15 m deep appeared in red Sherwood Sandstone near Ripon Railway Station (SE 3186 7260) in July 1834 (Tute 1868, 1870; Harrison 1892) and is still visible. A similar collapse occurred in 1860, breaching the Brotherton Formation at Ripon Parks (SE 3122 7328), and another affected the same formation at Hall Garth Ponds (SE 3184 7472) in 1939. At Hall Garth Ponds Mr J. Graham (pers. comm.) reported that "there was a loud noise and a large hole appeared and filled with water which bubbled and looked milky white. When the bubbling had ceased and the water cleared, trees were visible still rooted in the bottom of the hole". This hole is about 30 m in diameter. According to Mr N. Moore, a member of the local subaqua club, it is 7 m deep with rock (presumably all limestone) exposed in the sides; water flows from fissures in the rock and a tree is still rooted in the bottom.

Throughout much of the Ripon district thick superficial deposits, locally in excess of 20 m, blanket the bedrock. Subsidence hollows in these areas are nearly cylindrical when first formed but the sides rapidly slip to produce a conical depression. Corkscrew Pits (SE 320 731) are three such conical depressions. A small actively-subsiding hollow nearby (SE 3194 7317) has been repeatedly filled and is only about 1.5 m deep. The hollow has near-vertical sides, but if it were left unfilled they would become unstable.

The formation of a hollow in the drift-covered area (Figs 2 and 3) was illustrated on 3 August 1979 by the *Ripon Gazette* (Acrill Newspapers) which reported that on 28 July 1979 "A six-foot-deep circular pit appeared early on Saturday morning behind houses on Magdalen's Road" (SE 3170 7192). One resident described it as "like a minor earthquake". The Magdalen's Road subsidence hollow started as a slight downwarp and damaged two garages standing on the site. The garages were dismantled and soon afterwards the collapse occurred. The hollow was filled but subsided another metre in April and May 1980. The residents of the area were filling the hole during 1981, when slight slipping with associated tension gashes was evident on the western side. Left unfilled, slipping around the sides would have enlarged it, thus threatening the nearby houses; filling the hole is effectively controlling its size and preventing it from becoming a large conical depression.

On 1 February 1982 another hollow in a drift-covered area formed at Sharow (SE 3238 7182),

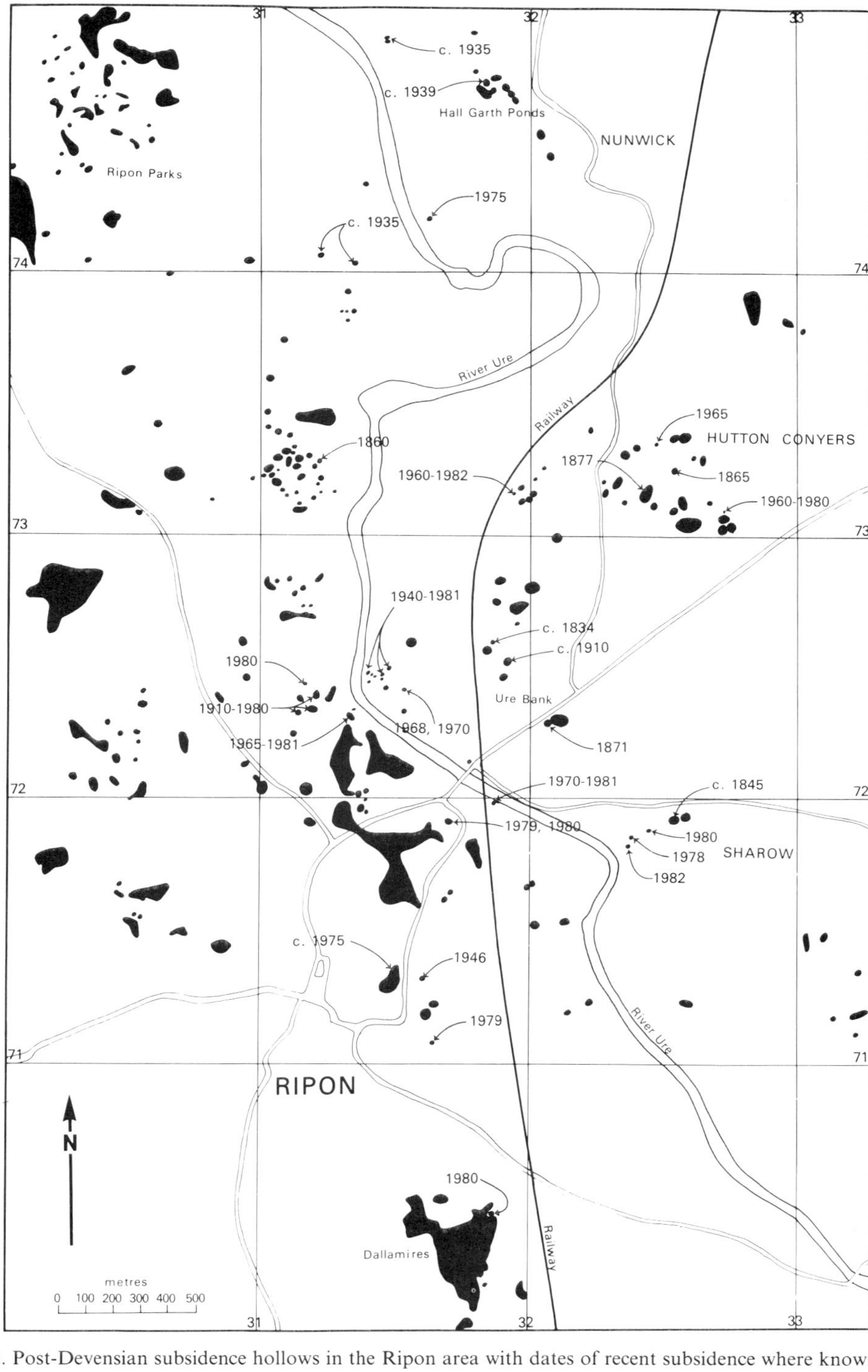

FIG. 2. Post-Devensian subsidence hollows in the Ripon area with dates of recent subsidence where known.

FIG. 3. (a) Subsidence hollow formed on 28 July 1979 behind houses on Magdalen's Road, Ripon (SE 3170 7192). Photo: Acrill Newspapers. (b) Subsidence hollow formed on 1 February 1982 at Sharow (SE 3238 7182). The hollow was 12 m in diameter and up to 9.7 m deep.

east of Ripon. The hollow was cylindrical, 12 m in diameter and up to 9.7 m deep (Figs 2 and 3). On 3 February 1982 it contained muddy water to within 0.8 m of the surface and incipient tension fissures had developed around the sides, which were actively caving in. This latest hollow appeared after a period of heavy rain, the field having been flooded a week or two earlier. When the flood abated a shallow puddle formed on the site of the subsidence hollow, but no other warning of impending collapse was observed.

Very few of the Ripon subsidence hollows are accurately dated, but by comparison with limestone karst areas (Newton 1977; Kemmerly 1980; Foose 1981) subsidence may be expected when there are changes in the water table level. Cameron (1879) recorded that small hollows or 'man holes' at Hutton Conyers, near Ripon, were formed mostly during very wet seasons.

The mechanisms controlling the subsidence

The solubilities of the rocks

Laboratory experiments on the dissolution rates of gypsum and limestone (Kemper *et al.* 1975; James & Lupton 1978; James & Kirkpatrick 1980); show that gypsum may be expected to dissolve in flowing water about 100 times faster than limestone. Limestone, in turn, is more soluble than dolomite. Chemicals present in the dissolving water also affect the solubility; waters rich in carbonic acid are aggressive to limestone (Sweeting 1966), and waters rich in $CaCO_3$ are capable of dissolving larger amounts of gypsum than pure water (Deer *et al.* 1962; Kempe 1972).

Field observations on gypsum and limestone dissolution support the experimental evidence. James *et al.* (1981) found that typical dissolution rates for gypsum by the River Ure at Ripon Parks were between 0.1 and 1.7 m per annum. By comparison Sweeting (1966) and Clayton (1981) considered dissolution in limestone of 0.01 m per annum to be a high value, achieved only by acid water drained from a peat bog.

Subsidence as a result of gypsum dissolution has been recorded in Texas (Olive 1957), Canada (Wigley *et al.* 1973), Germany (Hundt 1950; Herrmann 1964; Ströbel 1973), the Alps (Nicod 1977), Russia (Gorbunova 1977) and Newfoundland (Sweet 1977).

The gypsum-anhydrite transition

The western limit of the subsidence-affected area is the outcrop of the base of the Edlington Formation. The distribution of subsidence hollows indicates that the eastern limit, some 3 km away, approximates to the down-dip transition from gypsum to anhydrite in the Edlington and Roxby Formations. Within the subsidence belt this transition generally occurs at about 100 m, but if sufficient circulating water is available it is theoretically possible to extend to a depth of between 900 and 1200 m (Mossop & Shearman 1973).

The relationship between joints and subsidence

Observations of cave systems in limestone and gypsum indicate that their shape is largely controlled by bedding and joint orientations (Ford 1971; Kempe 1972; Waltham 1971; Jakucs 1977; Jackucs & Mezoso 1977). Furthermore, under phreatic conditions, roof dissolution can generate high caverns on sites where joints intersect (Waltham 1974). In the Ripon area the water flow in the gypsiferous beds is mainly phreatic and it is likely that the caverns that ultimately collapse to form subsidence hollows are located at the intersections of joints. Joint measurements in the Cadeby Formation and Sherwood Sandstone Group of the Ripon area (Fig. 4c) indicate that two major sets of joints cross at about 90°.

In much of the gypsum belt the subsidence hollows are apparently randomly scattered. However, hollows in the vicinity of Hutton Conyers, Ure Bank, Sharow and Ripon Golf Course have a linear or grid-like distribution (Fig. 2). The orientations of all the lines of three or more hollows are illustrated in Fig. 4(a), in which each segment is proportional to the total number of hollows on all the lines with that orientation. Because many of the hollows form a regular grid pattern (Fig. 4d,e,f) the rose diagram (Fig. 4a) shows four apparent trends. Many of the subsidence hollows are elongate or elliptical and a further rose diagram (Fig. 4b) shows the orientations of the long axes of 100 hollows near Ripon. The joint directions approximate to both the main lineations of hollows and the orientation of individual hollows (Fig. 4a–f), suggesting a close correlation between them. In Germany the location of subsidence hollows at the intersections of joints in gypsum was demonstrated by Ströbel (1973).

At Snape Mires (SE 285 860), east of Bedale, almost complete dissolution of the Permian gypsum, as proved by boreholes, has resulted in a large bedrock depression. This is about 8 sq km in area and bedrock levels are as much as 25 m below the drainage from the area. A fluvioglacial terrace which once covered the area has subsided and the depression is filled with laminated clay

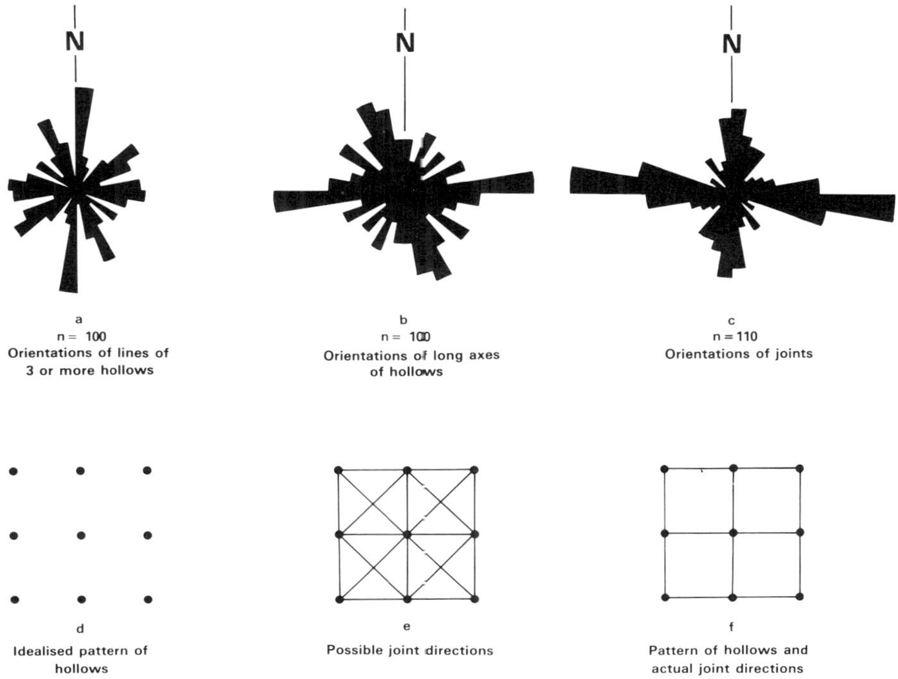

FIG. 4. Rose diagram and theoretical models showing the relationships between the orientations of lines of hollows, individual hollows and joints in the Ripon area.

deposited during the later part of the Late Devensian and during the Flandrian. Similar lacustrine deposits elsewhere in the Vale of York have a flat surface, but at Snape Mires they have a sharply uneven surface with a relief of up to 6 m. Between Rough Plantation (SE 2835 8525) and Flood Bridge (SE 2857 8752) the undulations take the form of N–S and E–W trending ridges and elongate hollows up to 1 km long (Fig. 5) and several turn through right angles. The probable sequence of events resulting in this morphology is shown in Fig. 5. The orientations of these subsidence ridges and hollows closely approximates to the joint directions found in the Permian rocks to the SW (Fig. 4c). It is likely, therefore, that the ridges represent the initial collapse of a joint-controlled cave system in the underlying gypsum (the pattern of which has been preserved by the drift infill) and the hollows coincide with areas where the remaining gypsum has subsequently been dissolved. The Snape Mires subsidence features are small drift-covered analogues of the joint-controlled subsidence troughs recorded from the gypsum plain in west Texas (Olive 1957).

Cavern collapse and breccia pipes

The volume of gypsum dissolved obviously affects the size and amount of the associated subsidence. The recent Ripon subsidences occur catastrophically, suggesting the sudden collapse of the roof of near-surface caverns. The gypsum beds, however, generally lie at depth and a cavity must propagate upwards by collapse, as a breccia pipe. A small exposure (SE 3129 7335) of brecciated Sherwood Sandstone in the Brotherton Formation outcrop near Ripon probably represents part of such a breccia pipe. Similar pipes above gypsum have been found in Texas (Eck & Redfield 1965) and the development of breccia pipes after the dissolution of evaporites was recorded for the Permian of Durham and Yorkshire (Smith 1972) and the Permo–Trias of Germany (Hundt 1950; Herrmann 1964; Prinz 1973).

Experience in mining areas shows that the collapse of a cavity generates an upward-propagating pipe, through overburden, for between four and 10 times the height of the cavity before breccia chokes the pipe (Piggott & Eynon 1978). Coal measures rocks have a high bulking factor (*op. cit.*), but Permo–Triassic mudstones and sandstones are soft and may be expected to compact more on brecciation, thus allowing their associated breccia pipes to penetrate greater thicknesses of strata. The size of the cavern also affects the propagation of the breccia pipes.

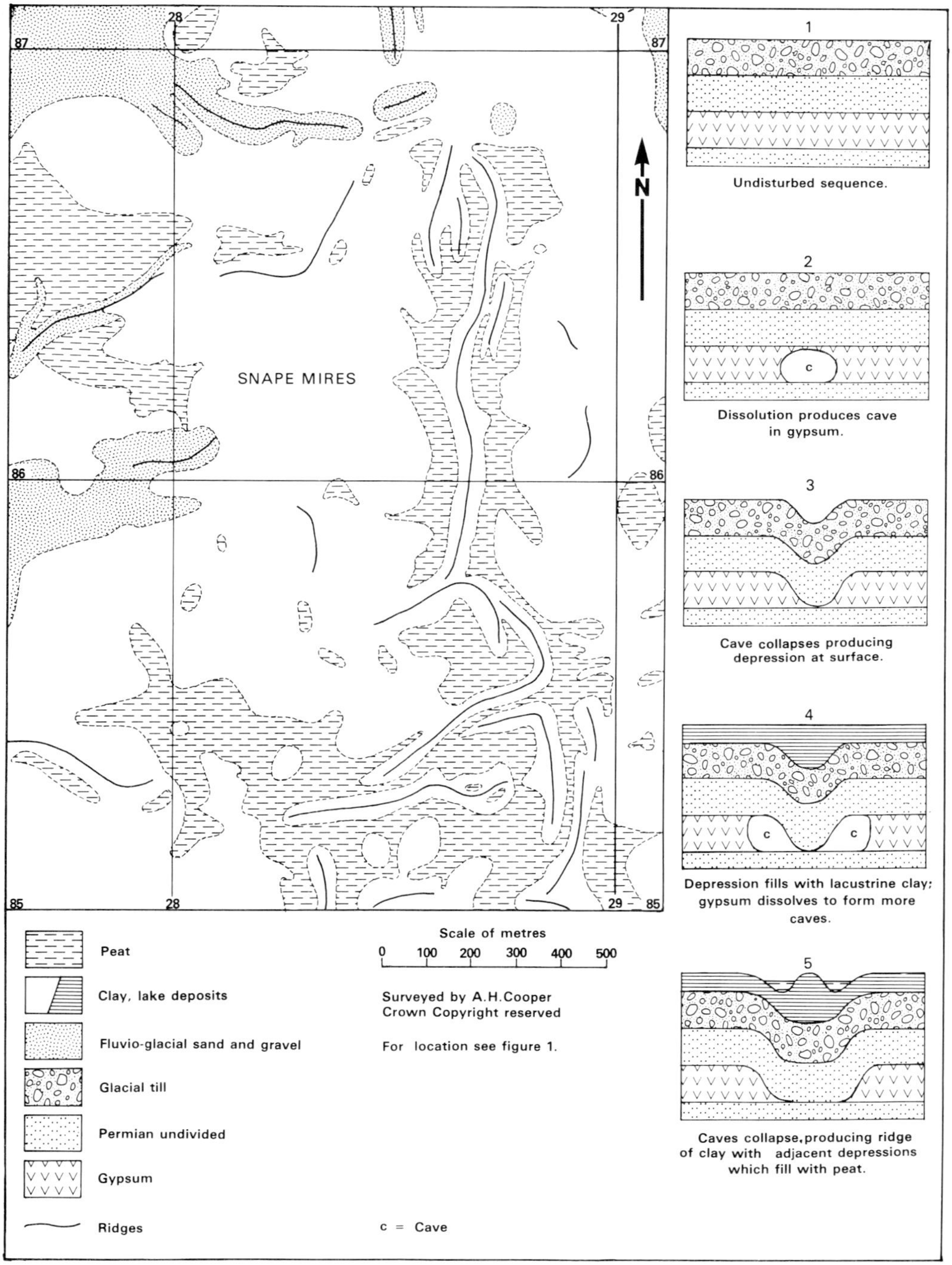

FIG. 5. The distribution and mode of formation of subsidence ridges in Snape Mires near Bedale, North Yorkshire. Geology based on provisional published maps of the British Geological Survey (NERC), crown copyright reserved.

Observations on mines where soft mudstones form the overburden (S. Penn, pers. comm.) suggest that mud flow away from the bottom of a breccia pipe can enhance the effect and a cavity may propagate upwards through overburden for more than 20 times its original height. If a propagation figure of 20:1 is applied to the Ripon–Bedale gypsum belt, it means that even at the eastern limit of the gypsum, collapse of cavities in excess of 5 m high may be expected to breach the surface. It would appear, therefore, that choking of breccia pipes is not the factor limiting the down-dip margin of this subsidence belt.

Water flow and the amount of subsidence

Water flow is a major factor controlling gypsum dissolution and subsidence. High water flow in the past has resulted in almost complete dissolution of the gypsum, as at Snape Mires. Nil water flow results in sulphate-saturation of the water in the rock and no dissolution occurs. A flow between the two extremes results in an active area.

Around Ripon, the most actively subsiding area at present, the volume of 35 subsidence hollows formed in the past 150 years is about 27,000 m^3. These hollows are scattered over an area of about 5 sq km, giving an average annual subsidence volume of about 36 m^3 per sq km. The actual rate of gypsum removal by groundwater must be greater than this, because of the choking of subsidence pipes. Many water wells in the area have very high concentrations of sulphates and the presence of such waters, in the valley gravels east of Ripon, points to a considerable down-dip movement of water. Calculations by Dr J. Aldrick (pers. comm.), based on assumptions of the infiltration and water flow, show that the possible amount of gypsum dissolved per square kilometre could be of the order of 120 m^3 per annum. In view of the assumptions used to make this calculation, a volume of collapse at surface of 36 m^3 per sq km per annum does not appear to be unreasonable and may indicate a smaller flow of water than was assumed.

In the Snape Mires area the presence of the large foundered area, filled with glacial and post-glacial lacustrine deposits indicates severe dissolution of gypsum during the later part of the Late Devensian and during the Flandrian. An approximate calculation shows that over an area of about 8 sq km a thickness of up to 25 m of gypsum has been dissolved in less than 13,000 yr ± 600 BP. This equates with an annual dissolution rate of at least 1,600 m^3 of gypsum per sq km. The area has abundant, commonly calcareous, springs and artesian water has been found in the Cadeby and Brotherton Formations. Even assuming generous water infiltration rates and a large catchment area for the groundwater, it is unlikely under present-day conditions that more than 1,000 m^3 of gypsum could be removed in a year. This fact, the lack of widespread active subsidence, and the presence of late and post-glacial deposits in the subsidence depression, suggest that most of the dissolution occurred when more water was available, possibly beneath or adjacent to the melting Devensian icesheet. Subsidence over gypsum in recently deglaciated valleys has been recorded in the Alps (Nicod 1977), and Sweeting (1974) attributed enlargement of Pennine cave systems to dissolution by the abundant water flow during the deglaciation.

Prediction of subsidence and remedial work

The long-term future of the actively subsiding areas is further subsidence and it is currently impossible to predict which places are likely to be unstable. However, in areas where the subsidence hollows have a regular grid-like or linear arrangement controlled by joints, it is likely that ground on lines between hollows is most at risk. The more numerous and closer the hollows, the greater likelihood there is of another hollow developing along that line. Land adjoining a subsidence hollow is also at risk because many of the hollows occur in groups of two or three.

If a subsidence hollow forms, the best course of action is to fill it with inert fill; this will prevent the sides from slipping and migrating outwards, thus stopping the hollow from becoming a large conical depression. Further subsidence of the fill in hollows, however, has occurred in several places. Filling the subsidence with toxic refuse should be avoided because pollution of local springs may occur. Disposal of surface water into hollows is also unwise as it may trigger off further subsidence. The effects of water abstraction from the gypsiferous formations is not known, but removal of large quantities of water high in sulphates may influence the subsurface dissolution of gypsum and so increase the likelihood of subsidence.

For very expensive structures, detailed site investigation is the only certain way to determine the stability of the ground. Such an investigation should involve drilling boreholes down to the Cadeby Formation (20–130 m) to determine the state of the gypsum beds, the presence or absence of cavities and the hydrogeological conditions in the adjacent strata. Down-hole and surface geophysical surveys may also be useful in defining the state of the gypsum. If cavities are found they may be filled and water flow through the gypsum thus stopped to prevent further dissolution; these measures are very expensive. For normal buildings, construction on rafts, subject to the advice of a qualified structural engineer, is probably the safest and most economical method.

ACKNOWLEDGMENTS: The author thanks the following for their assistance; the people of the Ripon–Bedale area for their recollections of subsidence, especially Mr R.C. Richardson, Mr J. Graham and Mr N. Moore; R.J. Aldrick, A.N. James, G.M. Davies, D.J.C. Mundy, S. Penn for useful discussion; E.G. Smith and I.C. Burgess for critically reading the manuscript. This paper is published by permission of the Director of the Geological British Survey (NERC).

References

ALDRICK, R.J. 1978. The hydrogeology of the Magnesian Limestones in Yorkshire between the River Wharfe and the River Aire. *Q. Jl eng. Geol. Lond.* **11,** 193–201.

CAMERON, A.G. 1879. Ripon swallow-holes. *Geol. Mag.* **6,** 575.

—— 1881. Subsidences over the Permian boundary between Hartlepool and Ripon. *Proc. Yorks. geol. polytech. Soc.* **7,** 342–51.

CLAYTON, K. 1981. Explanatory description of the landforms of the Malham area. *Field Studies*, **5,** 389–423.

DEER, W.A., HOWIE, R.A. & ZUSSMAN, J. 1962. Gypsum. In: *Rock Forming Minerals*, Vol. 5, 201–18. Longmans, London.

ECK, W. & REDFIELD, R.C. 1965. Engineering geology problems at Sanford Dam, Borger, Texas. *Eng. Geol. Sacramento*, **2,** 15–25.

EDWARDS, W., WRAY, D.A. & MITCHELL, G.H. 1940. Geology of the country around Wakefield. *Mem. geol. Surv. G.B.* London.

FOOSE, R.M. 1981. Sinkholes follow the pattern. *Geotimes*, **26,** 20–4.

FORBES, B.G. 1958. Folded Permian gypsum of Ripon Parks, Yorkshire. *Proc. Yorks. geol. Soc.* **31,** 351–7.

FORD, D.C. 1971. Geologic structure and a new explanation of limestone cavern genesis. *Trans. Br. Cave Res. Ass.* **13,** 81–94.

FORD, T.D. (ed.) 1977. *Proc. 7th int. Speliological Congress, Sheffield 1977*. British Cave Research Association, Sheffield.

GORBUNOVA, K.A. 1977. Morphology of gypsum karst. In: FORD, T.D. (ed.), (*q.v.*), 221–2.

HARRISON, W. (ed.) 1892. *Ripon Millenary. A record of the festival, also a history of the city arranged under its wakemen and mayors from the year 1400*. W. Harrison, Ripon.

HERRMANN, A. 1964. Gips-und Anhydritvorkommen in Nordwestdeutschland. *Silikat Jl*, **3,** 442–66.

HUNDT, R. 1950. *Erdfalltektonik*. W. Knapp, Halle (Saale).

JAKUCS, L. 1977. *Morphogenetics of Karst Regions*. Hilger, Bristol.

—— & MEZOSO, G. 1977. Genetic problems of the huge gypsum caves of the Ukraine. In: FORD, T.D. (ed.) (*q.v.*), 248–51.

JAMES, A.N. & LUPTON, A.R.R. 1978. Gypsum and anhydrite in foundations of hydraulic structures. *Geotechnique. London*, **28,** 249–72.

—— & KIRKPATRICK, I.M. 1980. Design of foundations of dams containing soluble rocks and soils. *Q. Jl eng. Geol. Lond.* **13,** 189–98.

——, COOPER, A.H. & HOLLIDAY, D.W. 1981. Solution of the gypsum cliff (Permian, Middle Marl) by the River Ure at Ripon Parks, North Yorkshire. *Proc. Yorks. geol. Soc.* **43,** 433–50.

KEMMERLY, P.R. 1980. A time-distribution study of doline collapse: framework for prediction. *Environ. Geol.* **3,** 123–30.

KEMPE, S. 1972. Cave genesis in gypsum with particular reference to underwater conditions. *Cave Sci. No. 49*, 1–6.

KEMPER, W.D., OLSEN, J. & DE MOOY, C.J. 1975. Dissolution rate of gypsum in flowing water. *Proc. Soil Sci. Soc. Am.* **39,** 458–63.

KENDALL, P.F. & WROOT, H.E. 1924. *Geology of Yorkshire*. Vienna.

LOWE, D.J. 1978. Farnham Cave, a rift cave in the Magnesian Limestone. *Trans. Br. Cave Res. Ass.* **5,** 23–8.

MAYHALL, J. 1878. *The annals of Yorkshire from the earliest period to the present time*, Vol. I. Johnson, Leeds.

MOSSOP, G.D. & SHEARMAN, D.J. 1973. Origins of secondary gypsum. *Trans. Instn. Min. Metall.* **82,** 147–54.

NEWTON, J.G. 1977. Induced sinkholes—a continuing problem along Alabama highways. *Publ. int. Ass. hydrol. Sci., Wallingford, No. 121*, 453–63.

NICOD, J. 1977. Deux karst du gypse remarquables des Alps occidentales. In: FORD, T.D. (ed.) (*q.v.*), 321–5.

OLIVE, W.W. 1957. Solution-subsidence troughs, Castile Formation of Gypsum Plain, Texas and New Mexico. *Bull. geol. Soc. Am.* **68,** 351–8.

PIGGOTT, R.J. & EYNON, P. 1978. Ground movements arising from the presence of shallow abandoned mine workings. In: GEDDES, J.G. (ed.) *Large ground movements and structures*, 749–80. Proceedings of the Conference held at The University of Wales Institute of Science and Technology, Cardiff 1977. Pentech Press, London.

PRINZ, H. 1973. Zur Entstehung von Einbruchsschloten und Korrisionkesseln über tiefem Salinarkarst. In: *Proceedings of a symposium on sink-holes and subsidence*, Hannover, [T2-D] 1–6. Deutsche Gesellschaft für Erd- und Grundbau. Essen.

SMITH, D.B. 1972. Foundered strata, collapse breccias and subsidence features of the English Zechstein. In:

RICHTER-BERNBURG, G. (ed.) *Geology of Saline Deposits*. Proc. Hannover Symp. (Earth Sciences 7), 255–69. Unesco, Paris.

—— 1974a. Permian. In: RAYNER, D.H. & HEMINGWAY, J.E. (eds) *The Geology and Mineral Resources of Yorkshire*, 115–44. Yorkshire Geological Society, Leeds.

—— 1974b. The stratigraphy and sedimentology of Permian rocks at outcrop in North Yorkshire. *J. Earth Sci. Leeds Geol. Ass.* **8,** 365–86.

SMITH, E.G., RHYS, G.H. & GOOSSENS, R.F. 1973. Geology of the country around East Retford, Worksop and Gainsborough. *Mem. geol. Surv. G B.* London.

STRÖBEL, W. 1973. Der Grundgips im Raum Stuttgart als Modell für Gipsauslaugung und Bildung von Erdfällen. In: *Proceedings of a symposium on sinkholes and subsidence*, Hannover, [T1-G] 1–8. Deutsche Gessellschaft für Erd- und Grundbau. Essen.

SWEET, G.A. 1977. Hydrogeology of gypsum karst in Newfoundland. In: FORD, T.D. (ed.) (*q.v.*), 390–1.

SWEETING, M.M. 1966. The weathering of limestones, with particular reference to the Carboniferous Limestone of northern England. In: DURY, G.H. (ed.) *Essays in Geomorphology*, 177–210. Heinemann, London.

—— 1974. Karst geomorphology in north-west England. In: WALTHAM, A.C. (ed.) *The Limestones and Caves of North-west England*, 46–78. David & Charles, Newton Abbot.

TUTE, REV. J.S. 1868. On certain natural pits in the neighbourhood of Ripon. *Geol. Mag.* **5,** 178–9.

—— 1870. On certain natural pits in the neighbourhood of Ripon. *Proc. Yorks. geol. polytech. Soc.* **5,** 2–7.

WALTHAM, A.C. 1971. Controlling factors in the development of caves. *Trans. Br. Cave Res. Ass.* **13,** 73–80.

—— 1974. The geomorphology of the caves of North-West England. In: WALTHAM, A.C. (ed.) *The Limestones and Caves of North-west England*, 79–105. David & Charles, Newton Abbot.

WIGLEY, T.M.L., DRAKE, J.J., QUINLAN, J.F. & FORD, D.C. 1973. Geomorphology and geochemistry of a gypsum karst near Canal Flats, British Columbia. *Can. J. Earth Sci.* **10,** 113–29.

A. H. COOPER, British Geological Survey, Windsor Court, Windsor Terrace, Newcastle upon Tyne NE2 4HE.

Germany

Environmental analysis of basin and schwellen facies in the lower Zechstein of Germany

J. Paul

SUMMARY: The Zechstein basin had a schwellen and basin topography, inherited from the late Variscan regime. The basinal facies of the lower Zechstein, mostly thin euxinic shales and limestones, are uniform throughout the basin. Different facies occur at the basin rim and on the schwellen. The schwellen facies of the Kupferschiefer varies from black and red shales to very fossiliferous dolomites. Reefs and other biogenic sediments dominate the shallow-water basin rim environments of the first- and second-cycle carbonates (Ca1 and Ca2). A regression during deposition of the upper part of the first-cycle carbonate led to emergence of the schwellen. This event can be recorded throughout the Zechstein basin. The deposition of great amounts of anhydrite upon the schwellen increased the relief between palaeohigh and basin. Subsequent sedimentation of thick halite blanketed the topography and filled the basin.

General structural and depositional framework

The Zechstein Sea probably invaded the Central European Basin from the north. This basin, extending from Great Britain to Poland, was the result of continued subsidence from the early Permian through to the end of the Triassic. The depositional setting of the Lower Permian, the Rotliegendes, was a complex inland drainage basin. Fluvial transport of detrital material from the rims of the basin toward the centre resulted in alluvial fan deposits and floodplains. Aeolian sand dunes and sabkhas formed in the centre of the basin (Glennie 1972; Plein 1978). The floor of this basin lay well below sea-level (Smith 1970, 1980).

The Zechstein basin was subdivided into separate sub-basins by schwellen or palaeohighs (Fig. 1). The schwellen became well developed during deposition of the Zechstein, although they were already in existence during the deposition of the Rotliegendes, when they were areas of non-deposition (Herrmann 1957). In general, the pattern of facies and the sequence of events of the Zechstein deposits is surprisingly similar throughout the basin from England to Poland.

Thickness and diversity of sediments and diversity of biota increase towards both the rims and schwellen. Progradational belts of thick (some 100 m) carbonates and anhydrite accumulated on the slopes.

The schwellen facies are good environmental indicators and can be used as gauges of hydrographic and sedimentary parameters. Clear changes in both water depth and position of the chemocline are indicated in sediments deposited on the schwellen whilst few or no variations occur in the basinal facies.

Kupferschiefer (T1)

The transgression of the Zechstein Sea was very rapid (Smith 1979, 1980). Reworking of the sediments deposited on pre-Zechstein surfaces, such as the dune sands, occurred only on a small scale.

The Kupferschiefer, or its English equivalent the Marl Slate, is a thin (<50 cm) euxinic black shale and laminated marl in the basinal facies. Thin laminae, each less than 100 μm thick, alternate between organic matter-rich layers, and white carbonate-rich layers. Benthic fauna, including trace fossils, are completely lacking, whereas nectonic and planktonic fauna are present. This basinal facies occurs uniformly from England to Poland and is the result of stratification in the Zechstein Sea. Within the waterbody a chemocline separated an oxygen-rich upper layer, with high plankton productivity, from a poisoned hydrogen sulphide-containing lower layer. The lamination is due to periodic (probably seasonal) variations in plankton productivity.

The schwellen facies of the Kupferschiefer is highly variable (Paul 1982a,b). In places it pinches out on schwellen whereas elsewhere it thickens to more than 2 m in small depressions (Fig. 2, locality 2). The carbonate content ranges to over 90% and is predominantly dolomite.

On the slopes of, or in, depressions on palaeohighs the lower part of the Kupferschiefer is represented by a very fossiliferous carbonate bed, which is called the Mutterflöz in Thuringia, the Productus Limestone in Hesse and the Border Dolomite in Poland (Fig. 2, locality 3). The most common fossils present are *Horridonia horrida*, other brachiopods, pelecypods, and crinoid ossicles. This carbonate bed is time-equivalent with the laminated black shale of the basin floor facies.

From HARWOOD, G.M. & SMITH, D.B. (eds), 1986, *The English Zechstein and Related Topics*, Geological Society Special Publication No. 22, pp. 143–147.

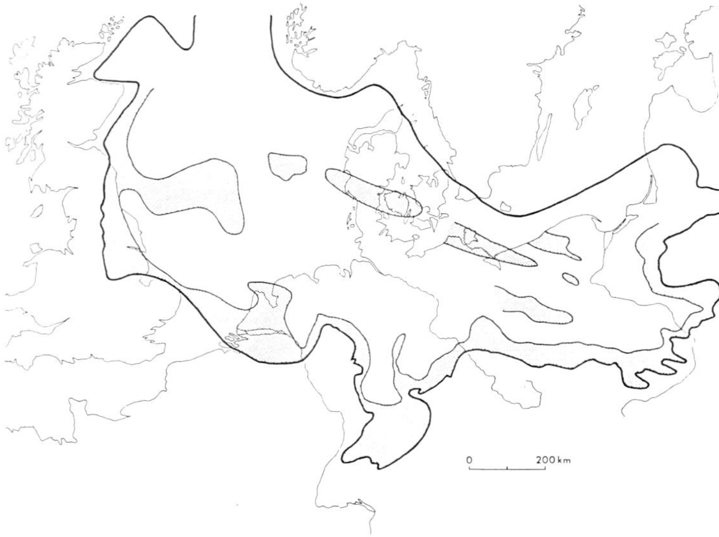

FIG. 1. The Central European Zechstein Basin. Thick line: supposed shoreline of the Werra Carbonate (Ca1). Stippled: areas of rim and schwellen facies. After Depowski (1978), Füchtbauer (1980), Smith (1980) and Ziegler (1981).

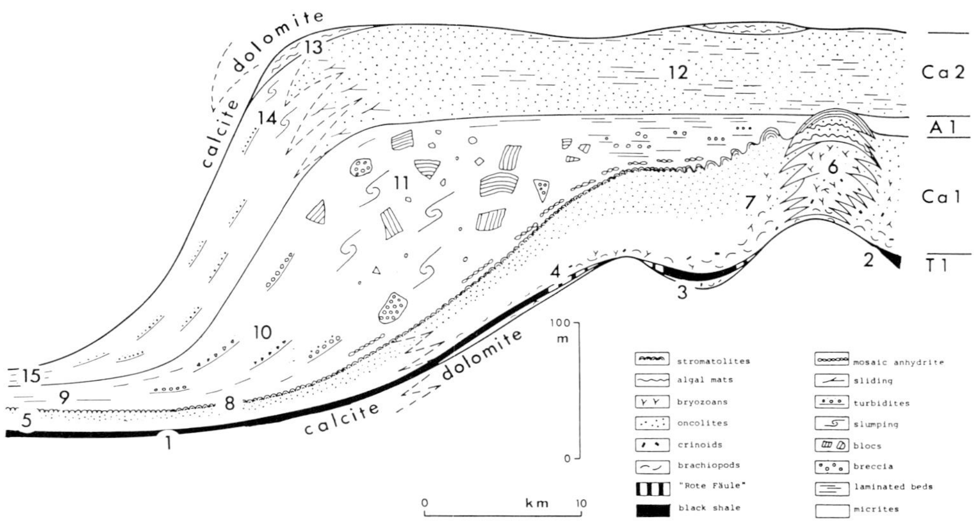

FIG. 2. Schematic cross-section of the lower Zechstein sediments from the basin to the top of a palaeohigh. Types of facies in the lower Zechstein: (1–4) Kupferschiefer (T1): 1, basinal facies, 2, schwellen facies, 3, Mutterflöz, 4, oxidized facies. (5–8) Werra Carbonate facies: 5, basinal facies, 6, reef facies, 7, biogenic facies, 8, algal stromatolite. (9–11) Werra anhydrite (A1) facies: 9, basinal facies, 10, turbidite facies, 11, slump facies. (12–15) Stassfurt Carbonate (Ca2) facies: 12, Hauptdolomit facies, 13, reef facies, 14, slope facies, 15, Stinkschiefer facies.

The Kupferschiefer of the schwellen facies is more a grey colour than black. The white carbonate laminae are broader than the black organic ones. Also, the organic matter content is lower than in basinal facies. Red colours, the so-called 'Rote Fäule', which are present especially on palaeohighs (Fig. 2, locality 4), indicate oxidizing conditions, both within the water column and the sediment.

On some schwellen slopes, bioclastic shell-beds

are interbedded with laminated marls. These shell-beds are the result of storms which swept the benthic fauna, present on the top of the topographic highs and above the chemocline, down into the adjoining depressions where euxinic conditions prevailed.

The Werra Carbonate (Ca1)

The laminated Kupferschiefer is abruptly overlain by bioturbated carbonates. The burrowed sediments indicate the onset of oxygenation throughout the water column. The benthic fauna spread from the palaeohighs, where they had persisted during Kupferschiefer deposition, to the basins. The carbonates of the first cycle show the greatest diversity of both sediments and biota. In the basinal facies the first cycle carbonate is a dark, laminated, marly micritic limestone (Füchtbauer 1968, 1972) with a thickness of only a few metres (Fig. 2, locality 5).

Thicknesses of schwellen and rim facies of the first cycle carbonate often exceed 100 m. Reefs are present on top of the palaeohighs (Fig. 2, locality 6). In Germany, these reefs are isolated reef knolls and seldom have diameters more than above 100 m (Kerkmann 1969; Füchtbauer 1980; Paul 1980). Reef distribution is strongly controlled by the underlying palaeotopography; all reefs were situated on topographic highs. Barrier reefs similar to those in the English Zechstein (Smith 1981) are not present. The reefs occur scattered in the Harz area and around the Thuringian Basin, but they are absent on the eastern border of the Rheinisches Schiefergebirge, although similar environments on palaeohighs were present.

Reef growth commonly started some metres above the base of the Zechstein sediments with basal coquina present between the reef and the sub-Zechstein surface. Reef development is similar throughout the basin (Kerkmann 1969; Smith 1981). In the lowermost part the framework is built by bryozoans of *Fenestella* and *Acanthocladia*-type with productids and other brachiopods also present. In the middle part of the reefs brachiopods are absent and bryozoans form the framework, with algal coatings around grains. In the upper part algal mats or stromatolites are the frame-builders; no other fossils were present, perhaps reflecting changes of salinity.

The schwellen facies is dominated by organisms. An abundant benthic fauna of brachiopods, bryozoans, and crinoids flourished; the main constituents of the sediments, however, were oncoids (Fig. 2, locality 7).

Recent studies (Smith 1980 and *this volume*; Peryt, *this volume*) show evidence for repeated oscillations of the sea-level throughout the Zechstein basin during deposition of the Werra Carbonate. During deposition of the lower part of the Werra Carbonate the transgression continued. This is shown at the basin margins and on the palaeohighs, where basement was progressively overstepped by the Werra Carbonate. Higher in the section, the carbonate sediments indicate a progressive shallowing, which culminated in retreat of the sea and formation of a palaeokarst (Paul, *this volume*). Overlying this horizon, a renewed transgression resulted in the deposition of coarse grainstones containing intraclasts up to several centimetres in diameter. Salinity was then near normal, as indicated by the presence of crinoid ossicles.

Although not yet certain, it is likely that most reefs were formed above the palaeokarst. Similar emergence with some erosion is known from the English Zechstein of Yorkshire with the Hampole Discontinuity at the base of the Hampole Beds (Smith 1968, 1980). Other examples of regressions within the Werra Carbonate (Ca1) are reported by Peryt (*this volume*) from Poland. Should further research show these to be time-equivalent, it may in future be possible to divide the first cycle carbonate (Ca1) into subcycles and hence correlate different parts of the Zechstein basin (Paul, *this volume*).

Near the top of the Werra Carbonate is a bed with columnar stromatolites (Fig. 2, locality 8), which occurs all over the NW-German sub-basin (Langer 1977; Richter-Bernburg 1982) and which may also be equivalent to the Trow Point Bed (Smith, *this volume*). Above this bed occur laminated algal mats thinner than 50 cm, followed by nodular anhydrite in a carbonate matrix and anhydrite.

The Werra Anhydrite (A1)

With the arid climate and the restricted circulation of the Zechstein Sea, salinity increased with eventual precipitation of calcium sulphate. The process of evaporation concentration commences in the near surface layers. Deeper regions of the basin have a low rate of evaporite formation due to mixing of the concentrated, near-surface fluids with the less-concentrated waters from deeper parts of the water column. Hence the deposition of calcium sulphate was much greater on the schwellen and basin margins than in the deeper basins. The thickness of the Werra Anhydrite increases from less than 50 m in the basin to 300 m on the schwellen slopes (Fig. 2).

The deposition of large thicknesses of unconsolidated evaporites led to instability on the slopes. Large-scale sliding (Fig. 2, locality 11), slumping, and debris flows took places down the slopes of the palaeohighs (Herrmann & Richter-Bernburg 1956). Turbidity currents (Fig. 2, locality 10), led to deposition of slope evaporites up to 50 km basinward of the source (Meyer 1977).

The basin facies anhydrite (Fig. 2, locality 9) is commonly laminated. The laminae reflect changing conditions for sulphate precipitation, the fall out of organic matter and detrital clay minerals. The laminae reflect changing matter and detrital clay minerals. Individual laminae, possibly of seasonal origin, can be correlated over large distances (Richter Bernburg, *this volume*). By contrast the anhydrite of the schwellen facies and the shallow-water basin margins shows chaotic bedding structures, the result of changing deposition, desiccation, and mass movements.

The difference in deposition of the anhydrite between the basin and palaeohighs emphasized the relief. The resultant large prograding areas broadened the former palaeohigh (Fig. 2). The presence of more minor transgressions and regressions during deposition of the Werra Anhydrite is not yet fully understood (Taylor 1980).

The Stassfurt Carbonate (Ca2)

The facies distribution of the Stassfurt Carbonate (Ca2), including the basin-palaeohigh relations, was recently documented in detail by Sannemann *et al.* (1978). In general, the pattern is similar to that of the first cycle carbonate. However, differences between schwellen and basins are more accentuated as a result of the anhydrite deposition and also many groups of organisms, including bryozoans, brachiopods, and crinoids, did not survive through the first Zechstein cycle.

The schwellen facies above the anhydrite platform is called the Hauptdolomit (Fig. 2, locality 12). It is a cross-bedded, shallow-water dolomite, mostly comprising oncolites, pisoids, and oolites. Small reefs (Fig. 2, locality 13), were composed of stromatolites and bivalves and were patchily developed along the edge of the schwellen. The slope facies (Fig. 2, locality 14) is dominated by slump and slide deposits, debris flows and turbidity currents.

The basinal facies of the Stassfurt Carbonate (Ca2) is, like the Werra Carbonate (Ca1), an euxinic laminated marly limestone, the Stinkschiefer or Stinking Shale (Fig. 2, locality 15).

The distribution of dolomite and calcite is similar to the Werra Carbonate; dolomite dominates the schwellen facies, calcite the basinal facies. The reasons for this, plus the interplay of dolomitization and dedolomitization, are not fully evaluated, although discussed recently (Clark 1980).

During deposition of the following Zechstein cycles the contrast between the schwellen and the basin facies decreased as a result of the large thicknesses of halite, which were deposited, in contrast to the anhydrite deposition, within the basins. By the end of deposition of the Zechstein sediments the entire area of the Zechstein basin was a flat trough, with the earlier topography filled with evaporites.

References

CLARK, D.N. 1980. The diagenesis of Zechstein carbonate sediments. In: FUCHTBAUER, H. & PERYT, T.M. (eds) (*q.v.*), 167–204.

DEPOWSKI, S. 1978. *Lithofacies Palaeogeographical Atlas of the Permian of Platform Areas of Poland.* Instytut Geologiczny, Warszawa.

FÜCHTBAUER, H. 1968. Carbonate sedimentation and subsidence in the Zechstein basin (northern Germany). In: MULLER, G. & FRIEDMAN, G.M. (eds) *Recent Developments in Carbonate Sedimentology in Central Europe*, 196–204. Springer, Berlin.

—— 1972. Influence of salinity on carbonate rocks in the Zechstein formation, north-western Germany. In: RICHTER-BERNBURG, G. (ed.) *Geology of Saline Deposits*. Proc. Hannover Symp. (Earth Sciences 7) 23–31. Unesco, Paris.

—— 1980. A stromatolitic bryozoan bioherm in the Zechstein. In: FUCHTBAUER, H. & PERYT, T.M. (eds) (*q.v.*), 233–42.

FUCHTBAUER, H. & PERYT, T.M. (eds). *The Zechstein Basin with Emphasis on Carbonate Sequences. Contr. Sedim.* **9.**

GLENNIE, K.W. 1972. Permian Rotliegendes of northwest Europe interpreted in the light of modern desert sedimentation studies. *Bull. Am. Ass. Petrol. Geol.* **56,** 1048–71.

HERRMANN, A. 1957. Der Zechstein am südwestlichen Harzrand. *Geol. Jb.* 1–72, Hannover.

—— & RICHTER-BERNBURG, G. 1956. Frühdiagenetische Störungen der Schichtung und Lagerung im Werra-Anhydrit (Z1) am Süd-Harz. *Z. dt. geol. Ges.* **105,** 689–703.

KERKMANN, K. 1969. Riffe und Riffbildungen im Zechstein von Thüringen. *Freiberger Forschungsh.* **C252**, 85 pp.

LANGER, M. 1977. Die Bedeutung der Algen für die Gliederung des Zechsteinkalks (Ca1) in der Beckenfazies und für die Korrelation mit den Zechsteinriffen. *Z. angew. Geol.* DDR, **23,** 83–7.

MEIER, R. 1977. Turbidite and Olisthostrome—Sedimentations phänomene des Werra-Sulfats (Zechstein 1) am Osthang der Eichsfeld-Schwelle im Gebiet des Südharzes. *Veröff. Zent. Inst. Phys. Erde*, **50,** 45 pp.

PAUL, J. 1980. Upper Permian algal stromatolite reefs, Harz Mountains (F.R. Germany). In: FUCHTBAUER, H. & PERYT, T.M. (eds) (*q.v.*), 253–68.

—— 1982a. Types of stratification in the Kupferschiefer. In: EINSELE, G. & SEILACHER, A. (eds) *Cyclic and Event Stratification*, 476–81. Springer-Verlag, Berlin.

—— 1982b. Zur Rand- und Schwellen-Fazies des Kupferschiefers. *Z. dt. geol. Ges.* **133,** 571–605.

—— (*this volume*). Stratigraphy of the Lower Werra Cycle (Z1) in West Germany (preliminary results).

PERYT, T.M. (*this volume*). Chronostratigraphical and lithostratigraphical correlations of the Zechstein Limestone in Central Europe.

PLEIN, E. 1978. Rotliegend-Ablagerungen im Norddeutschen Becken. *Z. dt. geol. Ges.* **129,** 71–97.

RICHTER-BERNBURG, G. (1982) Stratogenese des Zechsteinkalkes am Westharz. *Z. dt. geol. Ges.* **133,** 381–401.

RICHTER-BERNBURG, G. (*this volume*). Zechstein 1 and 2 anhydrites: facts and problems of sedimentation.

SANNEMANN, D., ZIMDARS, J. & PLEIN, E. 1978. Der basale Zechstein (A2–T1) zwischen Weser und Ems. *Z. dt. geol. Ges.* **129,** 33–70.

SMITH, D.B. 1968. The Hampole Beds—a significant marker in the Lower Magnesian Limestone of Yorkshire, Derbyshire and Nottinghamshire. *Proc. Yorks. geol. Soc.* **36,** 463–77.

—— 1970. The palaeogeography of the English Zechstein. In: RAU, J.L. & DELLWIG, L.F. (eds) *Third Symposium on Salt*, vol. 1, 20–3. Northern Ohio Geological Society, Cleveland.

—— 1979. Rapid marine transgressions and regressions of the Upper Permian Zechstein Sea. *J. geol. Soc. London*, **136,** 155–6.

—— 1980. The evolution of the English Zechstein basin. In: FUCHTBAUER, H. & PERYT, T.M. (eds) (*q.v.*), 7–34.

—— 1981. The Magnesian Limestone (Upper Permian) reef complex of northeastern England. In: TOOMEY, D.F. (ed.) *European Fossil Reef Models. Spec. Publ. Soc. econ. Paleont. Mineral. Tulsa*, **30,** 161–86.

—— (*this volume*). The Trow Point Bed—a deposit of Upper Permian marine oncoids, peloids and columnar stromatolites in the Zechstein of north-east England.

TAYLOR, J.C.M. 1980. Origin of the Werraanhydrit in the U.K. Southern North Sea—a reappraisal. In: FUCHTBAUER, H. & PERYT, T.M. (eds) (*q.v.*), 91–114.

ZIEGLER, P.A. 1981. Evolution of sedimentary basins in north-west Europe. In: ILLING, L.V. & HOBSON, G.D. (eds) *Petroleum Geology of the Continental Shelf of North-west Europe, Vol. 1, Geology*, 3–42. Applied Science, Barking.

JOSEF PAUL, Geologisch-Paläontologisches Institut und Museum, Georg-August Universität, Goldschmidt-Str. 3, 3400 Göttingen, Federal Republic of Germany.

Discussion contribution by D.B. Smith

(1) The distribution of the English Marl Slate is mutually exclusive of that of Cycle 1 grainstones and is closely controlled by basin-floor topography and the depth to an inferred pycnocline. It is interesting to note that Dr Paul considers wave-base to be a major factor affecting the distribution of the equivalent Kupferschiefer and it would be useful to learn if he believes that wave-base and the pyncnocline/chemocline were coincident and if he has any impression of the possible depth to wave base.

(2) Has Dr Paul speculated on the possibility that the sea-level fall that caused interstratal karst formation in the Cycle 1 carbonate in Germany might have been the same as that causing the Hampole Discontinuity in England?

Replies

(1) The facies of the Kupferschiefer is strongly affected by the pre-Zechstein topography and the position of the chemocline (redox discontinuity layer). The position of this chemocline changed during the deposition of the Kupferschiefer. At the beginning, the palaeohighs were above the chemocline in the aerated water, as can be deduced by a fosiliferous carbonate bed, the Mutterflöz. Later on, the chemocline rose to the wave base. This is demonstrated at steep inclined slopes, where laminated euxinic Kupferschiefer impinges at pre-Permian hills, whereas at the top of these hills they are bare of sediments. There is no evidence about the depth of the wave base. By comparison with recent marginal seas a depth of about 10–50 m may be estimated.

(2) In Germany as well as in England there are two cycles of rise and fall of the sea-level within the Zechstein Limestone. Therefore, it is sensible to regard these events as time-equivalent and to correlate the corresponding discontinuities. But T.M. Peryt has found in the Polish part of the basin three subcycles of transgression/regression. Hence further investigations are necessary to prove the existence and the significance of a third subcycle in the central and western part of the basin before a definite decision on correlation is possible.

Stratigraphy of the Lower Werra Cycle (Z1) in West Germany (preliminary results)

J. Paul

SUMMARY: The Zechstein Limestone (Ca1) of Western Germany has a clear stratigraphical subdivision. A lower unit (subcycle a) is separated from an upper unit (subcycle b) by a discontinuity, which is best exposed at the Solhops-Berg at the Harz Mountains and is here termed the 'Solhops Discontinuity'. The discontinuity was caused by an extensive regression of the Zechstein Sea which led to subaerial exposure of large areas of the Zechstein basin. Based on this discontinuity, a preliminary stratigraphy of the Lower Werra cycle in Germany is established and compared with that of the English Zechstein sequence.

Principles of Zechstein stratigraphy

The Zechstein basin was a restricted shallow marginal sea with only temporary connections with the open ocean. It was characterized by rapid marked changes of salinity that extinguished almost all life: only a few adaptable species survived in favoured ecological niches, or more likely, re-entered from the oceans. The history of life is, therefore, a continual decrease in the number of species.

As no useful biostratigraphy can be recognized, the subdivision of the Zechstein sequence is mainly lithostratigraphical; it is to the great credit of Richter-Bernburg (1955) that he established a classification based on cyclic changes of salinity; this classification has been generally accepted because the evidence of cyclicity is found almost everywhere in the Zechstein basin. Nevertheless, important problems of stratigraphy have yet to be solved; for example, the cyclic subdivision is difficult to apply near the margins of the Zechstein basin where locally there is no clear lithostratigraphical differentiation. There is also no generally accepted intracyclic subdivision that might allow more precise correlations throughout the Zechstein basin.

In the English part of the Zechstein basin there is a very clear partition of the first cycle shelf carbonate into a lower and an upper member (Smith 1968, 1980; Smith *et al.* 1974), but in West Germany there are only local formations in the marginal and swell areas; only tentative attempts have previously been made to correlate these local formations.

It may, therefore, be useful to apply the results of recent research suggesting that rapid and intense changes of the sea-level occurred during deposition of the Zechstein Limestone (Smith 1979; Paul, *this volume*; Peryt, *this volume*). These changes probably affected the whole basin and therefore may have been caused by eustatic movements rather than local tectonic events (Smith 1980). As the changes in the sea-level were apparently very rapid in comparison with the rate of sedimentation, it is possible to subdivide the Zechstein Limestone independently of facies and also to correlate these subcycles throughout the basin.

Regional settings

Five areas (Fig. 1, A–E) have been chosen to illustrate the existence of the subdivision and the correlation within the Lower Werra Cycle.

The Ibbenbühren and Hüggel areas

The Ibbenbühren and Hüggel areas, two isolated outliers of sedimentary rocks west and south of Osnabrück, are the most north-westerly outcrops of Zechstein deposits (Thiermann 1975; Harms 198[illegible]); Upper Carboniferous sandstones form the basement, which was part of the Rheinisches Schiefergebirge. Information from boreholes shows that Zechstein strata wedge out several kilometres south of the outcrops and a nearshore position may therefore be presumed. In the Ibbenbühren area the Kupferschiefer and the Zechstein Limestone are calcitic, whereas dolomites predominate at the Hüggel where a fossiliferous bed, the Mutterflöz (Paul, *this volume*) lies below the Kupferschiefer and indicates a higher palaeotopographical level. A discontinuity (Fig. 2) within the Zechstein Limestone separates a lower micritic unit with abundant slumps and slides from an upper unit composed of nodular limestones containing oncoids. The discontinuity surface has a relief, with depressions several centimetres deep; this could be a solution effect, but no other clear indications of karst features are detectable.

From HARWOOD, G.M. & SMITH, D.B. (eds), 1986, *The English Zechstein and Related Topics*, Geological Society Special Publication No. 22, pp. 149–156.

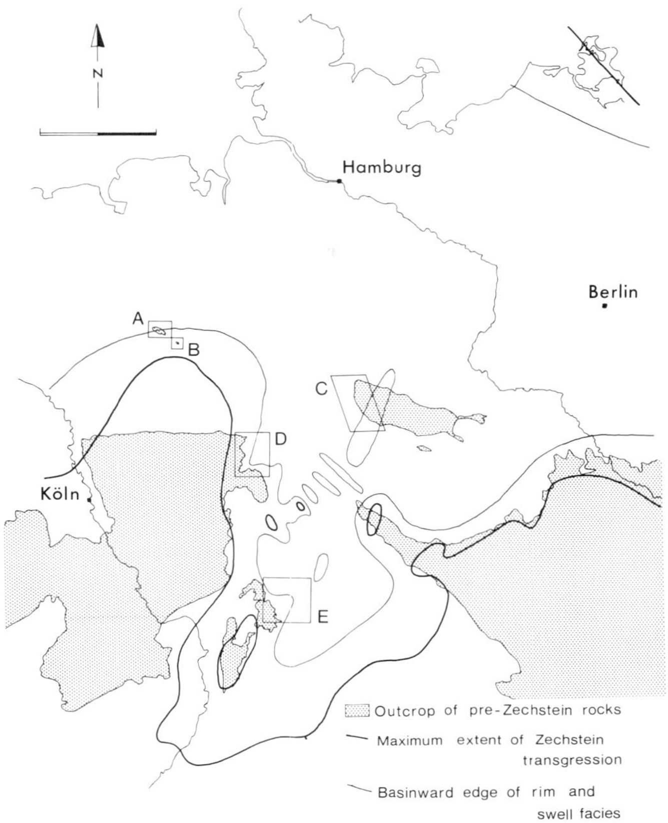

FIG. 1. The German Zechstein basin (after Füchtbauer 1972; Boigk & Schöneich 1974; Jung & Knitzschke 1976; Leifeld & Kulick 1981). A: Ibbenbühren, B: Hüggel, C: Harz Mountains, D: Waldeck, E: Spessart.

The Harz Mountains

A prominent palaeo-high, the Eichsfeld–Schwelle, crosses the Zechstein outcrops at the margin of the Harz Mountains; palaeoenvironments there range from lower slope to uppermost swell (Paul 1982a and *this volume*). The lower slope facies is best exposed at the Solhops-Berg at Seesen near the western extremity of the Harz Mountains, where the complete Cycle 1 carbonate (Ca1) is exposed (Fig. 3); the dominant feature of the 10 m thick Zechstein Limestone here is a palaeokarst surface in the upper part of the sequence (Figs 4 and 5). Underlying strata show indications of progressive shallowing of the water, with the proportion of oncoids increasing upwards as micritic limestones decrease. Above the discontinuity an intraformational breccia is evidence of a new transgression. Less than 1 m of oncoids were deposited and were succeeded by a very extensive stromatolite bed (Langer 1977; Richter-Bernburg 1982) that indicates a further phase of regression. Above this stromatolite bed, algal laminites and marls form an indistinct third subcycle or the transition to the Werra Anhydrite.

The position of the discontinuity at the top of the swell is not yet clear because turbulence in the shallow water caused many breaks in sedimentation; however there is a similar lithological sequence here of micritic dolomites overlain by oncolitic dolomites in the upper part of the Werra carbonate. The time of the initiation of reefs is unknown, but apparently was relatively late.

Several authors (Richter-Bernburg 1955; Herrmann 1956; Kerkmann 1969; Jordan 1979) separate the upper oncolitic part of the sequence from the Zechstein Limestone and term it the Werra-Dolomit, disregarding the fact that the lower part is also composed of dolomite. No exact definition of the subdivision had been made; for example Kerkmann writes ". . The Zechstein Limestone

FIG. 2. Zechstein Limestone. A discontinuity separates a micritic lower unit from an upper unit consisting of nodular limestones. Scale: 1 m. Quarry at Uffeln, near Ibbenbühren (area A).

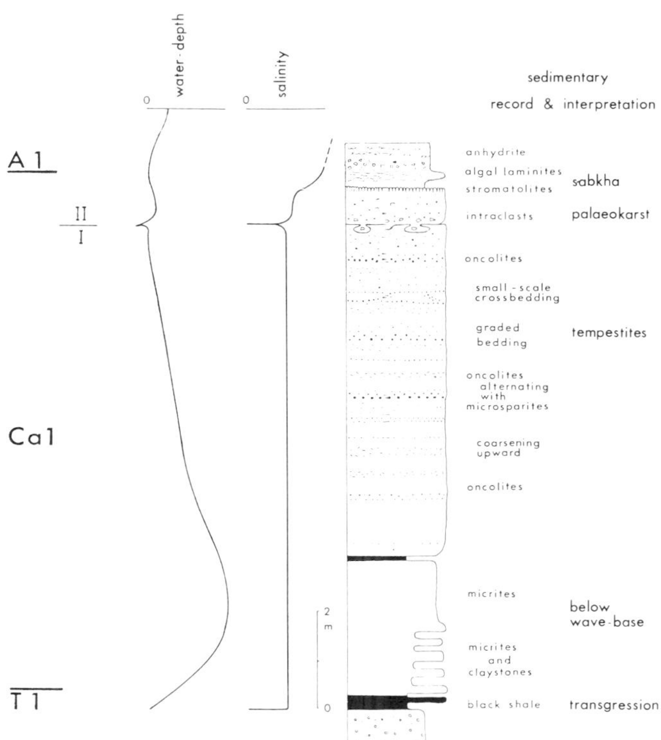

FIG. 3. Schematic section of the Zechstein Limestone at the lower slope of the Eichsfeld-Schwelle, near the Harz Mountains. Solhops (area C).

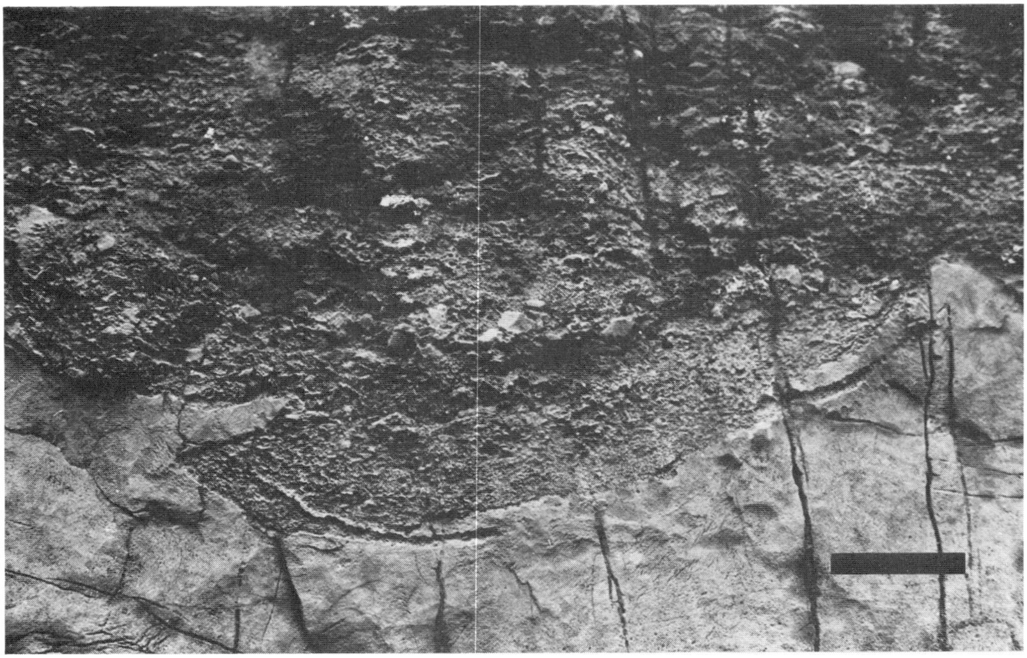

FIG. 4. Palaeokarst surface (discontinuity) with solution hollows, near top of Zechstein Limestone. Scale: 10 cm. Solhops-Berg, Harz Mountains.

FIG. 5. Intraformational breccia above the discontinuity passes upwards into oncolitic limestones. Solhops-Berg, Harz Mountains, same locality as Fig. 4.

passes over without a sharp boundary to the Werra-Dolomit." According to these authors this dolomite is stratigraphically equivalent to the Werra Anhydrite of the basinal facies.

The Waldeck area

The Zechstein outcrops of Waldeck are at the eastern margin of the Rheinisches Schiefergebirge. A small bay of the Zechstein Sea with moderately deep water was surrounded by more shallow areas and the facies were correspondingly differentiated; the assumed shoreline may have lain some 10 km west of the area.

Above the Kupferschiefer there are micritic limestones with an upwards-increasing content of oncoids. As in the other areas, there is a break in the sedimentary record, with hardgrounds and other indications of shallowing water. The overlying carbonate sequence is more or less wholly oncolitic. Only the palaeotopographically highest parts of the area were reached by the transgression during this period.

The upper part of the carbonate sequence is called the rim carbonate and its stratigraphical position is correlated by Richter-Bernburg (1950) and Kulick (1968) with the anhydrite (A1) of the basin. However, the presence of bryozoan patch reefs (Kulick 1968) and of stenohaline fossils throws doubt on this correlation because a hydrographical model combining the precipitation of sulphate in the basin and the presence of stenohaline fossils at the rims is very difficult to establish. I believe, therefore, that this unit is the upper subcycle of Ca1.

The Spessart Mountains

During Zechstein deposition the Spessart Mountains were a palaeohigh at the SW margin of the great bight between the Rheinisches Schiefergebirge and the Thuringian Forest; the highest parts were above sea-level. Because of this marginal position, stratigraphical relations are not clear. Micritic dolomites lie nearest to the shoreline and pass laterally into micritic limestones. A palaeokarst horizon may be a time-equivalent of the Solhops Discontinuity. Prüfert (1969) and Kowalczyk & Prüfert (1974, 1978) place this discontinuity at the top of the Zechstein Limestone. An unusually thick (*c.* 80 m) overlying sequence of marls and claystones in the slope facies north and east of the Spessart Mountains should be equivalent of the sequence Anhydrite 1 (A1) to Clay 3 (T3). A fossiliferous dolomite (Upper Carbonate Series of Prüfert) overlying the marls is thought to be the Plattendolomit (Ca3), but contains a very rich fauna of bryozoans, spiriferids and other stenohaline fossils that is very unusual for the Ca3 and would be unique in the Zechstein basin (Pattison 1970; Klapcinski & Karwowski 1981). It is more likely that this dolomite is the second subcycle of the Zechstein Limestone.

This interpretation is in good accordance with the findings of boreholes at the eastern margin of the Spessart-Schwelle, where there is also a thick sequence of marls and claystones between a lower and an upper carbonate below the Werra-Anhydrit.

Stratigraphic results

A preliminary correlation chart compares the lower Werra cycle of various areas in Germany with the English Zechstein (Table 1).

In all the areas investigated it is possible to recognize a clear subdivision of the Zechstein Limestone at a discontinuity which was caused by an extensive retreat of the sea. The lithological transition from a lower micritic to an upper oncolitic unit commenced earlier, probably with the beginning of the shallowing of the sea. In the first subcycle, dolomites are restricted to the upper parts of the swells, whereas in the second subcycle the dolomites spread over the former calcitic slope and basinal areas.

It seems that this discontinuity is a good marker horizon throughout the German part of the Zechstein basin; as it is best exposed and is clearly seen at Solhops in the Harz Mountains, I term it the 'Solhops Discontinuity'.

In England, the retreat of the sea and the resulting subdivision of the Zechstein Limestone has long been recognized. The retreat of the sea seems to have been very rapid in comparison with the rate of sedimentation and affected the whole basin; the discontinuity is therefore probably synchronous for stratigraphical purposes. The Hampole Discontinuity (Smith 1968) may therefore be correlated confidently with the Solhops Discontinuity and the Trow Point Bed (Smith, *this volume*) may be correlated with the stromatolite bed below the Werra Anhydrite in Central Europe.

However, according to Peryt (*this volume*) there were three subcycles of transgression and regression in the Polish part of the basin during Ca1 time. At Solhops a third cycle might be present but in the other areas no third cycle has yet been found. Further investigations in Germany as well as in England should prove whether or not there is a third subcycle and what significance it may have had on the evolution of the Zechstein Limestone.

TABLE 1. Preliminary correlation table of the lower Werra Cycle (Z1) in Western Germany. Ks—Kupferschiefer; Mf—Mutterflöz.

	A	B	C Harz Mountains		D	E Spessart			
A1	Ibbenbühren	Hüggel	lower slope	top schwellen	Waldeck	near-shore	slope	Yorkshire	Durham
Werra Carbonate (Ca 1): subcycle b	nodular limestone oncoids 30 m	oncolite 30 m	algal laminites stromatolites oncolite Solhops	stromatolitic reefs oncolite dolomite ? 40–80 m	oncolite limestone / dolomite 45 m	dolomite 40 m	dolomite –17 m marls & claystone	Cadeby Formation: Sprotbrough Member 0–40 m Hampole	Trow Point Bed Ford Formation 0–130 m
Werra Carbonate (Ca 1): subcycle a	micritic limestone 6 m	micritic dolomite 0–6 m	Discontinuity micritic limestone 9 m	oncolite micritic dolomite biomicrite	oncolite micritic limestone 25 m	dolomite 0–15 m	marl limestone 15 m	Cadeby Formation: Discontinuity oncolite Wetherby Member 0–60 m	Raisby Formation 30–70 m
Kupferschiefer	Kupferschiefer	Kupferschiefer / Mf	Kupferschiefer	Ks / Mf	Ks / Mf	Ks	Kupferschiefer / Weissliegendes	Marl Slate	Marl Slate
Basement	Upper Carboniferous sandstone		Rotliegendes	Carboniferous & Devonian	Carboniferous greywackes	Rotliegendes	Rotliegendes	Yellow Sands	

References

BOIGK, H. & SCHÖNEICH, H. 1974. Perm, Trias und älterer Jura im Bereich der südlichen Mittelmeer-Mjösen- Zone und des Rheingrabens. In: ILLIES, J.H. & FUCHS, K. (eds) *Approaches to Taphrogenesis*, 60–70. Schweizerbart, Stuttgart.

FUCHTBAUER, H. 1972. Influence of salinity on carbonate rocks in the Zechstein Formation, northwestern Germany. In: RICHTER-BERNBURG, G. (ed.) *Geology of Saline Deposits*, 23–31. UNESCO, Paris.

HARMS, F.J. 1981. Zur Geologie und Tektonik des Hüggel- und Silberberg-Gebietes bei Osnabrück (West-Niedersachsen). *Osnabrücker naturwiss. Mitt.* **8,** 19–62.

HERRMANN, A. 1956. Der Zechstein am südwestlichen Harzrand. *Geol. Jb.* **72,** 145–63.

JORDAN, H. 1979. Der Zechstein zwischen Osterode und Duderstadt (südliches Harzvorland). *Z. dt. geol. Ges.* **130,** 145–63.

JUNG, W. & KNITZSCHKE, G. 1976. Kupferschiefer in the German Democratic Republic (GDR) with special reference to the Kupferschiefer deposit in the southeast Harz Foreland. In: WOLF, K.H. (ed) *Handbook of Stratabound and Stratiform Ore Deposits*, **6,** 353–406. Elsevier, Amsterdam.

KERKMANN, K. 1969. Riffe und Riffbildungen im Zechstein von Thüringen. *Freiberger Forschhft.* **C252,** 85 pp.

KLAPCINSKI, J. & KARWOWSKI, L. 1981. Stratigraphical analysis of the Zechstein deposits in western Poland based on macrofauna studies. *Proc. int. Symp. Central European Permian, Warsaw, 1978*, 490–501. Wydawnictwa Geologiczne, Warsaw.

KOWALCZYK, G. & PRÜFERT, J. 1974. Gliederung und Fazies des Perms in der Wetterau. *Z. dt. geol. Ges.* **125,** 61–90.

—— & 1978. Exkursion F in das Oberrotliegende und den Zechstein am Rand von Spessart un Vogelsberg am 1. April 1978. *Jb. Mitt. oberrhein. geol. Ver.* N.F. **60,** 87–108.

KULICK, J. 1968. *Erläuterungen geologische Karte Hessen 1:25000, Bl. 4719, Korbach*, Wiesbaden.

LANGER, M. 1977. Die Bedeutung der Algen für die Gliederung des Zechsteinkalkes (Ca1) in der Beckenfazies und für die Korrelation mit den Zechsteinriffen. *Z. angew. Geol. DDR*, **23,** 83–7.

LEIFELD, D. & KULICK, J. 1981. Regionale Erkundung des Kupferschiefers in Hessen im Vorfeld industrieller Exploration. In: NEUMANN, R. & GUNTHER R. (eds) *Rahmenprogramm Rohstofforschung. Statusseminar 1981. Mineralische Rohstoffe*, 281–91 Statusbericht, Jülich.

PATTISON, J. 1970. A review of the marine fossils from the Upper Permian rocks of northern Ireland and northwest England. *Bull. geol. Surv. G.B.* **32,** 123–65.

PAUL, J. 1982a. Zur Rand- und Schwellen-Fazies des Kupferschiefers. *Z. dt. geol. Ges.* **133,** 571–605.

—— 1982b. Types of stratification in the Kupferschiefer. In: EINSELE, G. & SEILACHER, A. (eds) *Cyclic and Event Stratification*, 476–81. Springer-Verlag, Berlin.

—— (*this volume*) Environmental analysis of basin and schwellen facies in the lower Zechstein.

PERYT, T.M. (*this volume*). Chronostratigraphical and lithostratigraphical correlation of the Zechstein Limestone in Central Europe.

PRÜFERT, J. 1969. Der Zechstein in Gebiet des Vorspessarts und der Wetterau. Seine Stratigraphie, Fazies und Palaeogeography. *Sonderveröfftl. geol. Inst. Univ. Köln*, **16,** 186 pp.

RICHTER-BERNBURG, G. 1950. Zwei Beiträge zur Fazies, tektonik und Kupferführung des Zechsteins. I. Waldeck, II Nordsudeten. *Geol. Jb.* **65,** 145–212.

—— 1955. Stratigraphische Gliederung des deutschen Zechsteins. *Z. dt. geol. Ges.* **105,** 843–55.

—— 1982. Stratogenese des Zechsteinkalkes am Westharz. *Z. dt. geol. Ges.* **133,** 381–401.

SMITH, D.B. 1968. The Hampole Beds—a significant marker in the Lower Magnesian Limestone of Yorkshire, Derbyshire and Nottinghamshire. *Proc. Yorks. geol. Soc.* **36,** 463–74.

—— 1979. Rapid marine transgressions and regressions of the Upper Permian Zechstein sea. *J. geol. Soc. London*, **136,** 155–6.

—— 1980. The evolution of the English Zechstein basin. In: FUCHTBAUER, H. & PERYT, T.M. (eds) *The Zechstein Basin with Emphasis on Carbonate Sequences*, 7–34. *Contr. Sedim.* **9**.

—— (*this volume*). The Trow Point Bed—a deposit of Upper Permian marine oncoids, peloids and columnar stromatolites in the Zechstein of north-east England.

——, BRUNSTROM, R.G.W., MANNING, P.I., SIMPSON, S. & SHOTTON, F.W. 1974. *A correlation of Permian Rocks in the British Isles. Spec. Rep. geol. Soc. London. No. 5.*

THIERMANN, A. 1975. *Erläuterungen zur geologischen Karte Blatt 3611 Hopsten. Geol. Kte. Nordrhein-Westfalen 1:25000*, 214 pp.

JOSEF PAUL, Geologisch–Paläontologisch Institut und Museum Georg–August Universität, Goldschmidt-Str. 4, 3400 Göttingen, Federal Republic of Germany.

Addendum

Recent investigations proved the existence of three cyclic sea-level changes in the central part of the Zechstein basin similar to Poland. However, the second maximum and the second fall of the sea-level had only a limited range which are not discernible in the basin nor on the tops of the palaeohighs. We can thus trace only two cycles within the basin. Only on the slopes are there three cycles distinguishable.

Reference

PAUL, J. & PÖHLIG, 1984. Eustatische Spiegelschnankungen des präsalinaren Zechstein-Meeres. *Abstr. Geotajung 1984*, Hamburg, 114–5.

Zechstein 1 and 2 Anhydrites: facts and problems of sedimentation

G. Richter-Bernburg

SUMMARY: The distribution and lithology of the anhydrite rocks of Zechstein 1 and Zechstein 2 were significantly influenced by the contemporary geography of the basin. Thick 'pletholitic' shallow-water deposits with large-scale synsedimentary derangements lie near the margins and pass laterally, with steep boundaries, into thin laminated deep-water rocks on the basin floor. Five sub-cycles (I–V) have been recognized in cored boreholes in the Z1 Werraanhydrit throughout the German Zechstein basin and permit correlation with the Zechstein 1 sequence of the southern North Sea Basin; although not complete in every place, each sub-cycle generally comprises flaser anhydrite overlain by laminated anhydrite. The abrupt changes of facies at the contacts between the sub-cycles cannot be attributed to eustatic sea-level changes but are caused by changes of the climate, water currents and related effects; the periodicity must have resulted from factors originating outside the basin.

The following new adjectives are proposed, primarily for varieties of anhydrite although they may prove to be equally applicable to other sedimentary rock types:

'Pletholitic'—from the Greek 'plethos', an indefinable mass—applicable to thick anhydrite rock with a diffuse, cloudy or chicken-wire structure and fabric but with little overall stratification.

'Stratobolic'—from the Greek 'bolos', lump—anhydrite rock with extremely irregular small-scale and medium-scale bedding but with a strong sense of stratification *en masse*. This term includes 'flaser', which is mainly applied to stratobolic rocks with bedding of an exceptionally anastomosing character.

'Lamellitic'—from the German 'Lamellenanhydrit'—unusually uniformly and evenly laminated anhydrite, including some varved anhydrite. This term is proposed because the word 'laminated' is used in a range of senses and does not adequately stress the uniformity and continuity of bedding in some thinly and evenly layered anhydrite rocks.

It is well known that carbonate rocks display sharp changes in thickness and facies; during EZ82 this was abundantly demonstrated in Zechstein strata in the field and by several authors. The same is true for anhydrites although the scale is different. We know that the Z1 anhydrite is more than 400 m thick in places and is less than 10 m elsewhere. It is not surprising that the facies are completely different too; thus there are huge anhydrite masses that have a 'pletholitic' fabric (see Summary for definition) and, by contrast, there are anhydrite rocks that are fine-bedded or regularly laminated. These different sulphate rocks are the extreme types that we usually associate with 'shallow-water' or 'deep-water' environments respectively; this is undoubtedly correct in general, but in detail there are some unresolved problems and it is with these that this paper is mainly concerned. It is necessary, in discussing these problems, to recall some facts that have been published previously.

Basalanhydrit (Zechstein 2)

The so-called Basalanhydrit (A2), only 2–3 m thick, lies in Zechstein 2 between the Stinkschiefer (Ca2) and the thick Stassfurt rock salt (Na2); this halite can be about 500 m thick. The lower part of the A2 is finely laminated (varved) and consists of about 1000 layers that are very regular but of slightly varied thickness so that we can identify the individual layers by counting and measuring them. All these layers can be correlated throughout the German basin, which measures about 350 km W–E and about 320 km N–S; even individual peculiarities such as doubling, or ghost layers, distinguish the same bed everywhere. This consistency of the A2 is evidence of its deposition over the even and uniform floor of the Zechstein Sea, where no environmental events disturbed the quiet course of deposition for about 1000 yr. By counting and numbering the varves we get a time-scale, probably the most exact of all, which is important as a basis for the following findings: (1) each and every varve thickens towards the margin of the basin (Fig. 1); we may be justified in concluding that the conditions for the precipitation of $CaSO_4$ were fundamentally better in the shallower parts of the basin than in the deeper. (2) The deposition of the calcium sulphate started earlier near the edge of the basin than in the centre for the underlying Stinkschiefer (i.e. the carbonate phase of Zechstein 2), persists for 300

From HARWOOD, G.M. & SMITH, D.B. (eds), 1986, *The English Zechstein and Related Topics*, Geological Society Special Publication No. 22, pp. 157–163.

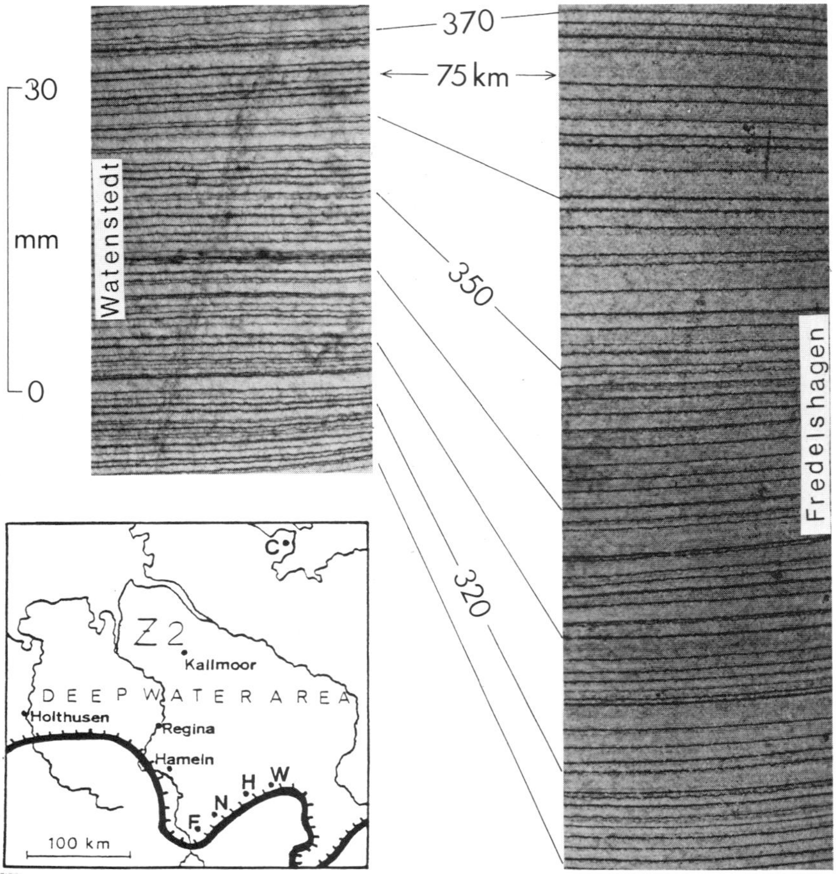

FIG. 1. Lamellitic (varved) Basalanhydrit, Zechstein 2, in cores of two boreholes NW of the Harz Mts. The congruent interval of varves 310–370 is 47 mm thick in the basin floor facies, but 86 mm thick closer to the southern border, where precipitation was more rapid. The map also shows the sites of boreholes depicted in Figs 2 and 3. C—Cismar; F—Fredelshagen; H—Hahndorf; N—Northeim; W—Watenstedt.

varves—or years—longer in the centre and passes into anhydrite towards the margin. (3) In a few places, the bedding of one or two varves is uneven, and this irregularity also distinguishes the same varves far away. The causal event, be it a storm or similar influence, had to be in exactly the same year—or moment.

The varved lower part of the Basalanhydrit has a sharp top and is overlain by flaser anhydrite or a similar 'stratobolic' sulphate (see Summary for definition). This rock type is usually included in pletholitic sediments which we generally suppose to be of shallow water origin.

The sharpness of the change in the sequence from lamellitic (see Summary for definition) to stratobolic anhydrite means that by counting the varves we can determine the year of this change: such counting reveals that in boreholes close to the basin margins this contact lies between varves 600 and 800 whereas in the inner part of the basin there are more than 1000 varves below the base of the flaser anhydrite (Fig. 2). The change from lamellitic to stratobolic anhydrite thus lies lower and was presumably earlier where the water was shallower than in the deeper more central areas.

The question arises of whether (a) the event responsible for the change from lamellitic to stratobolic anhydrite actually took place about 400 years earlier near the basin margin than in the centre or (b) this event occurred everywhere simultaneously but its effect has been disguised in marginal areas by eodiagenetic imposition there

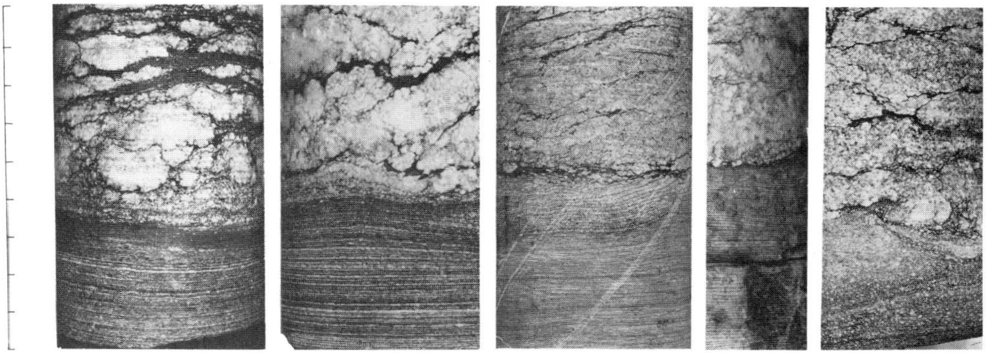

FIG. 2. Abrupt change from laminated lower part to stratobolic upper part of Basalanhydrit, Z2. The figures indicate the number of the youngest varve of the lower sequence. The scales are negligibly different. For key map see Fig. 1. Scale in centimetres.

of a flaser-type fabric on formerly laminated anhydrite. This question cannot satisfactorily be answered but a hint that at least some depositional changes were widespread and contemporaneous may be gleaned from a study of the cores of the Regina and Holthusen boreholes (sited 140 km apart) where a thin unit of stratobolic anhydrite lies above varve 480 (Fig. 3) although this feature is not found in surrounding boreholes; the brief change here from laminated to flaser anhydrite is petrographically similar to that at the top of the main lamellitic sequence.

In places the laminated rock type seems to be overwhelmed by the flaser type though it is partially preserved in stringers surrounding the lenses of white anhydrite; this may indicate replacement of laminated by flaser anhydrite. The thin stratobolic layer in the Regina and Holthausen could, alternatively, be regarded as a forerunner by about 500 years of the change in sedimen-

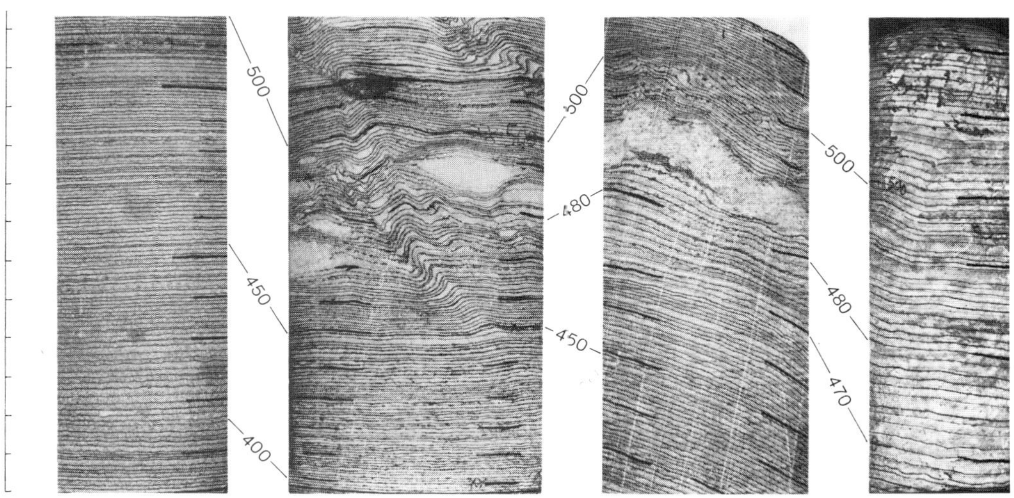

FIG. 3. Exceptional occurrence of flaser anhydrite within the varved part of Basalanhydrit, Z2. The interval of varves 480–490 shows a replacement by stratobolic fabric in only two boreholes (Holthusen and Regina). All other wells around present the normal varve sequence of the basin floor, as shown by the cores from the Kallmoor and Hameln boreholes. For key map see Fig. 1. Scale in centimetres.

tation at the top of the main laminated sequence and may have had a similar cause (see later). In seeking for this cause I will consider the evidence furnished by the fabric of the Z1 Werraanhydrit.

Werraanhydrit (Zechstein 1)

In the Werra Group, huge linear masses ('walls') of pletholitic anhydrite lie just inside the margin of the basin and follow the submarine ridges; they are commonly 200–300 m thick and locally exceed 500 m. From their geographical position and their stratigraphical relationship to older and younger rocks there can be no doubt that these masses formed in relatively shallow-water parts of the basin. They have a steep slope towards the main basin, comparable with the scarp of a barrier reef (Richter-Bernburg 1941, 1942, 1955, p. 612, 1972, p. 281 *et seq.*).

In the basin centre the Werraanhydrit is of the so-called deep water facies; it is only 40–50 m thick and very uniform through the large basin. It is composed of rhythmically alternating laminated and stratobolic anhydrite, and five sub-cycles (I–V) may be recognized (Richter-Bernburg 1955). Each sub-cycle starts with a coarse stratobolic (flaser or similar) anhydrite which is generally overlain by pearly anhydrite and this by lamellitic (mostly varved) anhydrite (Richter-Bernburg 1955, p. 628; von Hoyningen-Huene 1957; Jung 1958). The subdivision is present throughout the whole German Zechstein Basin, having been proved in many completely cored boreholes; some of these are summarized in Fig. 4.

It is of considerable importance that Taylor (1980) confirms comparable subdivisions from the southern North Sea where he distinguishes six sub-cycles, four of which (A–D) he assigns to a 'Lower' Werraanhydrit, and two (E, F) he assigns to an 'Upper' Werraanhydrit (or ?Hauptdolomit; but, I agree with Smith (1980) that this is still Zechstein 1). It seems reasonable to see a possibility for correlating the subdivisions between the North Sea and Germany. During EZ82 I briefly inspected this part of the cores from North Sea well 49/18–2 and can say that the rocks and sequence encountered are very similar to those in many complete cores from Germany. Other features common to the North Sea and German sequences are: (1) the occurrence of a thin deposit of oncoids and algal stromatolites at the top of the Cycle 1 carbonate (Smith 1974, and *this volume*; Richter-Bernburg 1982), (2) the presence of white nut-sized nodules of finely crystalline anhydrite in the upper few metres of Ca1 (Richter-Bernburg 1941; Taylor 1980), and (3) the presence of fairly clean, coarse structured, bright anhydrite of some decimetres thickness, succeeding the Zechsteinkalk with a sharp contact in exactly the same manner in the two places (Richter-Bernburg 1941, table 2; Taylor 1980, fig. 8a).

The sequences of the Werraanhydrit of basin facies are similar in the various boreholes in Germany and although details vary from place to place the individual cycles have the same successions of rock-types. This is equally true of the southern North Sea as described by Taylor (1980, p. 102) where he writes "details differ from one unit to the next but each retains its distinctive features across the basin..."

The whole basin extending from England through Germany and Poland, does not only share the same physicochemical and climatic environment but there is also a coincident run of sedimentation that responded to the same changes in the depositional conditions or the same supra-regional events.

Facies leaps

It has been noted above that each sub-cycle of the basin facies of the Werraanhydrit starts with a pletholitic bright anhydrite that succeeds the underlying laminite with a sharp contact and may indicate a rapid change of facies; this sharpness has been emphasized by the writer and also stressed by Taylor (1980). The relationship is comparable with that at the contact between the lower (varved) and upper (stratobolic) part of the Basalanhydrit in Z2, noted earlier. The significance of this is that the conditions of sedimentation which previously were dominant in the marginal wall or platform area, forming there pletholitic rocks, suddenly also dominated the basin floor at the moment when a new sub-cycle starts.

In searching for an explanation for this change we can exclude any local influence because of the ubiquity of the feature. The occurrence of stratobolic rocks in the interior of the basin, far away from the sediment input sources, argues against clastic accumulation as favoured by Schlager & Bolz (1977), although such accumulation could have occurred in the marginal belt. We have to take into account the following environmental facts:

(1) The thick mounds of Werraanhydrit are known from many large surface exposures not to form extended uniform and flat-lying deposits such as typify intertidal or supratidal rocks. On the contrary, we see stratigraphic anarchy caused by massive slumping and large-scale dislocations

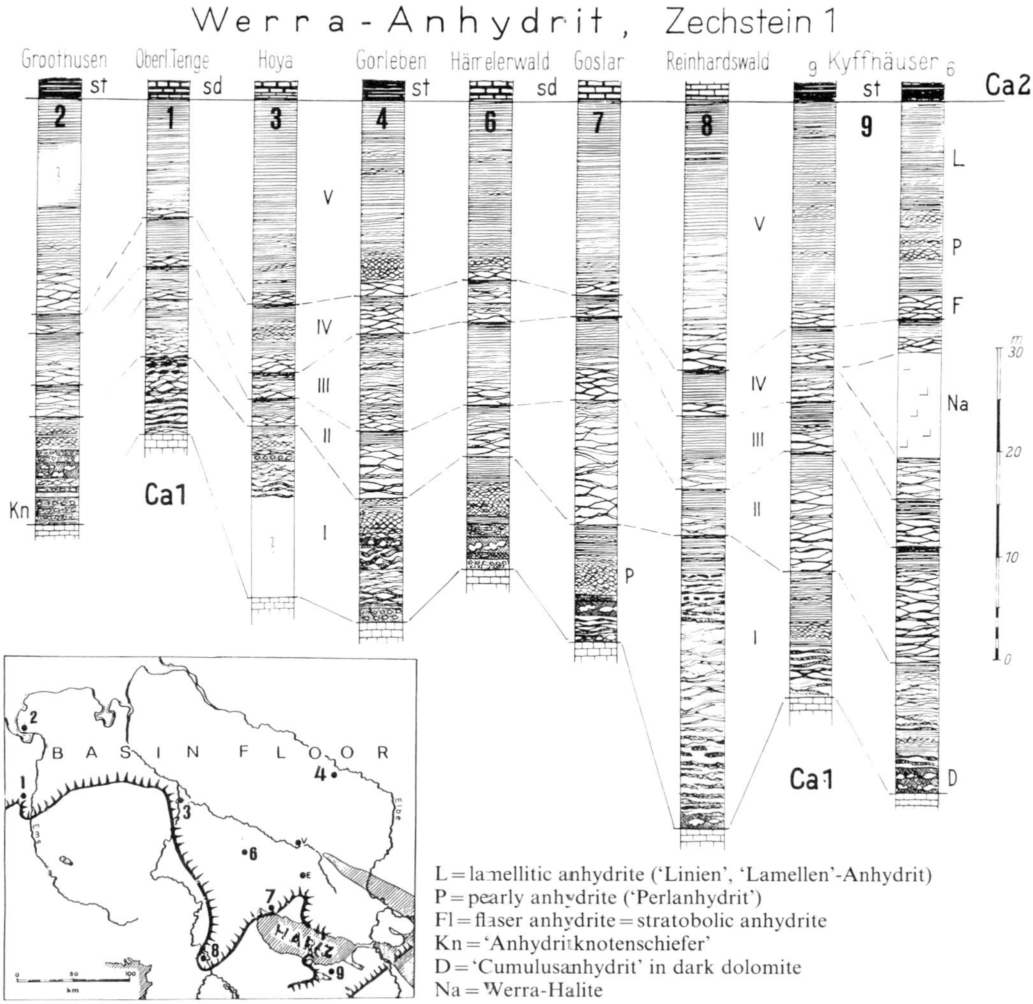

FIG. 4. Complete sequences of cored Werraanhydrit, Zechstein 1, showing the widespread subdivision into five sub-cycles. The site map also shows the position of the steep basinward slope of the Z1 anhydrite 'wall' (after Richter-Bernburg 1957a,b, 1972, p. 278).

which also involve regularly laminated anhydrite. Herrmann & Richter-Bernburg (1955) first explained these structures as synsedimentary deformations, though they are so large that they were taken to be tectonic; their origin by slumping has been confirmed by several later authors and excellent photographs of the disturbed strata were presented by Meier (1977).

(2) The so-called platform which is assumed to have lain landward of the anhydrite wall was surely not a smooth tidal flat, as it was during deposition of the succeeding Z2 shelf carbonates. In Z1, more than 200 m of halite was formed there (Richter-Bernburg 1964, p. 979), and the thicknesses of the anhydrite masses change remarkably sharply in accord with the contemporary water depth. I agree with Taylor (1980) that petrological study of the Werraanhydrit could readily lead to wrong conclusions on its mode of origin if the configuration of the whole basin is not also taken into account.

(3) The area with the thinnest Werraanhydrit is the classic Werra-Becken where about 300 m of halite with potash represents nearly the whole of Zechstein 1. Even there, though, the 'Anhydritknotenschiefer' contains intercalations of flaser anhydrite that are comparable with the stratobolic anhydrite in the sub-cycles of the basin facies.

(4) Wherever halite occurs in Z1 it is in sub-cycle IV, and this seems to be true also in North Sea well 49/18–2, as far as I could observe it; the salt is not associated with the greatest thickness of varved basin floor deposits as might be expected, but it occurs with stratobolic anhydrite. There are similar associations in Z2. Local details will be published separately.

It has to be assumed that in the Z1 basin the highest salinity was not encountered on the uniform seafloor but upon the large areas behind the barrier walls where the evaporation rate was greatest. This higher concentration resulted in faster deposition of $CaSO_4$ in the shallower water than in the deeper parts of the basin. Concurrently, gravity led to the movement of the heavier NaCl-brine to the still waters of the basin floor where little happened apart from the annual deposition of an average of 0.3 mm of $CaSO_4$ and of some remains of dead plankton (surely not algal mats). This tranquillity must have been abruptly disturbed at the moment when a new cycle started with a different facies.

It has therefore to be considered whether this invasion of 'shallow water' conditions into the 'deep water' dominion may have happened in connection with a eustatic sea-level fall. Vail *et al.* (1977) stated that cyclic sea-level changes with gradual rises and abrupt falls could conveniently explain facies cycles in Cretaceous to Tertiary sequences; perhaps similar sea-level changes could account for the cyclicity of the Werraanhydrit, with the stratobolic anhydrite forming during low stands and the gradual facies evolution to varved rocks accompanying a slow sea-level rise. The applicability of this interpretation depends on the amplitude of the eustatic movements. In assessing this applicability, the main factor is the general water depth of the Zechstein basin. This has to exceed the difference between the height of the anhydrite wall (150–300 m) and the thickness of the basin floor sediments (40–50 m), and is therefore 100–250 m if synsedimentary subsidence is not taken into account. The sediments upon the so-called platform require a continuous water cover which may be estimated at 5–10 m, though it was undoubtedly locally and regionally varied. To produce comparable 'shallow water' conditions on or near the basin floor, sea-level would have to fall at least 100 m and probably by much more.

During such low stands of sea-level large parts of the platform would dry up, but no real gaps in sedimentation or irrefutable proof of desiccation have yet been recognized in the many good exposures of pletholitic anhydrites; this suggests uninterrupted growth and a continuous water cover. Furthermore, if the stratobolic intercalations in the midst of the varved Basalanhydrit (Z2) indicate a sea-level fall, how should we explain the uninterrupted 'deep water' sedimentation all around?

I am not convinced, therefore, that the sharp changes of facies in Z1 and Z2 anhydrites are related to major sea-level oscillations. However, the five- or six-fold repetition of the same sequence of rock types requires explanation.

Rapid precipitation of $CaSO_4$, as in the pletholitic rocks, needs a greater concentration or a high rate of evaporation. Also, the proved association of flaser anhydrite and halite suggests a higher temperature than for laminated anhydrite. The sharpness of the facies change may therefore be caused by supersaturation of $CaSO_4$, resulting from rising temperature and by abruptly exceeding the critical concentration.

It must be mentioned that within the different anhydrite sequences, there are many facies changes of only centimetre-size; they have the same sequence of clean, white anhydrite above fine-bedded dark laminite, and are quite similar to the more major lithological changes. Generally the 'newcomer' has a very characteristic structure of nodular aggregates of tangentially-disposed acicular anhydrite. These crystalline nodules displace the lamination of the underlying varved beds, and are mostly involved in gliding deformations. It seems that these crystal assemblages have been built up in warmer water, subsided and drifted over the bottom. There is a certain similarity with the structures described as 'turbidity' products by Schlager & Bolz (1977, p. 603), and also with the ideas of von Gaertner (1932).

In summary, changes of currents and/or climate seem to be responsible for the lithological profile of the sub-cycles. The periodicity of the changes may be estimated as an order of 0.1–0.5 Ma each, and probably results from external effects.

References

DEAN, W.E., DAVIES, G.R. & ANDERSON, R.Y. 1975. Sedimentological significance of nodular and laminated anhydrite. *Geology*, **3**, 367–72.

GAERTNER, H.R. VON, 1932. Petrographie und paläogeographische Stellung der Gipse vom Südrande des Harzes. *Jb. Preuss. geol.* **53,** 655–94.

HERRMANN, A. & RICHTER-BERNBURG, G. 1955. Frühdiagenetische Störungen der Schichtung und Lagerung im Werra-Anhydrit (Zechstein 1) am Südharz. *Z. dt. geol. Ges.* **105,** 689–703.

HOYNINGEN-HUENE, E. VON, 1957. Die Texturen der subsalinaren Anhydrite im Harzvorland. *Z. Geol.* **6,** Beiheft 18, 3–48.

JUNG, W. 1958. Zur Feinstratigraphie der Werra-Anhydrite (Zechstein 1) im Bereich der Sangerhäuser und Mansfelder Mulde. *Z. Geol.* **7,** Beiheft 24, 3–88.

MEIER, R. 1977. Turbidite und Olisthostrome—Sedimentationsphänomene des Werra-Sulfats (Zechstein 1) am Osthang der Eichsfeld-Schwelle. *Akad. Wiss. Zentralinst. Phys. Erde*, **50,** 1–45.

RHYS, G.H. 1974. A proposed standard lithostratigraphic nomenclature for the Southern North Sea and an outline structural nomenclature for the whole of the (UK) North Sea. *Rep. Inst. geol. Sci. 74/8.* 14 pp.

RICHTER-BERNBURG, G. 1941. Paläogeographische und tektonische Stellung des Richelsdorfer Gebirges im Hessischen Raum. *Jb. Reichsstelle Bodenfschg.* **61,** 283–332.

—— 1942. Zur vergleichenden Stratigraphie des Zechsteins in Mitteldeutschland. *Z. "Kali"*, **12,** 3–16.

—— 1950. Zur Frage der absoluten Geschwindigkeit geologischer Vorgänge. *Die Naturw.* **37,** 1–8.

—— 1955. Über salinare Sedimentation. *Z. dt. geol. Ges.* **105,** 593–645.

—— 1957a. Zur Paläogeographie des Zechsteins. *I giassimenti gassiferi dell 'Europa occidentale. Accd. naz. Lincei.* 87–99.

—— 1957b. Isochrone Warven im Anhydrit des Zechstein 2. *Geol. Jb.* **74,** 601–10.

—— 1958. Die Korrelierung isochroner Warven im Anhydrit des Zechstein 2 (Zweiter Beitrag), *Geol. Jb.* **75,** 629–46.

—— 1960. Zeitmessung geologischer Vorgänge nach Warvenkorrelationen im Zechstein. *Geol. Rdsch.* **49,** 132–48.

—— 1963. Solar cycle and other climatic periods in varvitic evaporites. In: NAIRN, A.E. (ed.) *Problems in Palaeoclimatology*, 510–20. Proc. NATO Conf. Newcastle, London.

—— 1964. Entwicklungsmöglichkeiten der für die deutsche Wirtschaft bedeutsamen Erdgas-Lagerstätten. *Erdöl und Kohle*, **17,** 977–84.

—— 1972. Saline deposits in Germany. In: RICHTER-BERNBURG, G. (ed.) *Geology of Saline Deposits*, 275–82. Proc. Hanover Symp., 1968 (Earth Sciences 7), UNESCO, Paris.

SCHLAGER, R.W. & BOLZ, H. 1977. Clastic accumulation of sulphate evaporites in deep water. *J. sedim. Petrol.* **47,** 600–9.

SMITH, D.B. 1974. Permian. In: RAYNER, D.H. & HEMINGWAY, J.E. (eds) *The Geology and Mineral Resources of Yorkshire*, 115–44. Yorkshire Geological Society, Leeds.

—— 1980. The evolution of the English Zechstein Basin. In: FUCHTBAUER, H. & PERYT, T.M. (eds) *The Zechstein Basin with Emphasis on Carbonate Sequences*, 7–34. *Contr. Sedim.* **9.**

—— (this volume). The Trow Point Bed—a deposit of Upper Permian marine oncoids, peloids and columnar stromatolites in the Zechstein of NE England.

TAYLOR, J.C.M. 1980. Origin of the Werraanhydrit in the UK Southern North Sea. In: FUCHTBAUER, H. & PERYT, T.M. (eds) *The Zechstein Basin with Emphasis on Carbonate Sequences*, 91–113. *Contr. Sedim.* **9.**

—— & COLTER, V.S. 1975. Zechstein of the British Sector of the North Sea Basin. In: WOODLAND, A.W. (ed.) *Petroleum and the Continental Shelf of North-west Europe*, **1,** 240–63. Applied Science Publishers, Barking.

VAIL, P.R., MITCHUM, R.M. & THOMPSON, S. 1977. *Seismic Stratigraphy and Global Changes of Sea Level. Mem. Am. Ass. Petrol. Geol.* **26,** 83–97.

G. RICHTER-BERNBURG, D 3000 Hannover 1, Haarstrasse 8, Federal Republic of Germany.

Zechstein salt correlation: England–Denmark–Germany

G. Richter-Bernburg

SUMMARY: A correlation of the Zechstein strata of England and Germany is presented. It is based on strong similarities in the distribution, thickness and lithology (including lithological succession) of the various evaporites in the two sequences and, particularly, on an upwards change in colour from mainly grey to mainly red between the second and third cycles; this change probably indicates a basin-wide change in the oxidation state of the Zechstein Sea. By these criteria, it is suggested that strata of the Don and Aislaby groups of England together are equivalent to the deposits of the Werra Cycle in Germany and that those of the Teesside Group are correlatives of the German Stassfurt sequence.

Any attempt to correlate the Zechstein salt formations in England (including offshore), Denmark and Germany soon discloses the limits of our knowledge. The reason is the scarcity of complete core sections on the English side and the halotectonic disturbance of the salt sequences in the diapirs of North Denmark (Jutland), although complete cores do exist there. However, it is possible to attempt a correlation based on our knowledge of the German Zechstein salt sequence because of the wealth of detailed observations in many potash and rock salt mines. I have personally had the opportunity to study many fully cored boreholes near Whitby in North Yorkshire as well as in Jutland (Suldrup and others) and Germany, and offer the following facts for consideration.

The Leine Folge and its equivalents

The most obvious fact is the similarity of the 'younger' halite group in all the areas mentioned and its contrast with the 'older' groups. Thus, the Sneaton Halite of the Staintondale Group in England is about 40–50 m thick and resembles the German Leine-Salz (Na3) in many points: (1) the lower third has the same orange (bright- to red-orange) colour, the same mainly clean halite, and a bedding type which is, in Germany, called Liniensalz (Na3 alpha-beta-gamma) because of the presence of evenly-spaced clearly-defined dark anhydritic films (Richter-Bernburg 1955); (2) the overlying halite is, in Yorkshire, more coarsely crystalline and diffusely striped by alternations of brownish-orange and bright-orange colour. This rock-type is known in Germany as Bändersalz (Na3 delta-epsilon) and also occurs in a corresponding stratigraphical position in the Suldrup boreholes in Denmark (Richter-Bernburg 1962).

The Stainton Potash of North Yorkshire, situated in the midst of the Stainton Halite member, is a group of red sylvinite beds—more or less rich in KCl—with a fairly high content of red-brown clay filling the interstices between the salt crystals. This description would also be valid for the Riedel potash bed which is exposed and exploited in NW German mines. This has not only the same stratigraphic position but, in the whole Na3 sequence, is the level at which mud first contaminates the previously purely saline sedimentation. All younger beds of Z3 in Germany are blended with more or less abundant clay and are called Tonflockensalz as a rock type and Tonmittel-Salz as the stratigraphic member. The equivalent salts from here and from the upper third of the Sneaton Halite are identical. In contrast, the clay-rich rock salts have not yet been noted in the Danish boreholes where the Anhydritmittel-Zone (Na3) is like that in Germany.

Altogether the Na3 seems to be much thicker in Denmark and Germany than in Yorkshire. However, in all these places the same type of sharp and regular bedding and the same orange colour are present in the lower half of Na3, and do not exist in any other Zechstein salt member. A further similarity is that extremely white pure halite occurs at the top of both the Sneaton Halite and Na3 (in which it is called the Reines Grenzsalz); each occurrence is immediately overlain by red clay-rock.

The Sherburn Anhydrite is 6–10 m thick in England and overlies the thin carbonate rock of the Upgang Formation (Smith 1974); it contains random carbonate intercalations. Although the thickness of the Hauptanhydrit (A3) of Germany is about five times greater than the Sherburn Anhydrite, its structure is similar although its base is only patchily calcareous. Cross-bedding is typical of a distinct stratigraphic level in the German A3 (Kosmahl 1967) and also occurs in the Sherburn Anhydrite in England (Smith 1980, p. 26). Meadows of primary gypsum crystals,

From HARWOOD, G.M. & SMITH, D.B. (eds), 1986, *The English Zechstein and Related Topics*, Geological Society Special Publication No. 22, pp. 165–168.

later replaced by anhydrite, form parts of A3 in both countries; in England the crystals are larger in some beds than in Germany and replacement by halite is widespread (Stewart 1951b).

In Denmark, this anhydrite is not well recorded, but it too seems to have only a thin underlying carbonate; this is surprising because the fine clastic basal member of this cycle is very well developed with a dark grey to black silty and even sandy shale about 0.5 m thick (Richter-Bernburg 1962). This member in Denmark is clearly the equivalent of the Grauer Salzton (T3) of Germany, but is very unlike the basal unit of the Staintondale Group which is the red Carnallitic (or Rotten) Marl.

The lowest 1–2 m of the Carnallitic Marl is commonly a tea-green claystone, that passes up into a uniform brown-red pelitic sediment of 10–25 m which contains some carnallite in pores and fissures, as a relic of the original brine. This rock resembles the Roter Salzton (T4) of Germany but it must be recalled that the Grauer Salzton (T3) is also red in its basal facies in an extensive southern coastal belt and is grey only where it is exclusively a basinal rock: a red mudstone must be expected as close to the shoreline as Yorkshire is situated.

From the foregoing it seems that the Scandinavian landmass was shedding grey detritus into the Jutland trough whilst the southern and south-western continent was under highly arid conditions of weathering and supplied red sediments to its surroundings; the red colour of the Carnallitic or Rotten Marl by no means proves that is the equivalent of the Roter Salzton (T4) of Germany.

In summary we may state that the Staintondale Group, as far as it can be observed in completely cored sequences, has by far the most similarities with the rocks of the well-known Leine-Cycle (Z3) of Germany and some of its equivalents in Denmark; it does not readily compare with other groups of Zechstein strata.

The Stassfurt Folge and its equivalents

The correlation outlined above requires that the Teesside-Group should be the counterpart of the Stassfurt-Cycle (or Z2). There is no problem in viewing the famous German potash bed Flöz Stassfurt as the equivalent of the Boulby Potash of Yorkshire; the deposit is indeed petrologically different but the facies of this member is so sharply variable, even in the same mine, that we should not be surprised that it has a somewhat different character near Whitby. Fundamentally, the potash seam (local name 'Veggerby') in Denmark is similar to that in the northernmost boreholes in Germany where the Mg-salts carnallite and kieserite are predominant. It must be remembered that the deposition of these chemically highly sensitive salts changes with the slightest change in the environmental conditions, much more so than with halite; we cannot expect identical sediments in places so far apart. This notwithstanding, the climate was favourable over the whole basin for the precipitation of potash salts at the end of Z2.

The onshore profiles of the English Zechstein do not have the true basinal character known in Germany and Jutland where the Z2 halite is estimated to be about 500 m thick. The rock salt of the Teesside Group, the Boulby Halite, is only about 50–100 m thick onshore but thickens sharply eastwards (Smith 1974; Taylor & Colter 1975). As far as can be seen in Boulby Mine and in complete cores, it consists of colourless halite blended with much grey anhydrite and mudstone so that it appears dirty at some levels. In the lower part, beds of anhydrite are intercalated with the grey halite; in the upper part, the rock salt is pure and snow-white in places and brownish or brown elsewhere. In this connection it is of interest that the uppermost Z2 halite in both Germany and Denmark includes very pure white halite and a rocksalt with a very distinctive dark-brown gleam—just the same colours that are present in the uppermost Boulby Halite. That is not a convincing argument for correlation but is one additional coincidence.

These small conformities in the facies should not be dismissed lightly; they reflect environmental conditions that affected sedimentation and probably indicate climatic events. They are, despite their wide extent through the Zechstein basin, contemporaneous and therefore indicate correlation. No petrographic similarity can be observed between the Leine halites in Germany (including their equivalents in Denmark) and the Boulby Halite in England.

When the Boulby Halite is regarded as the equivalent of the German Stassfurt Halite (No. 2), the Billingham Anhydrite should be correlated with the Basalanhydrit (A2). The Billingham Anhydrite is about 15–20 m thick, much thicker than the Basalanhydrit of the basin floor, but near the coast of the southern continent the latter thickens to 25 m and more. The subdivision into a lower part interbedded with carbonate (12 m) and an upper part with halite intercalations as described by Stewart (1951a, p. 446), can also be observed in some 'shallow water' places in Germany; its stratigraphic identity as A2 is more than probable.

Correlation of the Billingham Anhydrite with the Basalanhydrit requires that the Seaham Formation (and the equivalent Brotherton Formation in Yorkshire) is the correlative of the Hauptdolomit/Stinkschiefer (Ca2); all members older than the Seaham Formation would belong to Zechstein 1. This first cycle is influenced by the proximity of the coast and is very variable in its carbonate and sulphate facies (see references in Smith 1980). The Fordon Evaporites (Stewart 1949) and the Edlington Formation would all be of Z1 age.

Conclusions

Although the colour of salt rocks cannot be regarded as proof of stratigraphic correlation it must be taken as evidence of an important change of the redox potential during Zechstein deposition; thus, after white, colourless and grey halites and dark pyritic and/or bituminous mudstones, the change is marked by the incoming of pink or orange or red salt rocks together with almost red pelites. This change characterizes only the contact between Zechstein 2 and Zechstein 3, indicating that it was then that the change from a reducing to an oxidizing environment occurred throughout the basin, including the shelf areas. In England this change lies between the Teesside and Staintondale groups (Table 1).

TABLE 1. Correlation of Zechstein strata in England and Germany–Denmark

ENGLAND Smith *et al.* (1974)			GERMANY Richter-Bernburg (1955)		
Eskdale Group	Roxby Formation	Z4	Roter Salzton T 4 etc.		Allerfolge
STAINTONDALE GROUP	Sneaton Halite (including Sneaton Potash near middle)	Z3	Tonmittel-Salz Anhydritmittel-Salz Bandersalze Na3 Liniensalze	K-Floz 'Riedel'	LEINE-FOLGE
	Sherburn Anhydrite		Hauptanhydrit A3		
	Upgang Formation		Dolomit (= Plattendol. Ca3)		
	Carnallitic or Rotten Marl		Grauer Salzton T3		
TEESIDE GROUP	Cover Beds Boulby Potash	Z2	Deckhalit Na2r Kali-Floz Stassfurt K2 (='Veggerby')		STASSFURT-FOLGE
	Boulby Halite		Stassfurt-Halit Na2		
	Billingham Anhydrite		Basalanhydrit A2		
	Seaham Formation		Hauptdolomit Stinkschiefer Ca2		
AISLABY GROUP	Fordon Evaporites Kirkham Abbey Formation	Z1	Oberer Werra-Anhydrit A1	Werra-Halit Na1 +K1	WERRA-FOLGE
DON GROUP	Hayton Anhydrite		Unterer		
	Cadeby Magnesian Limestone Formation Marl Slate		Zechsteinkalk Ca1 Kupferschiefer		

References

KOSMAHL, W. 1967. Grauer Salzton und Hauptanhydrit des Zechsteins in Nord-westdeutschland. *Jb. geol.* **84,** 367–406.

Nederlandse Aardolie Maatschappij B.V. & Rijks Geol. Dienst, 1980. Stratigraphic nomenclature of The Netherlands. *Verh. K. ned. geol. mijnb. Genoot.* **32,** 77 pp.

RHYS, G.H. 1974. A proposed standard lithostratigraphic nomenclature for the southern North Sea and an outline structural nomenclature for the whole of the (UK) North Sea. *Rep. Inst. geol. Sci. 74/8*, 14 pp.

RICHTER-BERNBURG, G. 1955. Stratigraphische Gliederung des deutschen Zechsteins. *Z. dt. geol. Ges.* **105,** 843–54.

—— 1962. Geologische Stellungnahme (1961) zum Bohrprogramm auf Kalisalze bei Suldrup/Jutland. In: *Kaliboringerne ved Suldrup*, **1,** 70–114. Danm. Geol. Unders., Kobenhavn.

SMITH, D.B. 1974. Permian. In: RAYNER, D.H. & HEMINGWAY, J.E. (eds) *The Geology and Mineral Resources of Yorkshire.* 111–44. Yorkshire Geological Society, Leeds.

—— 1980. The evolution of the English Zechstein Basin. In: FUCHTBAUER, H. & PERYT, T.M. (eds) *The Zechstein Basin with Emphasis on Carbonate Sequences, Contr. Sedim.* **9,** 7–34.

——, BRUNSTROM, R.G.W., MANNING, P.I., SIMPSON, S. & SHOTTON, F.W. 1974. *A Correlation of Permian Rocks in the British Isles. Spec. Rep. geol. Soc. London*, **5,** 45 pp.

STEWART, F.H. 1949. The petrology of the evaporites of the Eskdale No. 2 boring, east Yorkshire. Part I. The Lower Evaporite Bed. *Mineralog. Mag.* **28,** 631–75.

—— 1951a. Part II, The Middle Evaporite Bed. *Mineralog. Mag.* **29,** 445–75.

—— 1951b. Part III, The Upper Evaporite Bed. *Mineralog. Mag.* **29,** 557–72.

TAYLOR, J.C.M. 1980. Origin of the Werraanhydrit in the U.K. Southern North Sea—a reappraisal. In: FUCHTBAUER, H. & PERYT, T.M. (eds) *The Zechstein Basin with Emphasis on Carbonate Sequences, Contr. Sedim.* **9,** 91–113.

—— & COLTER, V.S. 1975. Zechstein of the British Sector of the North Sea Basin. In: WOODLAND, A.W. (ed.) *Petroleum and the Continental Shelf of North-west Europe*, **1,** 240–63. Applied Science, Barking.

G. RICHTER-BERNBURG, D 3000 Hannover 1, Haarstrasse 8, Federal Republic of Germany.

Discussion contribution by D.B. Smith

I too examined the 15 boreholes on which Professor Dr Richter-Bernburg based his proposed correlation and also more than 30 other cored boreholes drilled through EZ3 and EZ4 evaporites in the same area. My views on the correlation of the sequences proved in these boreholes with that of the Zechstein in Germany differ from that now proposed and have already been published (Smith *et al.* 1974).

Powerful new evidence supporting this earlier correlation has since become available from seismic exploration in the southern North Sea and from wire-line logs and the cores of boreholes drilled there; these show that the Z3 carbonate beneath the Z3 evaporites is continuous across and around the south and west parts of the Zechstein Basin and becomes the EZ3 carbonate in England. The Z3 carbonate and the EZ3 carbonate, moreover, have almost identical biotas which differ substantially from those of Cycle 1 and Cycle 2 carbonates. On this evidence, and notwithstanding any similarities of colour and clay content, the Z3 evaporites of Germany seem most unlikely to be equivalent to the Fordon Evaporites of England; the latter, furthermore, cannot be the EZ1 evaporite, for the Hartlepool/Hayton Anhydrite of England can readily be traced without break across the North Sea into the Werraanhydrit of the Zechstein sequence in Holland and Germany (see, e.g. Taylor 1980).

Discussion contribution by H.A. Van Andrichem Boogeart

The Coppershale (T1) and the Grey Salt Clay (T3) are the most conspicuous and continuous marker beds in log correlation of the Zechstein sequence in The Netherlands. They can be followed from Germany, via The Netherlands into the English offshore, and thence into the English onshore (Rhys 1974).

In addition, the Z3 Carbonate Member (Plattendolomit, Ca3) is very continuous along the margins of the basin and is clearly still present in areas where the evaporites and other carbonates have completely or almost completely wedged out. With these observations in mind, a log correlation of appropriate onshore and offshore wells in and around the Whitby area should readily indicate to which Zechstein cycles the formations in the Whitby area must be assigned.

Poland

On the Zechstein Copper Shale lithofacies and palaeoenvironments in SW Poland

S. Oszczepalski

SUMMARY: Microfacies and geochemical studies on the Copper Shale in the Sudetic Foreland of SW Poland show the strong influence of palaeorelief on the distribution of lithofacies. Spatial variability in microlithofacies and lithofacies indicate sedimentation in the following palaeoenvironments: (1) offshore—deep shelf, characterized by the black shale lithofacies (30–50 cm thick) and shallow shelf, where the Copper Shale is either absent or is represented by red shale lithofacies (up to 60 cm thick), and (2) nearshore, characterized by grey shale lithofacies (40–100 cm thick or more). Shorewards the Copper Shale is replaced by carbonate and/or terrigenous rocks of the first Zechstein cycle.

Areas dominated by erosion during the Saxonian may have no Copper Shale, it may be of Red Shale lithofacies or its place may be taken by basal parts of the Zechstein Limestone. This suggests the existence of elevated areas where the oxidized (Rote Fäule) facies occurred.

The Copper Shale sea was shallow, mud-dominated and epeiric, with oscillations of the redox boundary and controlling the evolution of sedimentation and the facies diversity of the basal part of the Zechstein Werra cyclothem.

Basal parts of the Zechstein have been intensively studied for a long time in connection with the search for copper deposits. The studies were primarily focused on the Copper Shale/Kupferschiefer/Marl Slate which is a black to dark grey or red-brown clay-carbonate rock that is generally 30–50 cm thick; it is characterized by marked shaly parting and a generally high content of organic matter, pyrite and other sulphides, and by local mineralization with iron oxides. The rocks usually rest on clastic Weissliegendes sandstone, Rotliegendes, or Basal Limestone (Mutterflöz) and, in a few places, directly on earlier rocks. Upwards, the Copper Shale passes into the Zechstein Limestone (Zechsteinkalk) (Figs 1–3).

In the last 50 years, studies of basal Zechstein rocks concentrated on polymetallic mineralization as well as on lithology, petrography and geochemistry (Hirst & Dunham 1963; Rentzsch 1964; Wedepohl 1964; Konstantynowicz 1965; Wyzykowski 1965; Jarosz 1968; Salski 1968; Rydzewski 1969; Haranczyk 1970; Jung & Knitzschke 1976; Turner *et al.* 1978; Oszczepalski 1979). In addition, much attention was paid to geochemical facies and their relations to regions of the richest copper mineralization (see review in Jung & Knitzschke 1976). These studies suggested that the distribution of mineralization and geochemical facies is dependent on the relief of the Zechstein seafloor. However, the depth of the Copper Shale sea is still the subject of controversy (see Rentzsch 1964; Wedepohl 1964; Brongersma-Sanders 1971; Oszczepalski 1980; Smith 1980).

The paper presents characteristics of the Copper Shale in the Fore-Sudetic region of Poland and a preliminary reconstruction of its palaeogeographic facies based on the results of microfacies and geochemical analyses.

Lithofacies of the Copper Shale

The lithology of the Copper Shale is generally in the mudrock-carbonate spectrum, differing in the local proportion of individual components. The Copper Shale in the basal part of the Zechstein was originally differentiated by its major primary rock properties such as texture and lamination, treating the mode of parting as an auxiliary feature (see Lundegard & Samuels 1980). For this reason the formation was not distinguished at many places, despite the presence of clayey and (locally) organic-rich rocks in the appropriate part of the lithostratigraphic column (e.g. Figs 2 and 3). The shaly rocks commonly pass laterally and vertically into limestone-dolomite rocks that are themselves shaly (the term shale, as used here, denotes a rock with most grains less than 0.062 mm diameter and generally with more than 35% of terrigenous components; this definition is consistent with the subdivision of fine-grained sedimentary rocks of Lundegard & Samuels (1980) and Potter *et al.* (1980).

Detailed studies of the Copper Shale have made it possible to differentiate a number of microfacies on the basis of the content and ratios of clay minerals, carbonates and organic matter and of the predominant sedimentary structures. In the two-component nomenclature of microfa-

From HARWOOD, G.M. & SMITH, D.B. (eds), 1986, *The English Zechstein and Related Topics*, Geological Society Special Publication No. 22, pp. 171–182.

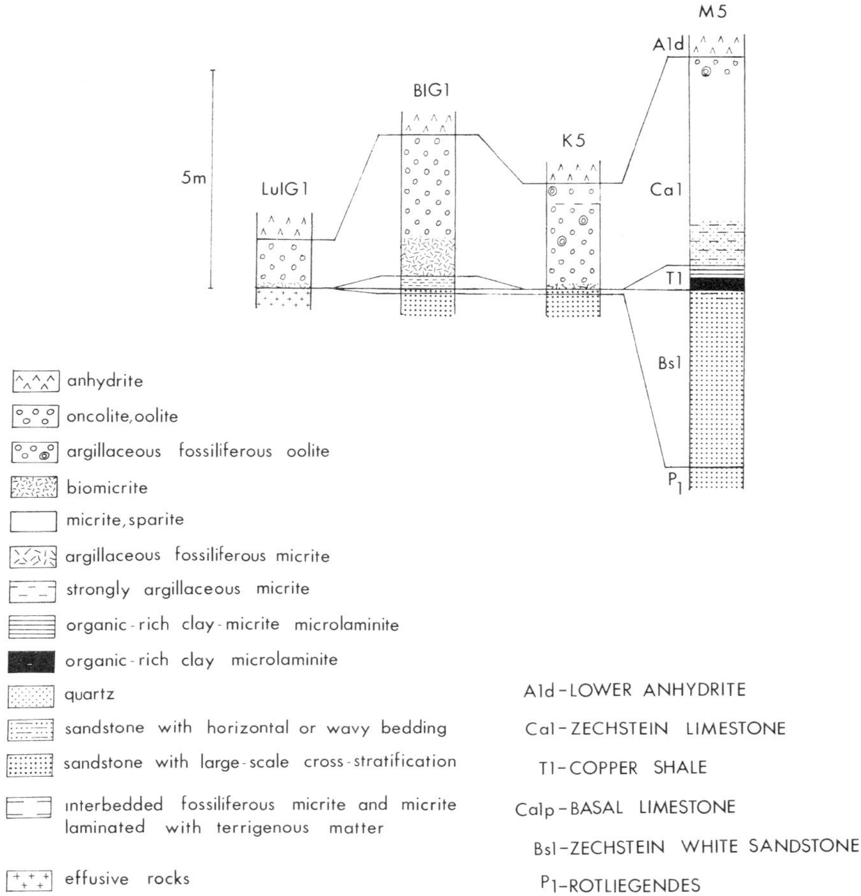

FIG. 1. Correlation of the basal Z1 deposits of the offshore zone of the Copper Shale; see Fig. 6 for location of boreholes.

cies used here, the rock name refers to the two main components in a given rock; single-component rock names have been given to microfacies in which organic matter is absent or is present in trace amounts only.

The analysis of variations in the distribution and proportions of the main microfacies has made it possible to differentiate three major lithofacies of the Copper Shale: (1) black shales, (2) grey shales and (3) red shales (Table 1).

Black shales

Copper Shale rocks of this lithofacies are black to dark-grey fissile clay rocks with a low carbonate content. They are regarded as typifying the Copper Shale in places where sulphide mineralization is most intense. Smooth, commonly slickensided shale surfaces often bear impressions of *Lingula* and fish.

The basic components of rocks of this facies include terrigenous material such as clay minerals, quartz, micas, rock debris and feldspars, carbonates, organic matter and metal sulphides (Table 1). Clay minerals are represented almost exclusively by illite, accompanied by negligible amounts of kaolinite and montmorillonite (*see also* Haranczyk 1970). Quartz and other terrigenous components are dispersed throughout the rock, only locally forming flat lenticular bodies or laminae (Fig. 4a). Carbonates, both dolomite and calcite, occur as laminae or lenses 15–40 μm thick and are of micrite or, in places, sparite; some carbonates are dispersed in clay-organic matter.

The black shale lithofacies is characterized by a high share of organic carbon (generally 2–10%). Organic matter here occurs in the form of brownish irregular streaks and laminae of kerogen (94–96%) and bituminous matter (4–6%). Bitumens were found to yield isoprenoids and relatively

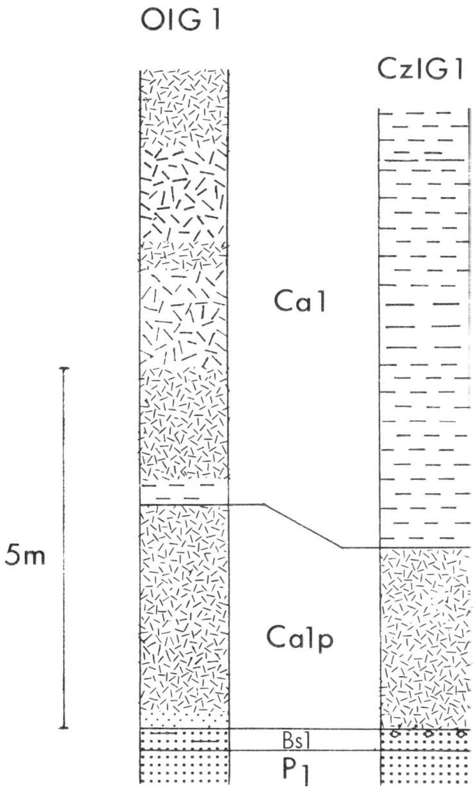

FIG. 2. Correlation of the basal Z1 deposits of the nearshore zone of the Copper Shale; see Fig. 6 for location. Key as for Fig. 1.

large amounts of metal porphyrin compounds of nickel, and vanadium and isoprenoids. Odd-numbered paraffinic chains from the interval *n*-C15 to *n*-C21 were found to predominate in the composition of the *n*-alkanes.

Parallel lamination visible with the naked eye is typical of this lithofacies. Microscopic studies show the laminae to be less than 0.1 mm in thickness and to form part of lithological rhythms also involving clay, organic matter and carbonates. Clay-organic microlaminites are characterized by the predominance of parallel planar but laterally discontinuous laminae (Fig. 4a). This variety commonly displays various deformational structures (Salski 1968); in the microfacies with a similar proportion of clay-organic and carbonate matter, parallel wavy lamination with discontinuous laminae is more common.

Interpretation

Sediments of this lithofacies were formed in a low-energy environment, presumably below effective wave base. This is suggested by the predominance of clay-organic microlithofacies with parallel lamination. Such deposits are typical products of sedimentation in low-energy environments, where bottom currents are too weak to sculpture the flat seafloor and can transport only the finest particles. The high concentration of organic matter also suggests sedimentation in an environment of limited water circulation.

Lamination of sediment reflects changes in the rate of sedimentation and/or biological productivity. It should be added here that high productivity in the Zechstein Sea, by using up the available supply of oxygen, led to reducing conditions at the bottom. Benthic fauna and bioturbation are almost completely missing here, indicating that the Eh = 0 interface was high above the seafloor (Byers 1977). Anoxic sedimentary conditions are also shown by the geochemical indices: increased content of Cu, Mo, Ni, Pb, S and C_{org}, increased concentration of isoprenoids regarded as 'geochemical fossils', and the presence of porphyrin (Didyk *et al.* 1978). In turn, the predominance of odd-numbered *n*-alkanes of medium molecular weight over those of high molecular weight compounds in the distribution of *n*-alkanes indicates the presence of autochthonous organic material in the sediment, insignificant admixture of terrigenous organic components and the presence of phytoplankton and procaryotic micro-organisms as precursors of bitumens (Tissot & Welte 1978).

There is growing evidence that sedimentary structures are preserved and the proportion of organic matter in sediment is high under dominantly anoxic bottom conditions where the concentration of oxygen is below 0.1 ml l^{-1}. Such conditions kill almost all benthic organisms (Byers 1977) and anaerobic bacteria are the only organisms reworking the organic matter. The reduction of sulphates by bacteria leads to the release of large quantities of H_2S to bottom waters (Tissot & Welte 1978); this results in the intense creation of metal-organic compounds, their sorption and chemisorption by organic particles and clay minerals, and precipitation of metal sulphides through the reaction of metal compounds and metal-organic complexes with organic reductors and H_2S (Saxby 1976).

In the Copper Shale basin, these processes were accompanied by early-diagenetic dolomitization of carbonate mud; the magnesium presumably came from pore water as well as from organic compounds (see Irwin *et al.* 1977). Neomorphic processes and carbonization of organic matter during later stages of diagenesis led to the des-

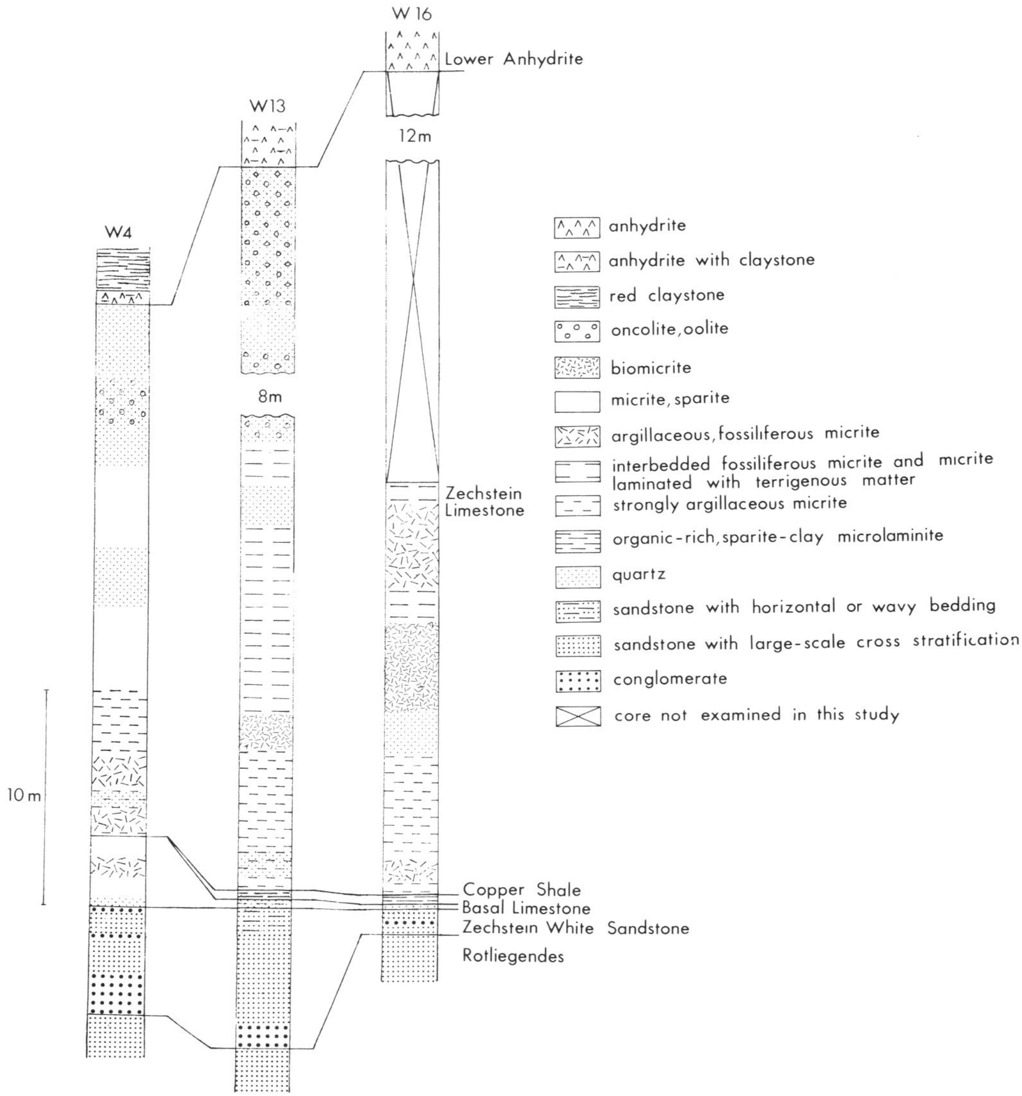

FIG. 3. Correlation of the basal Z1 deposits of the nearshore zone of the Copper Shale; see Fig. 6 for location of boreholes.

truction of metal-organic structures and the migration of metals and, therefore, to late diagenetic mineralization (e.g. Haranczyk 1970; Oszczepalski 1979).

Grey shales

Rocks of this lithofacies of the Copper Shale are usually dark-grey mud shales with platy to flaggy parting and a varied but usually fairly high content of carbonate and organic matter. Plant debris and micro- and macrofaunal remains occur here in subordinate amounts. An especially abundant faunal assemblage in the North-Sudetic Syncline comprises bivalves, brachiopods, foraminifers and bryozoans (Gunia 1962; Krason 1964). In the same region there are also marked changes in the colour of rocks; they are paler coloured and lower parts of the shaly sequences commonly bear reddish spots and are locally underlain by red clayshale in the SE part of the Syncline (Niskiewicz 1967). Towards the NW, the deposits tend to be darker and red spots and streaks are arranged parallel to the bedding

TABLE 1. Summary of observed and interpreted features of lithofacies of the Copper Shale

		Black shales	Grey shales	Red shales
Characteristics	Lithology	Organic-rich clay-shales	Organic-rich mud-shales	Clay-shales
	Colour	Black, dark-grey	Dark grey	Red, green, grey
	Parting	Fissile	Platy	Papery to fissile
	Stratification	Parallel	Lenticular	Fabric, colour
	High-energy sedimentary structures	Absent	Rare	Common
	Fossils	Rare	Rare	Common locally
	Terrigenous components	45–55%	30–50%	40–70%
	Carbonates	20–30%	30–50%	15–40%
	Organic carbon	2–10%	2–7%	<1%
	Porphyrins	High	High	Low
	Isoprenoids	High	High	Low
	Sulphides	Very common	Very common	Rare
	Iron oxides	Absent	Rare	Very common
	Microlithofacies	Organic-rich clay microlaminite	Organic-rich sparite-clay microlaminite	Clay microlaminite, sandy
Inferred environment	Sedimentary environment	Offshore, deep shelf	Nearshore	Offshore, shallow shelf
	Environmental energy	Low turbulence	Low turbulence, occasional weak currents	Moderate turbulence, occasional currents and rare storms
	State of oxidation	Anoxic	Anoxic	Oxic

(Krason 1964; Konstantynowicz 1965; Oszczepalski 1979).

Clay minerals and carbonates predominate in the grey shales. In comparison with the black shale lithofacies, the rocks are impoverished in organic matter relative to non-clay terrigenous material of silt and sand grade. Clay material consists of illite, locally accompanied by kaolinite. Carbonates are present in large quantities (Table 1), being represented by dolomite in the SE Fore-Sudetic Monocline and by calcite in the North-Sudetic Syncline; they generally form thick lenses or discontinuous laminae less than 0.5 mm thick, of microsparite or sparite, and in places, pelletoids or small intraclasts (Oszczepalski 1979).

Organic matter is still abundant in the grey shales (2–7% C_{org}), occurring in the form of streaks and irregular laminae between carbonate-rich laminae (Fig. 4b). Bitumens are mainly represented by *n*-alkanes (predominantly of odd-numbered *n*-alkanes in the *n*-C15–*n*-C21 interval), isoprenoids and organometallic compounds. Kerogen comprises a relatively large proportion of the carbonized organic remains.

Lamination in this lithofacies is mainly of the flat lenticular type (Fig. 4b) or, in places, irregular planar parallel. Thick carbonate lenses and laminae predominate in the rhythms, and are separated by thin wavy clay laminae that are enriched in organic matter and are locally microstylolitic. Sparite-clay microlaminites are characterized by irregular planar or wavy parallel or coarsely lenticular lamination. Laminae up to 10 mm thick of clastic material have been noted, as in W13 borehole. Microfacies with a predominance of clay over carbonate components are usually characterized by flat-lenticular lamination. Thin-section analyses show that skeletal carbonate grains are here represented by foraminifers (mainly of the genera *Agathammina* and *Nodosaria*) and fragments of bryozoans. Non-skeletal organic remains are present in sparite lenses (Z. Walenczak, *pers. comm.*); the remains resemble carbonate grains originating from the activity of photosynthetic bacteria and/or algae (e.g. Krumbein & Cohen 1977).

Interpretation

Rocks of this lithofacies originated in an environment similar to that of the black shale lithofacies, but shallower and closer to the shore. This is

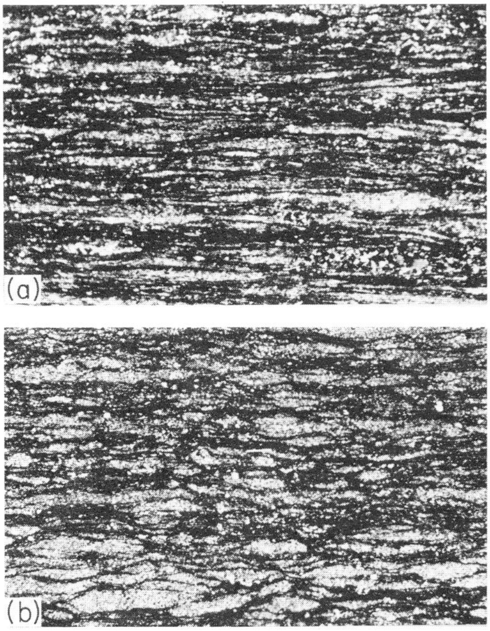

FIG. 4. (a) Clay-shale. Organic-rich clay microlaminite of the black shale lithofacies. Scattered flat silt and carbonate lenses. M9 borehole, ×30. (b) Mud-shale. Organic-rich sparite-clay microlaminite of the grey shale lithofacies. W17 borehole, ×30.

indicated by the presence of microlaminites and by the nature and abundance of the organic remains. The relatively high proportion of terrigenous organic remains and clastics in the laminae suggests proximity to land and the main sedimentary structures may partly have originated as small-scale ripples caused by episodic weak bottom currents (see Reineck & Singh 1980).

The boundary between oxidation and reduction zones was presumably above the surface of the bottom sediments, as suggested by the almost complete lack of benthic fauna and related sedimentary structures and the increased proportion of S, C_{org}, porphyrins and isoprenoids.

The finely-dispersed organic remains in the carbonate laminae suggest intense stabilization of carbonate mud and lithification by cyanobacterial populations (Krumbein & Cohen 1977). The micro-organisms may be responsible for the formation of wavy sparite laminae, which suggests repeated extinction of phytoplankton and the development of microbenthos, leading to alternation of sparite, cryptalgal and clay-sapropelic laminae.

Red shales

The red shales are clay-shales with a distinct fissile and platy (or locally papery) parting, a low content of carbonates and negligible organic matter. They comprise alternating rhythmic sequences of brown-red or grey to grey-green, 0.5–3 cm layers ('rhythmites', see Reineck & Singh 1980). Fissile planes are generally smooth and glitter markedly in grey-green layers but are rough in the more carbonate-rich brown-red layers; in places they bear impressions of shells and fish scales, as in M24 borehole.

Shales of this facies are a mixture of clay material and fine clastics, mainly micas and quartz grains of silt size (Fig. 5a); illite is accompanied by small amounts of chlorites, kaolinite and spherical glauconite grains. Carbonates are mainly represented by calcite but their proportion is generally only 15–40%; they are dispersed in the form of micrite (or, in places, sparite or micritized grains) in clay material or are concentrated in weak laminae. Clay-carbonate material is commonly coloured brown by haematite pigment. Laminar and lenticular concentrations of iron oxides are fairly common (Fig. 5b).

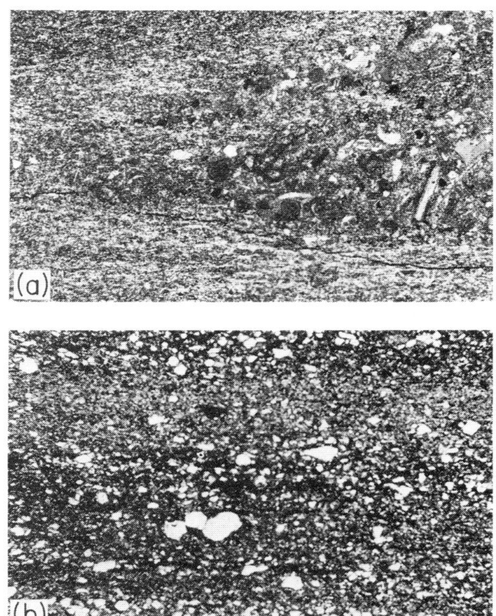

FIG. 5. (a) Clay-shale. Clay microlaminite of the red shale lithofacies, with fabric lamination, interlaminated with sandy biomicrites. UIG1 borehole, ×30, crossed polars. (b) Sandy mud-shale of the red shale lithofacies, with fabric and colour lamination (abundant iron oxides). K12 borehole, ×30.

Organic matter occurs in very small amounts (generally below 1%) in rare thin bituminous streaks. Bitumens include here trace amounts of *n*-alkanes, porphyrins and isoprenoids.

Clay forms plates arranged on and parallel to parting surfaces (fabric lamination, Fig. 5a). The macroscopic differences that permit differentiation of colour rhythms are confirmed by microscopic studies; the grey and grey-green varieties were found to be almost exclusively formed of clay material with fabric lamination and an insignificant admixture of haematite pigment. Clay microlaminites are characterized by colour lamination and, where enriched in micrite and clastic or carbonate granular material, a weak irregular to streaky lamination. Small-scale cross-bedding, flute casts and fill structures are found locally (Oszczepalski 1980), as are intercalations of sandy biomicrite. Skeletal grains present in the latter include thick-tested *Agathammina* and *Ammodiscus* and fragments of bivalves, brachiopods, echinoderms, bryozoans and algae (Fig. 5a).

Interpretation

Features such as the diverse assemblage of sedimentary structures, the local marked abundance of sand-size grains, biomicrite intercalations, lateral passage to argillaceous pelmicrites, and local enrichment in carbonate grains indicate sedimentation in an environment of generally low but episodically varying energy. This is further supported by geochemical data, which indicate the weakly oxidizing character of the environment with the Eh$=0$ surface lying close to the seafloor. The low content of organic matter and the lack of porphyrins suggest that early diagenetic processes took place in an oxidizing environment in which the degradation of organic matter, including chlorophyll compounds (the parent matter for porphyrins and isoprenoids) was fairly rapid (Didyk *et al.* 1978; Tissot & Welte 1978). This resulted in a marked enrichment of the sediments in iron oxides, the stability field of which depends on high values of Eh (Berner 1971).

Local oxide pseudomorphs after framboidal pyrite, and thin organic streaks, suggest the possibility of burial and fossilization of part of the organic matter; the preservation was primarily due to rapid calcitization which precluded compactional expulsion of the organic matter from the sediment. The processes of early diagenetic cementation and dolomitization were impeded here because of strongly oxidizing bottom water, the prevalence of aerobic bacteria and the oxidation of organic matter (drop in pH) and related loss of magnesium (Mitterer 1971; Irwin *et al.* 1977). It seems that platy minerals became compactionally ordered and pore space reduced at a very early stage of diagenesis, a process that could markedly reduce the rate of late-diagenetic processes.

Copper Shale palaeogeography

The tectonic and sedimentary processes active during sedimentation of the Rotliegendes fluvial and aeolian deposits exerted a decisive influence on the development of palaeorelief of the pre-Zechstein surface (Pokorski 1978). The major palaeogeographic units from the end of the Saxonian in the Sudetic Foreland include the Szprotawa-Highland (Wyzykowski 1965; Pokorski 1978). The uplands were subjected to intense degradation in the Saxonian, supplying detrital material for infilling the Zielona Gora and other depressions (Fig. 6).

Transgression of the Zechstein Sea, related to subsidence at about the time of the Rotliegendes Zechstein transition, was marked by resedimentation of Rotliegendes deposits almost throughout the Zechstein basin (Krason 1964; Wyzykowski 1965; Peryt 1976). According to some authors (e.g. Krason 1964; Smith 1970, 1980), the transgression was by rapid flooding of a basin originally situated below sea-level. However, there is evidence for slow transgression (Peryt 1976) and, therefore, for the existence of some areas (especially those characterized by predominant erosion in the Saxonian) above sea-level in the initial marine phase of the Zechstein Sea (Fig. 6); islands are inferred to have persisted in the area of the Szprotawa Elevation throughout deposition of the Zechstein Limestone (Oszczepalski 1980; Peryt 1981).

Comparison of a number of sequences of basal parts of the first cyclothem and of relevant sequences of microfacies permits the palaeogeographic differentiation of the Copper Shale into an offshore zone, deep and shallow shelf and a nearshore zone or belt (Fig. 6).

Offshore zone

Deep shelf

In the deep shelf area the Copper Shale is of black shale lithofacies with organic-rich clay microlaminites at the base and clay-micrite laminae at the top. Its thickness ranges from 30 to 50 cm, only locally increasing to about 100 cm.

In the lithostratigraphic column typical of this zone (M5 borehole, Fig. 1), the Copper Shale rests on Weissliegendes rocks which generally pass downwards into the clastic Rotliegendes via

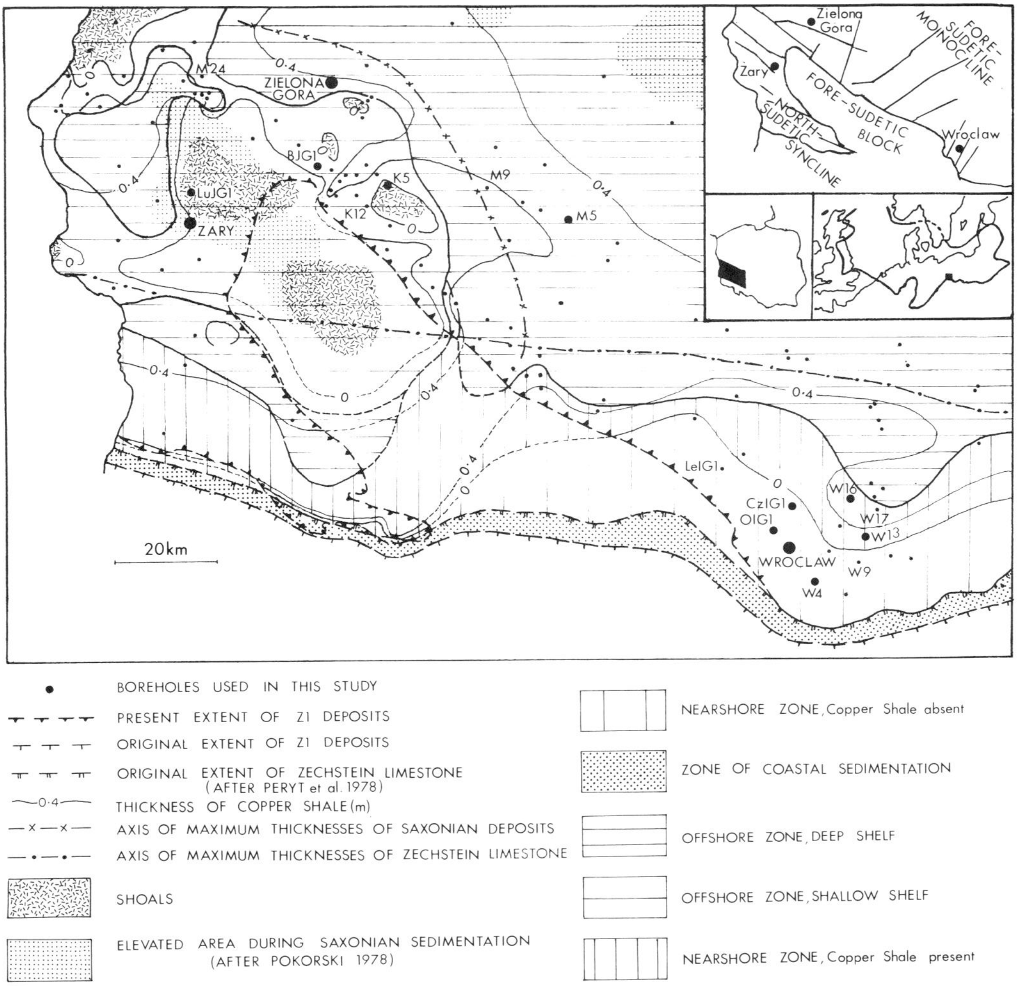

FIG. 6. Postulated palaeogeographical zones of the Copper Shale in Poland. In the North-Sudetic Syncline, data are compiled after Gunia (1962), Krason (1964) and Oszczepalski (1979).

passage beds, so-called Grauliegendes. The Copper Shale is locally separated from the Weissliegendes by a Basal Limestone of argillaceous micrites and biomicrites a few cm thick (Salski 1968; Peryt 1976). The Zechstein Limestone also is varied in thickness and lithology (Peryt 1978). In the central part of the zone it is mainly represented by 4–10 m of argillaceous micrites with oncolites at the top, but in the southern part it is up to about 120 m thick (Fig. 6) and comprises rocks typical of the barrier zone.

Taking into account the thickness, typical microfacies successions and geochemical characteristics of the Copper Shale (e.g. Oszczepalski 1980, fig. 5) it may be judged that the black shale-type lithofacies was deposited in the deepest part of a mud-dominated shelf, in an anaerobic environment characterized by clearly marked lamination, a wealth of organic matter and by a lack of benthos and bioturbation.

Shallow shelf

On elevations in the shallow part of the shelf the Copper Shale is absent and here the Zechstein Limestone rests on Weissliegendes or Rotliegendes or older bedrock (LuIG1 and K5 boreholes, Fig. 1). The Zechstein Limestone is rather thin (1–5 m) and comprises sandy biomicrites and micrites, some rich in glauconite in the lower part and in well-sorted oncolites rich in radial ooids, algal envelopes and sessile tubular foraminifers in the upper part. The sequences are also characterized by low geochemical differentiation, locally intense bioturbation and the abundance of red spots (Rydzewski 1969; Oszczepalski 1980, fig. 5).

Close to shoals, the Copper Shale is represented by thin (at most 30 cm thick) rocks of the red shale lithofacies, resting on the Weissliegendes or Rotliegendes or, in places, a few centimetres of sandy biomicrites.

Depressions in the shallow shelf are characterized by somewhat greater thickness of the Copper Shale (30–60 cm) and the Zechstein Limestone is of 4–10 m of fossiliferous micrites with oncoids at the top. In some areas, basal parts of the Z1 carbonate are of argillaceous or sandy micrites with bioclasts or of pelmicrites; the structural and textural features of these rocks show that they are part of the Zechstein Limestone (Oszczepalski 1980).

The above data suggest sedimentation of the shaly-carbonate series in a shallow part of the sublittoral zone, within the range of effective wave base. The locally high content of sand, marked admixture of non-skeletal carbonate grains, the presense of biomicrite intercalations and of small-scale high-energy sedimentary structures indicate higher energy of the sedimentary environment than in the black shale zone, suggesting the possibility of episodic reworking of sediment by storm-created bottom currents (Reineck & Singh 1980) and the transport of material from neighbouring shoals.

Nearshore zone

In this zone, the Copper Shale is of the grey shale lithofacies, 40–100 cm thick, locally thickening to about 6 m in the North-Sudetic Syncline (Gunia 1962; Krason 1964). The rocks are interlaminated with clastics, and skeletal grains (mainly *Agathammina* and *Nodosaria* and, in places, fragments of shells, bryozoans and plants) attain rock-forming importance here.

Basal parts of the Zechstein rocks here are part of the Weissliegendes, usually with conglomerates at the base and sandstones at the top (Fig. 3). In some places these sandstones yield abundant faunas of bivalves and brachiopods (Gunia 1962, figs 2–3; Krason 1964) and oncoids and skeletal grains (mainly *Glomospira* and sessile tubular foraminifers, Fig. 2). The overlying Basal Limestone rocks are locally separated from the Weissliegendes by a black shale layer a few centimetres thick (Salski 1968) or micrites interlaminated with terrigenous material (as in borehole W9). The Basal Limestone forms a constant lithostratigraphic horizon, ranging in thickness from 30 cm in the northern part of this zone to about 1–3 m in the south (Figs 2 and 3) and exceptionally up to 7 m (Krason 1964). It is formed of biomicrites and sandy micrites.

Southward in this zone, the Copper Shale wedges out or its place is taken by terrigenous Zechstein rocks (Fig. 6). The Zechstein Limestone is represented by shaly argillaceous micrites with lamination at the base disturbed by bioturbation (as in the LeIG borehole), passing upwards into interbedded fossiliferous micrites and micrites laminated with terrigenous material and with bryozoan biomicrites with interlayers of oncolites in the north (Peryt 1978; Oszczepalski 1979) and by sandy oolites in the south (Fig. 3).

Comparisons and conclusions

The microsequences of the Z1 shaly-carbonate suggest relatively high differentiation of the paleorelief of the floor of the Zechstein basin as well as a small-scale relief. The seafloor topography had a decisive influence on the energy of the sedimentary environments, on the degree of oxidation of bottom waters, on circulation and, therefore, on the distribution of lithofacies of the Copper Shale.

In central parts of the Zechstein Basin, the major phase of marine transgression was preceded by a gentle ingression (Plumhoff 1966; Peryt 1976) which explains the local transitions and interdigitation of the Rotliegendes and Zechstein. Also, in England, the uppermost parts of dune deposits in the Yellow Sands were subjected to reworking and redeposition during this transgression (Smith & Francis 1967; Bell *et al.* 1979). In nearshore areas and those subjected to uplift, the Zechstein sequence begins with a transgressive conglomerate (Krason 1964; Jordan 1969) or there is a sedimentary cycle with conglomerates at the base (Zary and Wroclaw regions, Fig. 3).

At the beginning of Zechstein sedimentation, deposition of clays and muds was presumably limited to the deepest parts of the basin. Rapid growth of phytoplankton in the depressions soon resulted in stratification of the water and the deposition of organic matter there. A thin layer of argillaceous rock found locally at the junction of the Weissliegendes and Zechstein (Fig. 2) may be an equivalent of the shaly deposits formed at the beginning of the transgression in the marginal part of the basin. Shorewards, the Copper Shale deposits laterally pass into layers rich in benthic fauna (i.e. Basal Limestone—see also Gunia 1962; Krason 1964; Jordan 1969); this relationship has also been recognized by Paul (1982), according to whom the Basal Limestone schwellen deposits represent a time equivalent of the lower part of the Copper Shale. This suggests that there was a uniform pattern of sedimentation in the whole southern part of the Zechstein Basin, as well as the possibility of a synchronous improvement of environmental conditions (downward shift of the redox boundary?) that was responsible for the spread of benthos in nearshore shoals. The upper invertebrate bed recorded in County Durham, England, may correspond to that event (Bell *et al.* 1979), and there are also some traces of shallowing in the upper part of the Basal Limestone in the North-Sudetic Syncline (Gunia 1962).

The major phase of transgression appears to be synchronous throughout the Zechstein Basin (Smith 1980). It widened the area of sedimentation of shaly deposits and resulted in partial overflooding of islands in the basinal zone (Oszczepalski 1980; Peryt & Wazny 1980). The rise of sea-level and increase in the supply of nutrients led to high productivity of phytoplankton, eutrophication of the basin and deposition of large amounts of organic matter. Because of a rise of the redox boundary, anoxic waters transgressed over biogenic banks and black shales began to form there also (see Fig. 7). Closer to shore, sedimentation proceeded under dysaerobic conditions where an oxygen content below 1 ml l^{-1} is shown by the intense reworking of sediments by a soft-bodied infauna (Byers 1977). The wedging out of the Copper Shale takes place 5–20 km from the inferred shoreline (Fig. 6).

In shallow parts of the shelf, large amounts of fine-grained terrigenous sediment were supplied as Rotliegendes rocks continued to be eroded. Clay minerals with iron oxide coatings were deposited in areas adjoining shoals where oscillations in bottom water aeration and/or the supply of organic reductors occurred (Berner 1971; Reineck & Singh 1980). This oscillation resulted in red-green rhythmites. Such rocks are also known from the nearshore zone (Niskiewicz 1967). The rocks known as 'Rote Fäule' are regarded as shallow-water equivalents of black copper-bearing shales and their distribution reflects that of redox potential in seafloor sediments during sedimentation and diagenesis (Rentzsch 1964; Konstantynowicz 1965; Rydzewski 1969, 1978; Jung & Knitzschke 1976).

The next stage in sedimentation was related to a decline in organic productivity, presumably caused by oligotrophism of waters. This could have resulted in a downward shift of the oxic/anoxic interface and the spread of benthic fauna throughout the Zechstein Basin (Productus-bank—see also Krason 1964; Smith & Francis 1967; Jordan 1969; Peryt 1978; Paul 1982). Subsequently, along with the predominance of regressive conditions and the initiation of a zonal pattern in the sedimentation of the Zechstein Limestone, progradation of the carbonate platform began (Taylor & Colter 1975; Peryt 1978; Smith 1980).

The interpretation presented here, as well as sedimentary wedging out of the Copper Shale in the GDR (Jung & Knitzschke 1976) and England (Smith 1970; Turner *et al.* 1978), supports the hypothesis of marked influence of the bottom topography on the sedimentation of the Copper Shale, suggesting a relatively small sea depth and large-scale circulation strongly impeded by elevations in central parts of the basin (e.g. Peryt & Wazny 1980). The elevations were situated within the wave base which implies marked aeration in relevant parts of the basin. In the deep shelf, the redox interface was presumably close to the effective wave base, in disagreement with the model of a deep stratified basin presented by Byers (1977). However, Byers (1977, p. 7) admits that the redox boundary may be only a few metres

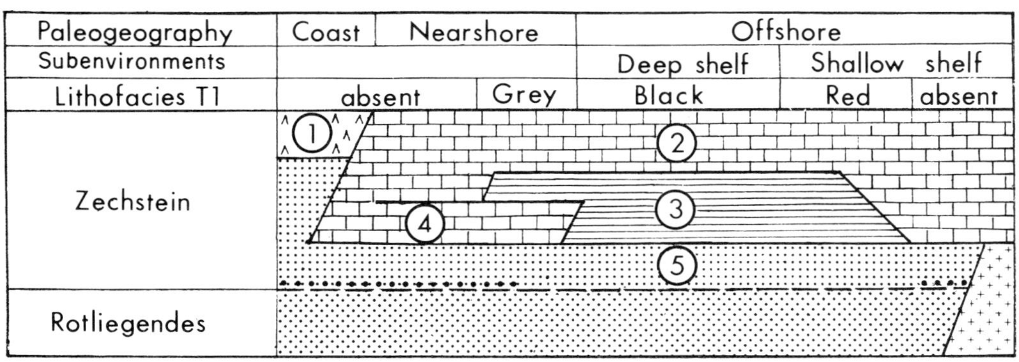

FIG. 7. Simplified correlation chart of basal Z1 lithostratigraphic units in SW Poland. Key as for Fig. 1.

below sea-level under conditions of impeded vertical water exchange and limited circulation. In modern seas the wave base is generally at a depth of 30–40 m and, under storm conditions, even close to 60 m in the Persian Gulf (Purser 1973). By analogy it may be assumed that the sedimentary basin of black shales was not much deeper than 40–60 m whereas the grey shale lithofacies was related to shallow nearshore areas (see also Bell *et al.* 1979). This interpretation is consistent with the view of Rentzsch (1964; see also Jung & Knitzschke 1976), according to whom the sea was not deeper than 50–100 m. According to others (Wedepohl 1964; Haranczyk 1970; Brongersma-Sanders 1971; Smith 1980; Paul 1982), the basinal facies of the Copper Shale originated in the deeper parts of an epicontinental sea (200–300 m deep) and are euxinic. In the light of data presented here, the more-shallow-water interpretation of sedimentary environment of the Copper Shale seems more likely.

ACKNOWLEDGMENTS: The studies were carried out on cores made available through the courtesy of the Instytut Geologiczny and Zjednoczenie Gornictwa Naftowego i Gazownictwa. Sincere thanks are expressed to Drs A. Rydzewski and F. Ekiert for guidance and discussions, and to Dr T.M. Peryt and D.B. Smith for critical reviews of the manuscript and valuable comments. Thanks are also due to Drs M. Podemski and Z. Walenczak for fruitful discussions. X-ray analyses were made by Dr M. Stepniewski and analysis of organic matter by Dr J. Calikowski and Z. Rzepkowska.

References

BELL, J., HOLDEN, J., PETTIGREW, T.H. & SEDMAN, K.W. 1979. The Marl Slate and Basal Permian Breccia at Middridge, Co. Durham. *Proc. Yorks. geol. Soc.* **42,** 439–60.

BERNER, R.A. 1971. *Principles of Chemical Sedimentology.* McGraw-Hill, New York.

BRONGERSMA-SANDERS, M. 1971. Origin of major cyclicity of evaporites and bituminous rocks: an actualistic model. *Mar. Geol.* **11,** 123–44.

BYERS, C.W. 1977. Biofacies patterns in euxinic basins: a general model. In: COOK, H.E. & ENOS, P. (eds) *Deep-water Carbonate Environments. Spec. Publ. Soc. econ. Palaeont. Mineral, Tulsa,* **25,** 5–17.

DIDYK, B.M., SIMONEIT, B.R.T., BRASSEL, S.C. & EGLINTON, G. 1978. Organic geochemical indicators of palaeoenvironmental conditions of sedimentation. *Nature, London,* **272,** 216–22.

FUCHTBAUER, H. & PERYT, T.M. (eds) 1980. *The Zechstein Basin with Emphasis on Carbonate Sequences. Contr. Sedim.* **9.**

GUNIA, T. 1962. Cechsztyn synkliny leszczynieckiej. *Biul. Inst. geol.* **173,** 57–114.

HARANCZYK, C. 1970. Zechstein lead-bearing shales in the Fore-Sudetian monocline in Poland. *Econ. Geol.* **65,** 481–95.

HIRST, D.M. & DUNHAM, K.C. 1963. Chemistry and petrography of the Marl Slate of S.E. Durham, England. *Econ. Geol.* **58,** 912–40.

IRWIN, H., CURTIS, C.D. & COLEMAN, M.L. 1977. Isotopic evidence for source of diagenetic carbonates formed during burial of organic-rich sediments. *Nature, London,* **269,** 209–13.

JAROSZ, J. 1968. Charakterystyka mineralogicznopetrograficzna zloza "Lubin". *Rudy Metale Nieżel.* **13,** 625–34.

JORDAN, H. 1969. Zur Biostratigraphie und Fazies des Zechsteins in Germanishen Becken, unter besonderer Berucksichtigung des Thuringer Beckens. *Freiberger Forschhft.* **C245,** 27–45.

JUNG, W. & KNITZSCHKE, G. 1976. Kupferschiefer in the German Democratic Republic (GDR) with special reference to the Kupferschiefer deposit in the Southeast Harz Foreland. In: WOLF, K.H. (ed.) *Handbook of Stratabound and Stratiform Ore Deposits,* **6,** 353–406. Elsevier, Amsterdam.

KONSTANTYNOWICZ, E. 1965. Mineralizacja utworow cechsztynu niecki polnocnosudeckiej (Dolny Slask). *Prace Kom. Nauk Geol.* PAN. Oddz. w Krakowie, **28.**

KRASON, J. 1964. Podzial stratygraficzny cechsztynu polnocnosudeckiego w swietle badan facjalnych. *Geol. Sudetica,* **1,** 221–55.

KRUMBEIN, W.E. & COHEN, Y. 1977. Primary production, mat formation and lithification: contribution to oxygenic and facultative anoxygenic cyanobacteria. In: FLUGEL, E. (ed.) *Fossil Algae,* 37–56. Springer-Verlag, Berlin.

LUNDEGARD, P.D. & SAMUELS, N.D. 1980. Field classification of fine-grained sedimentary rocks. *J. sedim. Petrol.* **50,** 781–86.

MITTERER, R.M. 1971. Influence of natural organic matter on $CaCO_3$ precipitation. In: BRICKER, O. (ed.) *Carbonate Cements,* 252–58. Johns Hopkins University Studies in Geology, **19,** Baltimore.

NISKIEWICZ, J. 1967. Podzial cechsztynu w rejonie Nowego Kosciola (Dolny Slask). *Przeglad geol.* **15,** 268–72.

OSZCZEPALSKI, S. 1979. Mineralizacja polimetaliczna spagowych osadow cechsztynu w rejonie Weglinca—Zar. *Kwart. geol.* **23,** 563–78.

—— 1980. Paleogeography, sedimentation and mineralization of the Z1 carbonate series (Zechstein) in the western part of the Fore-Sudetic Monocline (western Poland). In: FUCHTBAUER, H. & PERYT, T.M. (eds) (*q.v.*), 307–23.

PAUL, J. 1982. Zur Rand-und Schwellen-Facies des Kupferschiefers. *Z. dt. geol. Ges.* **133,** 571–605.

PERYT, T.M. 1976. Ingresja morza turynskiego (gorny perm) na obszarze monokliny przedsudeckiej. *Rocz. Pol. Tow. geol.* **46,** 455–65.

—— 1978. Sedimentology and paleoecology of the Zechstein Limestone (Upper Permian) in the Fore-Sudetic area, western Poland. *Sediment. Geol.* **20,** 217–43.

—— 1981. Cechsztyn w otoczeniu bloku przedsudeckiego. *Kwart. geol.* **25,** 75–91.

PERYT, T.M., PIATKOWSKI, T.S. & WAGNER, R. 1978. Paleogeographical map of the Zechstein Limestone (Ca1). In: DEPOWSKI, S. (ed.) *Atlas litofacjalno-paleogeograficzny permu obszarow platformowych Polski.* Inst. Geol., Warszawa.

PERYT, T.M. & WAZNY, H. 1980. Microfacies and geochemical development of the basin facies of the Zechstein Limestone (Ca1) in western Poland. In: FUCHTBAUER, H. & PERYT, T.M. (eds) (*q.v.*), 279–306.

PLUMHOFF, F. 1966. Marines Ober-Rotliegendes (Perm) im Zentrum des nordwest-deutschen Rotliegend-Beckens. *Erdöl Kohle*, **19,** 713–20.

POKORSKI, J. 1978. Paleogeographical map of the end of Saxonian. In: DEPOWSKI, S. (ed.) *Atlas litofacjalno-paleogeograficzny permu obszarow platformowych Polski.* Inst. Geol. Warszawa.

POTTER, P.E., MAYNARD, J.B. & PRYOR, W.A. 1980. *Sedimentology of Shale.* Springer-Verlag, New York.

PURSER, B.H. (ed.) 1973. *The Persian Gulf.* Springer-Verlag, New York.

REINECK, H.E. & SINGH, I.B. 1980. *Depositional Sedimentary Environments.* Springer-Verlag, New York.

RENTZSCH, J. 1964. Der Kenntnisstand uber die Metall- und Erzmineralvereilung im Kupferschiefer. *Z. angew. Geol. DDR*, **10,** 281–8.

RYDZEWSKI, A. 1969. Petrografia lupkow miedzionosnych cechsztynu na monoklinie przedsudeckiej. *Biul. Inst. geol.* **217,** 113–58.

—— 1978. Facja utleniona cechsztynskiego lupku miedzionosnego na obszarze monokliny przedsudeckiej. *Przegl. geol.* **26,** 102–8.

SALSKI, W. 1968. Charakterystyka litologiczna i drobne struktury lupkow miedzionosnych monokliny przedsudeckiej. *Kwart. geol.* **12,** 855–73.

SAXBY, J.D. 1976. The significance of organic matter in ore genesis. In: WOLF, K.H. (ed.) *Handbook of Stratabound and Stratiform Ore Deposits*, **2,** 111–33. Elsevier, Amsterdam.

SMITH, D.B. 1970. Permian and Trias. In: HICKLING, G. (ed.) *Geology of Durham County, Trans. nat. Hist. Soc. Northumb.* **41,** 66–91.

—— 1980. The evolution of the English Zechstein basin. In: FUCHTBAUER, H. & PERYT, T.M. (eds), (*q.v.*), 7–34.

—— & FRANCIS, E.A. 1967. Geology of the country between Durham and West Hartlepool. *Mem. geol. Surv. G.B.* 354 pp.

TAYLOR, J.C.M. & COLTER, V.S. 1975. Zechstein of the British sector of the southern North Sea Basin. In: WOODLAND, A.W. (ed.) *Petroleum and the Continental Shelf of North-West Europe*, 240–63. Applied Science Publishers, Barking.

TISSOT, B. & WELTE, D.H. 1978. *Petroleum Formation and Occurrence.* Springer-Verlag, New York.

TURNER, P., VAUGHAN, D.J. & WHITEHOUSE, K.I. 1978. Dolomitization and the Mineralization of the Marl Slate (N.E. England). *Miner. Deposita*, **13,** 245–58.

WEDEPOHL, K.H. 1964. Untersuchungen am Kupferschiefer in Nordwest-Deutschland. *Geochim. cosmochim. Acta*, **28,** 305–64.

WYZYKOWSKI, J. 1965. Die Lithologie des Zechsteins in der Vorsudetischen Monoklinale und das Auftreten der Roten Faule. *Freiberger Forschhft.* **C193,** 247–59.

S. OSZCZEPALSKI, Instytut Geologiczny, 00–975, Warszawa, ul Rakowiecka 4, Poland.

Comments on the genesis and structure of the copper-polymetallic ore deposit of the Foresudetic Monocline, SW Poland

J.B. Tomaszewski

SUMMARY: The copper-polymetallic ore deposit is situated within the Foresudetic Monocline near its boundary with the Foresudetic Block. The copper-mineralized zone extends through a series of dolomites, dolomite/clay shales and white sandstones of the lowermost Zechstein and, possibly, the uppermost part of the Rotliegendes. Mineral compositions, the manner of their occurrence and the diverse internal structure of this stratified deposit all indicate a multistage origin. Synsedimentary mineralization took place in a reducing environment present both during deposition of the shales and dolomites and during sedimentation of the white sandstones. Black concretions containing large amounts of organic carbon, FeS_2 and FeO occur in the upper part of these sandstones. Mineralization also took place throughout diagenesis, mainly in the sandstones but also in the shales and, to a lesser degree, the dolomites.

Sedimentation was on an unstable shelf within the neritic zone. It is considered that natural marine aqueous solutions were enriched in metal ions (mainly Cu, Pb, Zn) by hot brines flowing out into the sedimentary basin through deep fissures, such as those in the Middle Odra fault zone, situated near the deposit.

The copper-polymetallic ore deposit of the Foresudetic Monocline is situated near Lubin, Sieroszowice and Glogow (Fig. 1) near the tectonic boundary between the Foresudetic Block (Tomaszewski 1963a). The extent of the deposit, particularly in the down dip direction, has not yet been established and geological reconnaissance work is still in progress. The deposit is stratified and is contained within certain lithological-stratigraphical units. Other similar occurrences of copper mineralization have been discovered within the Foresudetic Monocline (Fig. 1). Many authors have discussed various problems connected with these deposits, including Tomaszewski (1962a, 1963b), Konstantynowicz *et al.* (1963), Oberc & Serkies (1968), Harańczyk (1970) and Rydzewski (1976). In this paper the author presents further data on this deposit; his views are based on results obtained during many years of research. The aim is to draw attention to some problems, formerly imperceptible in studies based on incomplete data.

Geological outline of the region

Three rock complexes are present in the regional structure (Tomaszewski 1962b) (Fig. 2):

(1) The Cenozoic complex, which is tectonically undisturbed, includes Quaternary and Tertiary sediments and forms the cover of both the Foresudetic Monocline and Forsudetic Block. The Quaternary and Tertiary deposits, 300–400 m thick, are of continental origin and are present as loams, sands, clays and siltstones with seams of Tertiary brown coals.

(2) The Permo–Triassic complex was divided into blocks by Alpine faulting and forms the Foresudetic Monocline, a fragment of the west European Palaeozoic platform. The Triassic sediments, up to 800 m thick, contain continental Lower and Middle Bunter sandstones, with clay intercalations, Roethian carbonate-anhydrite deposits and Muschelkalk carbonate-argillaceous marine sediments. Zechstein marine sediments occur in four sedimentary cycles; these are incomplete in this marginal zone of the Zechstein basin (Tomaszewski 1962b, 1966). The sequence developed is shown in Table 1. The underlying Rotliegendes contains continental red sandstones about 300 m thick. These sandstones are bleached near the top, forming the Weissliegendes with variable, 5–60 m, thickness.

(3) The Proterozoic–Palaeozoic complex was intensely folded and metamorphosed during Variscan movements and forms the crystalline basement of the Foresudetic Monocline and the Forsudetic Block. It contains schists and gniesses with some granitic intrusions.

Sediments associated with the ore deposit

The Rotliegendes and Weissliegendes were formed in a continental environment. In contrast, the white Zechstein sandstones (P1) were deposited during the initial Zechstein transgression.

From HARWOOD, G.M. & SMITH, D.B. (eds), 1986, *The English Zechstein and Related Topics*, Geological Society Special Publication No. 22, pp. 183–194.

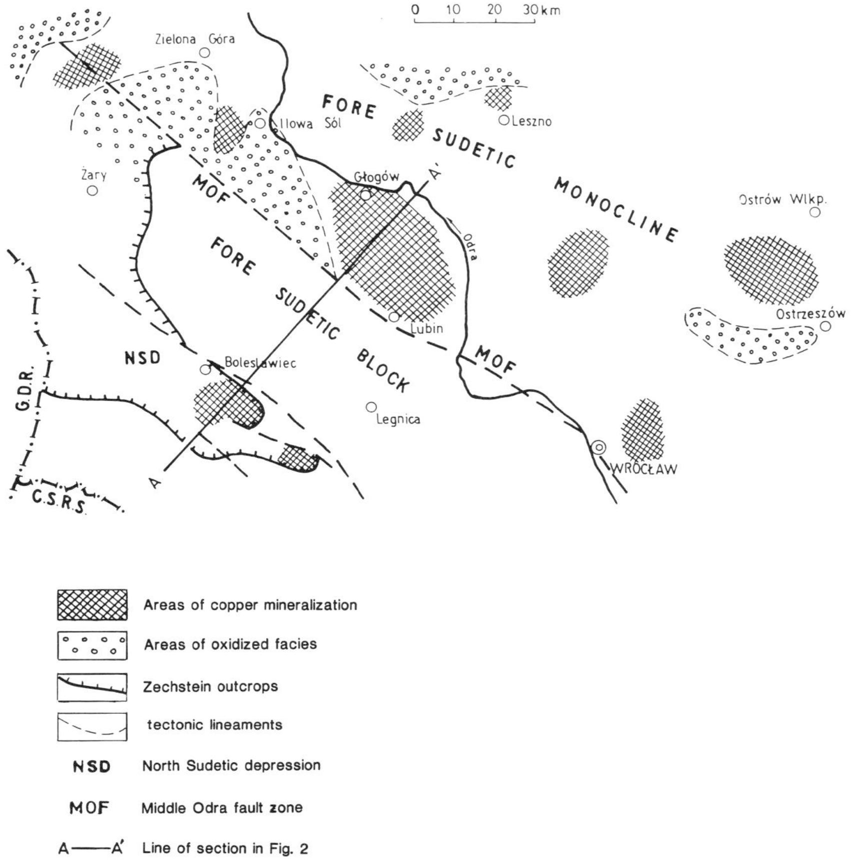

FIG. 1. Sketch map of the Foresudetic region showing the areas of copper mineralization and occurrence of oxidized facies.

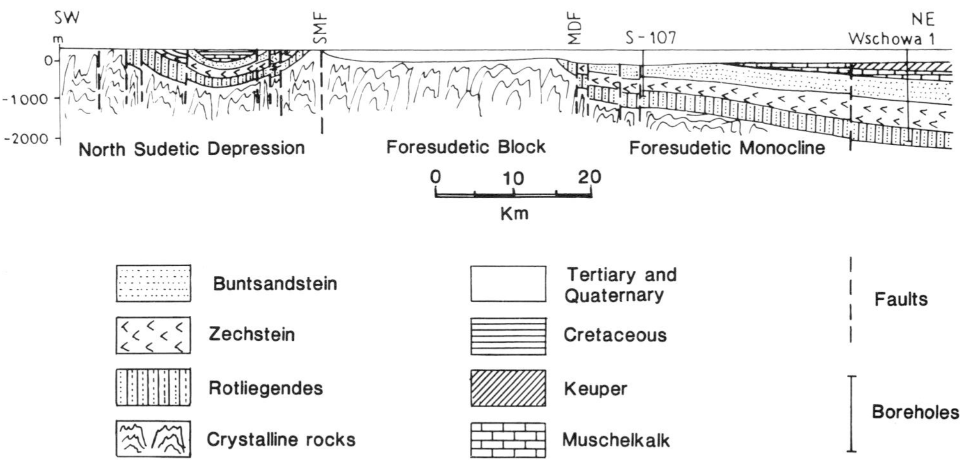

FIG. 2. Geological section through the Foresudetic region.

TABLE 1. Zechstein sediments present in the Lubin area.

Lithology	Abbreviation	Approximate thickness (m)	Cycle
Red clay shales with a thin layer of anhydrite (A4)	L4	40	PZ 4
Anhydrites	A3	30	
Grey clay-stones with lenses of (W3) dolomites at the top	L3	5	PZ 3
Anhydrites	A2	40	PZ 2
Dolomites	W2	10	
Upper anhydrites	A1g	40	
Rock salt	Na1	0·70	
Lower anhydrites	A1d	50	
Dolomites and limestones with Cu mineralization at base	W1	60	PZ 1
Copper-bearing shales, locally with the border dolomite (Wgr) at their base	L1	0·5	
White sandstone with Cu mineralization	P1	0·5–5	

The presence of black concretions, reworked in places, containing organic carbon, pyrite and FeO argues for the existence of a reducing environment in certain areas of the basin during their deposition (Tomaszewski 1978). Reducing conditions persisted in the shallow epicontinental sea during deposition of the succeeding copper-bearing shales (L1) and dark grey limestones and dolomites (W1). Subsequent Zechstein sediments evidence periodic evaporation of the basin (Table 1) with the red claystones (L4), a regressive sequence, constituting the transition from the Zechstein to Lower Triassic intracontinental sedimentation.

Tectonic relationships

The original structure of the Foresudetic Monocline, which dips 2°–6° towards the SE, was complicated by later Alpine faulting (Fig. 2). Major faults trend NW–SE and east to west, with minor north to south trends. The faults transect both Monocline and ore deposit; faulting was thus after mineralization. Intense faulting is present in the Middle Odra Fault Zone, near the boundary between Block and Monocline (Tomaszewski 1963c). Away from the Block margin faulting is less intense (Fig. 2).

Characteristics of the deposit

The deposit is stratified; copper-polymetallic mineralization is spatially and, to a certain extent, genetically connected with the copper-bearing shales, dolomites and white sandstones. The deposit exhibits large variability in:

(1) the thickness of mineralized beds (0.5–16 m);

(2) the concentration of copper in the mineralized beds, both horizontally and vertically;

(3) the mineralized lithology; mineralization occurs in dolomite/clay shales, dolomites and white sandstones.

There is no continuity of mineralization within any one sediment type across the area; mineralization is present within seven major lithologies (Konstantynowicz *et al.* 1963) (Fig. 3). Most mineralization is contained in dolomites and limestones in the western part of the deposit, in the copper-bearing shales in the centre and in sandstones in the SE (Fig. 4). Other mineralized lithologies are present over smaller areas.

Mineralization within the deposit is diverse. Major ore minerals are chalcocite, bornite and chalcopyrite with galena, sphalerite and pyrite. Covellite, tetrahedrite, tennantite, enargite, arsenopyrite, stromeyerite, safflorite, cobaltite, nicke-

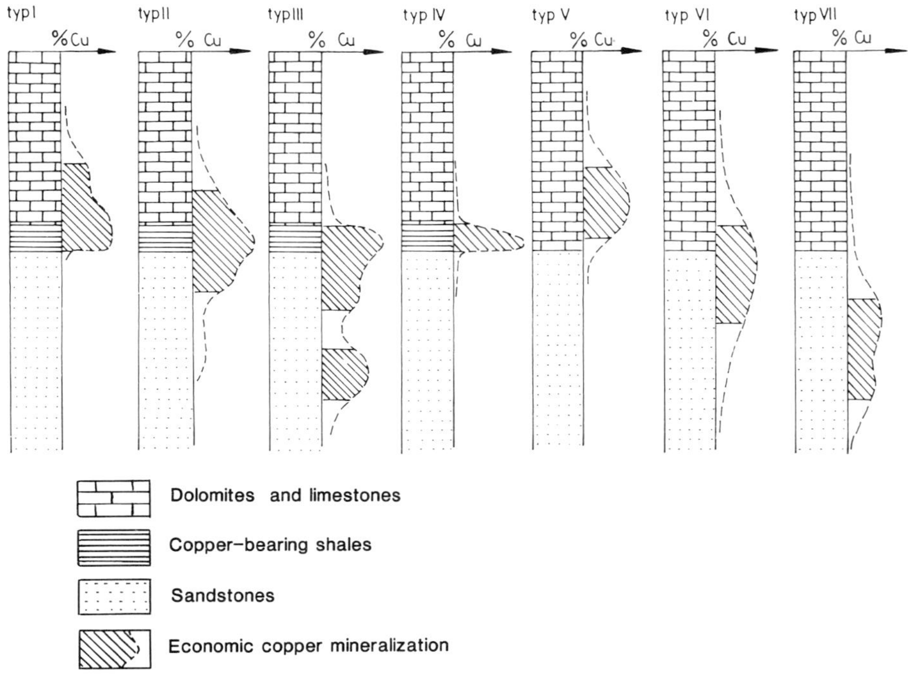

FIG. 3. Vertical distribution of copper mineralization within the different lithologies.

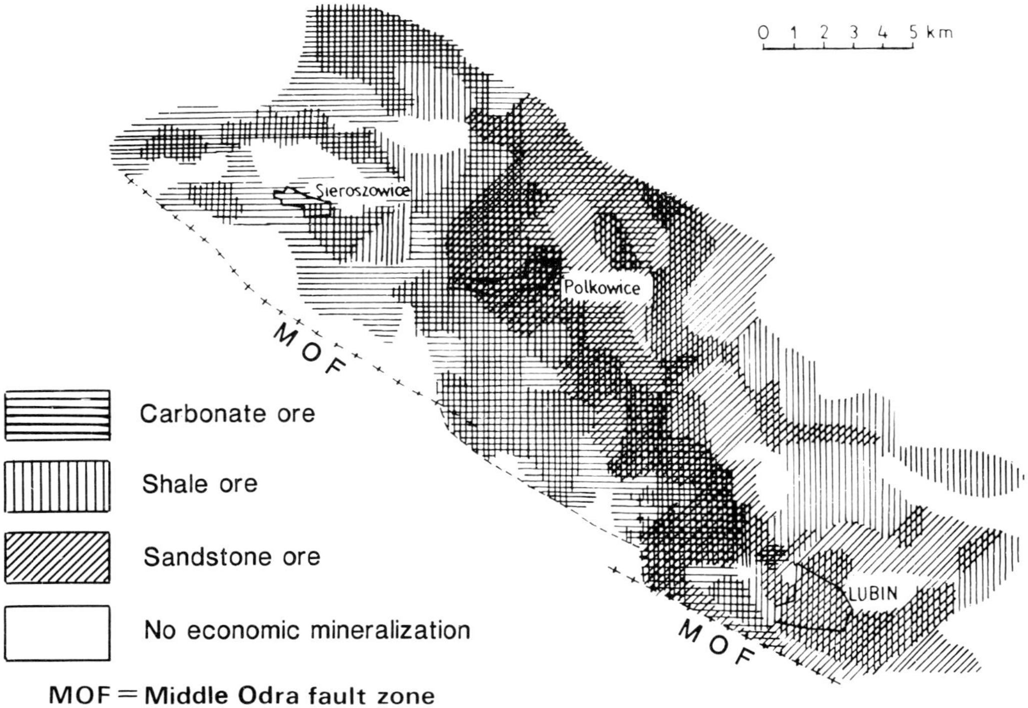

FIG. 4. Lateral distribution of lithologies containing the deposit.

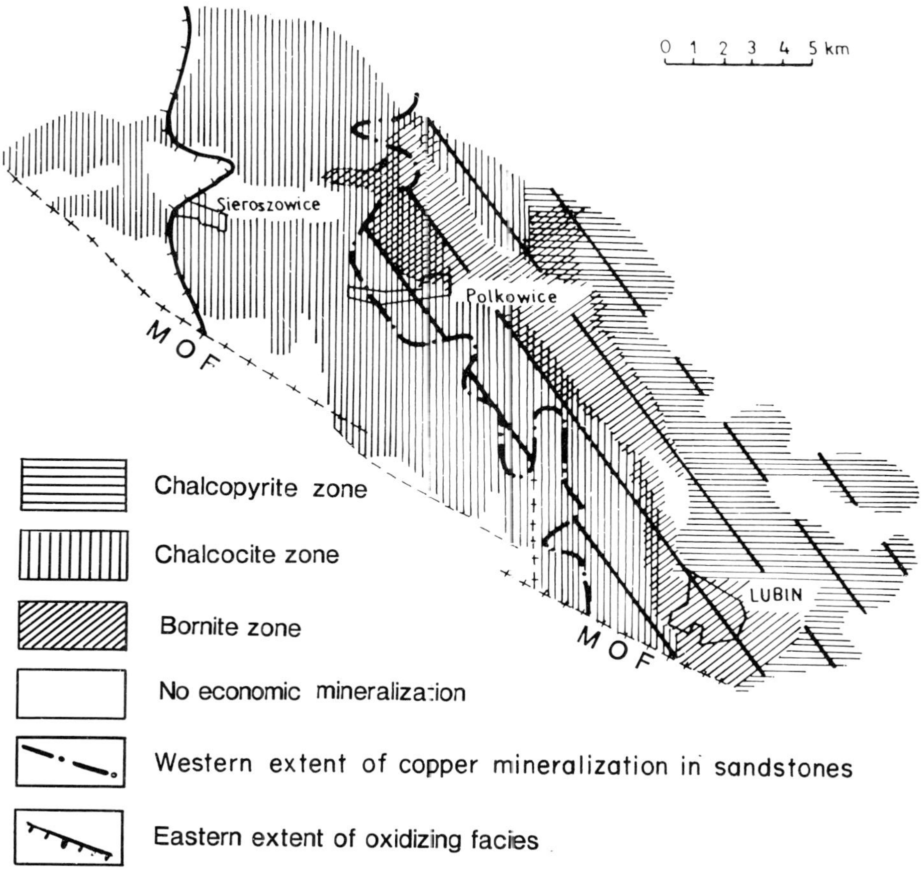

FIG. 5. Lateral distribution of chalcocite, bornite and chalcopyrite zones.

lite, molybdenite, hematite, magnetite and others are minor ore minerals (Konstantynowicz *et al.* 1963). There is a striking horizontal zonation of the major copper, zinc and lead minerals (Fig. 5) plus a less obvious vertical zonation (Fig. 6). The horizontal zonation was governed by reduction/oxidation boundaries (Fig. 5), the position of the Zechstein shorelines plus sulphide solubilities.

Ore minerals occur in three typical forms within the deposits:

(1) dispersed fine particles (from several to tens of micrometres in diameter) present in many places in streaks superimposed on the foliation of the copper-bearing shales (Fig. 7);

(2) elongated lenses and laminae (up to several millimetres thick) concordant with the foliation of the shales, plus thin veinlets (of similar thickness) which transect the mineralized dolomites shales and sandstones in various directions (Figs 7 and 8);

(3) larger concentrations (mineral aggregates), of variable shape (usually not exceeding several millimetres in diameter), present in mineralized sandstones and dolomites.

The dispersed mineralization (in the shales) or in the forms of small irregular concentrations (in the sandstones) was probably synsedimentary. Larger concentrations of mineralization, laminae and veinlets are epigenetic, either diagenetic or catagenetic, as they involve replacement of the host rock by the minerals; Fig. 9 shows one such example. Near Lubin, there is a local occurrence of two mineralized zones (one above the other) in sandstones separated by a barren zone (some tens of centimetres to over 1 m thick). The copper-ore veinlets occur only within the mineralized zone and are absent at the top and at the base of the deposit.

The mineralized zone cuts across the different lithologies within the deposit (Fig. 10). This may result from different physico-chemical conditions in different parts of the deposit during sedimen-

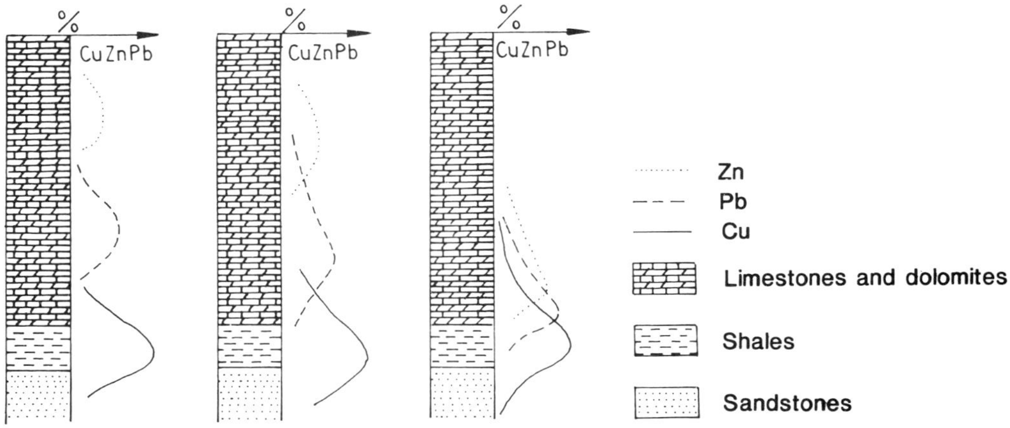

FIG. 6. Vertical zonation of the mineralization.

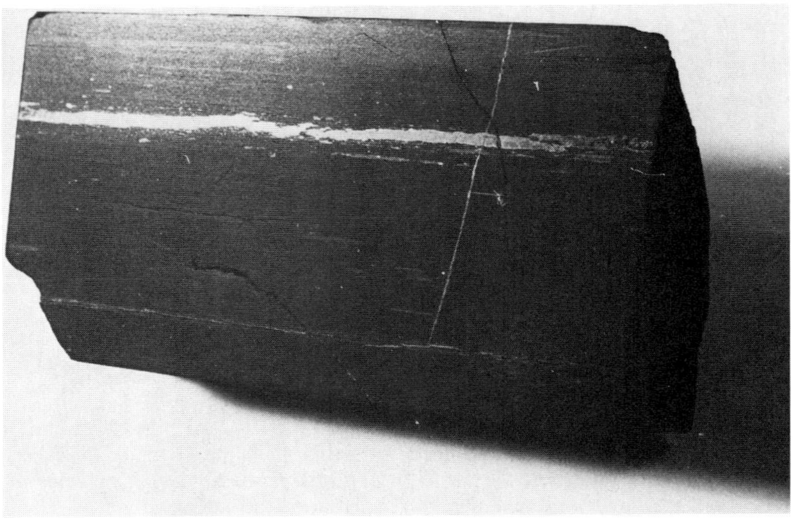

FIG. 7. Copper-bearing dolomite/clay shale with dispersed synsedimentary mineralization and diagenetic veinlets. Sample is 8 cm long.

tation or from movement of the ore minerals during diagenesis. There is a direct relationship between the thickness of the white sandstones and that of the mineralized copper-rich zone (Fig. 10).

Black oval concretions occur in the top two metres of the white sandstones. These sandstones, and some concretions, were reworked during the Zechstein transgression and are therefore synsedimentary (Oberc & Tomaszewski 1963). They indicate the existence of reducing conditions in Zechstein sediments both during and immediately after the transgression. These reducing conditions enabled the synsedimentary precipitation of copper sulphides within the P1 sandstones. Analysis of the concretions shows organic material (up to 0.6% organic carbon), reduced iron (up to 1.8% FeO) and pyrite (up to 4.1%), all indicating continuation of reducing conditions during and after concretion growth. This continuation of reducing conditions could have stimulated precipitation of dispersed copper sulphides both within and away from the nodules. In places, sulphides replace cement between sand grains. The richer, diagenetic mineralization was superimposed upon the low grade synsedimentary one under favourable conditions. The diagenetic mineralization in the sandstones may have resulted from compaction and subsequent lithifi-

FIG. 8. Highly mineralized white sandstone. Diagenetic chalcocite mineralization forms veinlets and cements. Sample is 7 cm high.

FIG. 9. White sandstones with diagenetic chalcocite mineralization following laminations. Sample is 10 cm across.

cation of the copper-rich carbonate-argillaceous Zechstein sediments and the expulsion from these of metal-ion-bearing aqueous solutions into the underlying porous coarse-grained sandy sediments. The Zechstein copper-bearing shales were rich in sapropelic organic matter. Anoxic breakdown of the sapropel caused the reduction of sulphates present in the seawater to sulphides. Continued anoxic decomposition of organic matter in the shales during diagenesis could have been a further source of continued reducing conditions in the underlying white sandstones and, in certain areas, may have been responsible for the synsedimentary mineralization found in their upper part.

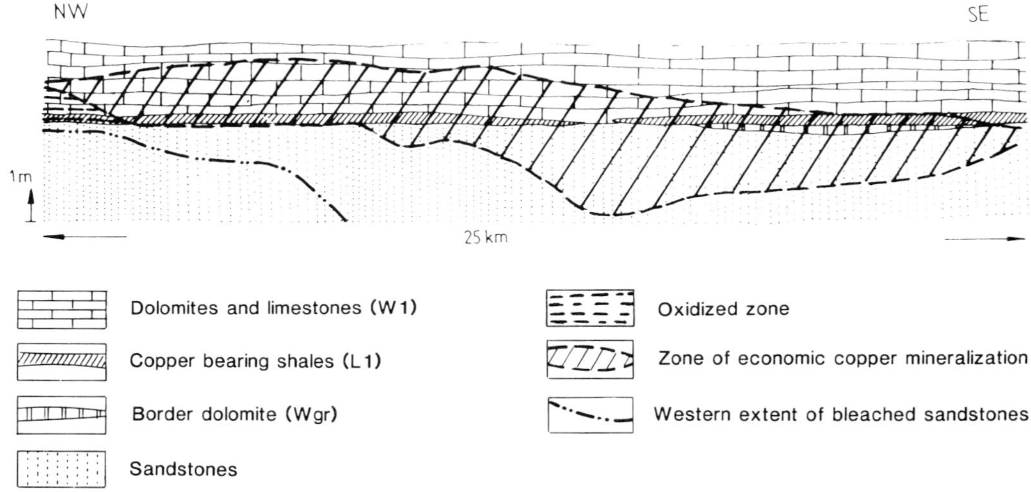

FIG. 10. Diagram of vertical extent of mineralization in longitudinal section through deposit.

This is reinforced by local decrease of copper mineralization in areas of gypsum-anhydrite cements within the sandstones which is found in the western and north-western part of the deposit (Tomaszewski 1963c) (Fig. 11). The western boundary of mineralization within the white sandstones does not parallel that of the Foresudetic Block and Monocline and is more or less conformable to the chalcocite, bornite and chalcopyrite mineralization zones (Fig. 5). Mineralization within the sandstones is at a maximum where the white sandstone is thickest (Fig. 12).

As well as copper mineralization there are subordinate lead and zinc ores plus a group of the

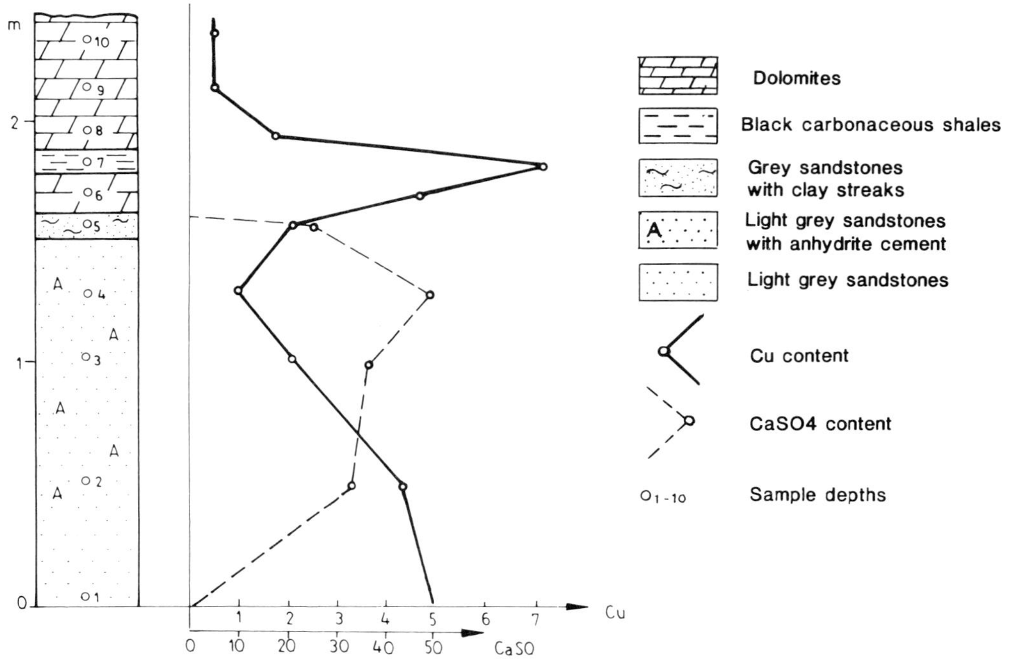

FIG. 11. Decrease of copper mineralization in anhydrite-cemented grey sandstones.

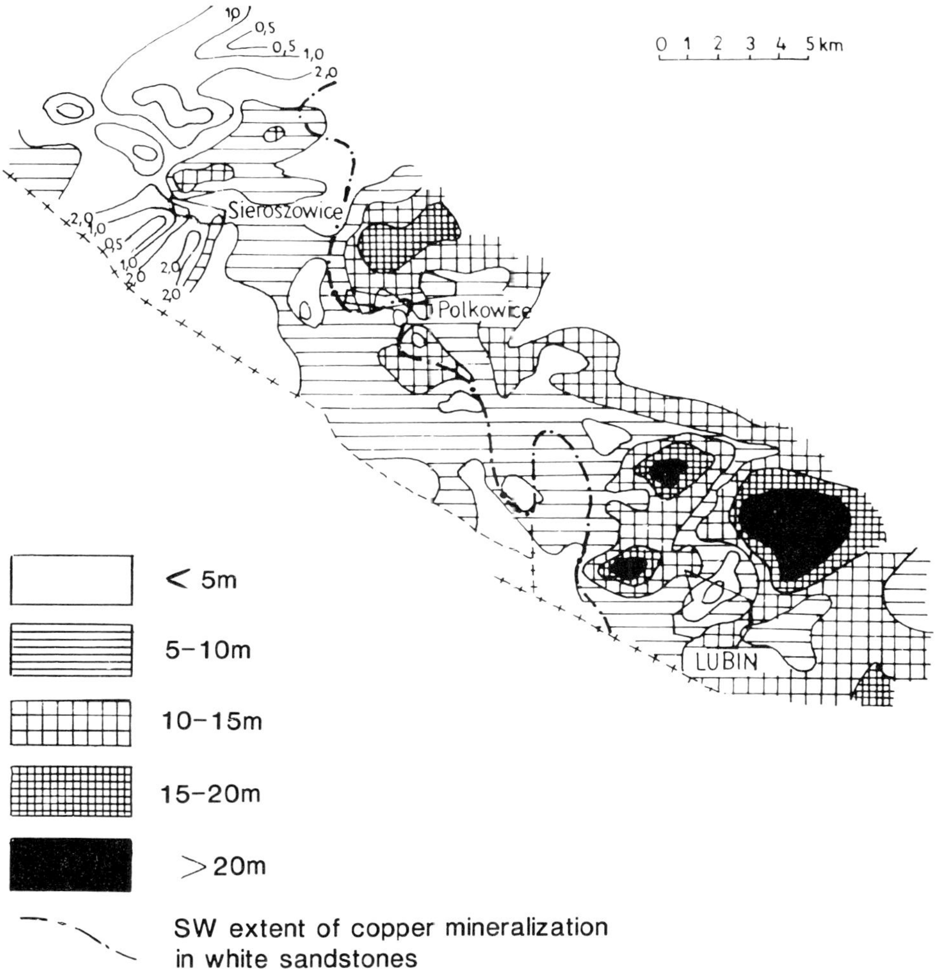

FIG. 12. Isopach map of total white sandstones; (P) Zechstein plus Weissliegendes.

accompanying metals (e.g. Ag, Co, Ni, Mo, V, REE) present in trace amounts. Higher lead and zinc sulphide concentrations occur only in certain parts of the deposit. Zinc mineralization is commonly poor and dispersed. Lead mineralization is much more pronounced and usually found in carbonate rocks; lead is rare in shales and only exceptionally present in sandstones. This argues against Harańczyk's views (1970) on lead-bearing shales. The relative maxima of the Cu, Pb and Zn metals show different vertical relationships throughout the deposit (Konstantynowicz *et al.* 1963) (Fig. 13). Increased contents of lead occur in the NE of the deposit, and form an elongated zone which strikes NW–SE (Fig. 14) parallel with the trend of the bornite and chalcopyrite mineralization zone (Fig. 15). Lead mineralization (galena) is economically important. Where the dolomites and sandstones are copper-bearing only small amounts (2–10 times less than in shales) of Ag, Co and Mo occur. This also supports synsedimentary formation of the main ore mass within the shales with diagenetic mobilization of sulphides from shales to sandstones and dolomites; this may have involved loss of some trace elements. The horizontal mineral zones present in the Mansfeld deposit, e.g. copper in the marginal parts, lead in the deeper, intermediate parts and zinc in the deepest parts of the basin are not present in the area of the Foresudetic Monocline.

An essential feature of this deposit is the lack of genetic and spatial relationships of the mineralization with volcanism, magmatism and fault

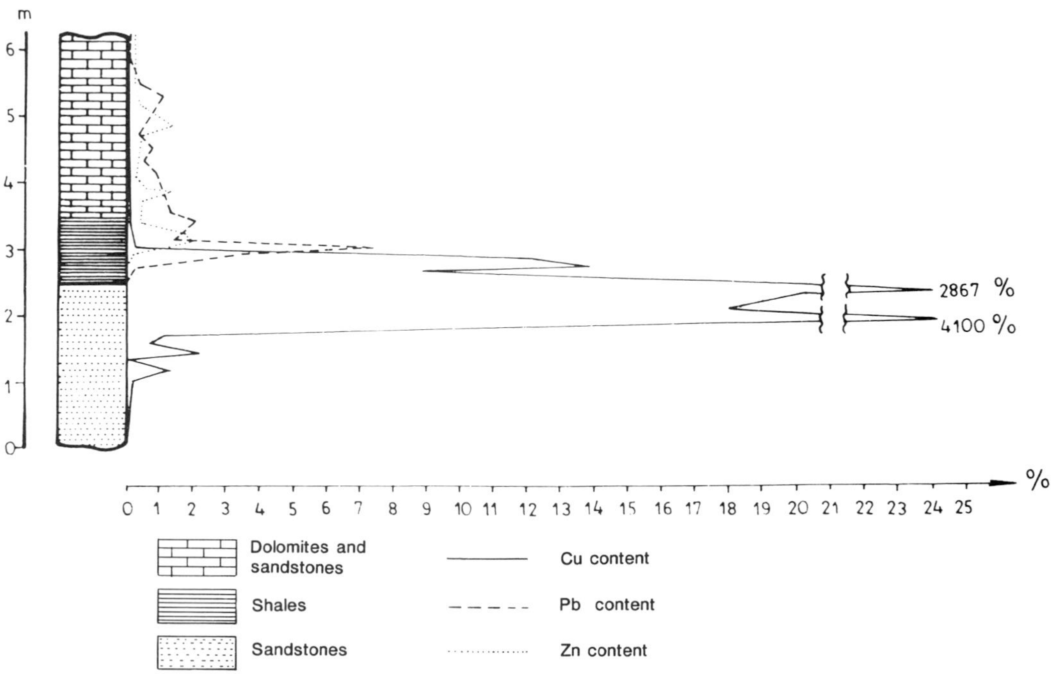

FIG. 13. Vertical distributions of copper, lead and zinc mineralized zones in the deposit.

tectonics. This deposit is present over a vast horizontal area with a relatively small thickness; the ratio of length to its width and thickness is 10,000:4,000:1. Copper mineralization within sediments of reducing facies is present in the immediate neighbourhood of barren red-spotted oxidized facies sediments (Fig. 1). These oxidized sediments may have formed in coastal zones of the basin and in elevated parts of the seafloor within wave-base, where high energy water movement caused their oxidization. This deposit can be distinguished from other types of stratified deposits associated with copper-bearing shales as a separate copper-polymetallic ore deposit of the 'MP' type (MP=monoklina przedsudecka, or Foresudetic Monocline).

The genesis of the deposit

These stratified deposits originated in the Zechstein basin in areas mostly near elevations in the seafloor, or near the margins of the basin where there were calm depositional conditions and a reducing environment. It is difficult to explain the source of the amounts of metals (tens of millions of tonnes), deposited over small areas (a range of several hundred square kilometres), in relatively thin rock strata (a range of several metres) and in a very short, geologically speaking, time span of the copper-bearing shales sedimentation (a dozen or so thousand years). It is also difficult to give an explicit interpretation of the mechanism of their accumulation. Natural concentrations of metal ions in seawater could not have been the only source of metals and there must have been considerable enrichment, perhaps by metalliferous thermal solutions which flowed out into the basin through deep fissures (Bischoff 1969). If this were so, hydrothermal activity must have been synchronized with sedimentation of the copper-bearing shales and thus there must have existed a loose genetic relationship with post-Variscan post-magnetic activity. Using the evidence discussed in this paper, the multistage genetic history of the deposit may be summarized:

(1 a) Synsedimentary precipitation of metal sulphides during deposition of the white sandstones (P1) where suitable reducing conditions existed;

(b) synsedimentary precipitation in a reducing environment of the main mass of copper sulphides during the deposition of the (L1) copper-bearing shales. Primary concentrations of metals in seawater plus metalliferous thermal solutions flowing into the basin through deep fissures constituted the source of the metals;

(c) synsedimentary precipitation of metal sul-

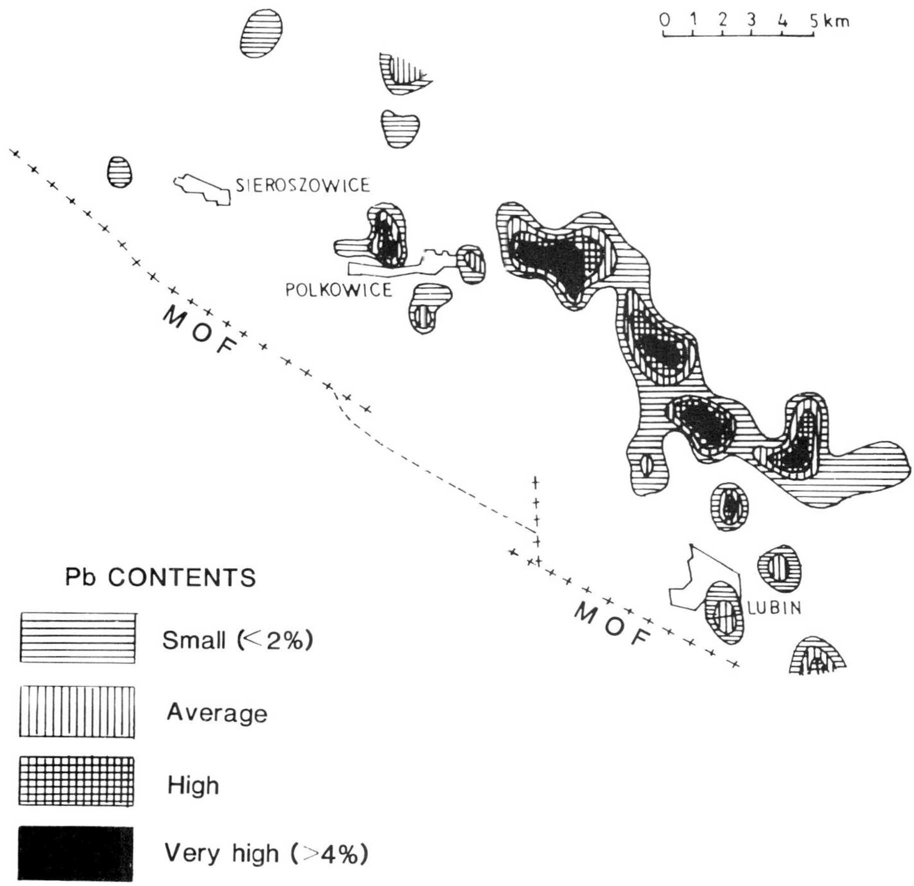

FIG. 14. Distribution of lead mineralization in the copper mineralized zone.

phides, with enrichment in more soluble Pb and Zn sulphides, during the initial sedimentation of the succeeding dolomites and limestones;

(2) Gradual diagenesis of the sediment. Compaction of the copper-bearing shales containing the main mass of synsedimentary mineralization plus expulsion of metalliferous solutions into the porous white sandstones coupled with diagenetic alteration of organic matter. Resultant reducing conditions enhanced sulphide mobility. Formation of some diagenetic sulphide mineralization;

(3) In the catagenetic, or late diagenetic, stage, metal sulphides as well as calcite, barite and gypsum formed veinlets, perhaps from remobilization, and further concentration of minerals. The most intense catagenetic events took place in granular clastic sediments and terminated their cementation.

The stages distinguished above are connected and may overlap to a large extent.

Conclusions

Geological and genetic features of the stratified deposits of the Foresudetic Monocline may be summarized as follows:

(1) Geotectonic position: on margins of platforms, deposition of sediments during transgression following the accumulation of red, clastic continental deposits, lack of relationship between mineralization and tectonics;

(2) Type of basin: unstable shelf, where rate of deposition compensates subsidence, plus cyclic sedimentation; acceleration of sedimentation, decrease of subsidence and consequent shallowing and regression of the sea towards the end of a major period.

(3) Palaeogeographic conditions: bays and depressions, usually near structural highs or elevations of the seafloor;

(4) Sedimentary conditions; neutral to alkaline

marine environment, restricted water circulation, zero redox potential surface within the water column and hence negative redox potential at the sediment–water interface;

(5) Facies: neritic shallow sea with deposition of carbonate-terrigenous (argillaceous) sediments, connection with evaporites (sulphates and chlorides).

(6) Multistage deposit-forming process: (a) synsedimentary sulphide precipitation; (b) diagenetic sulphide concretions, pseudomorphs and pervasive mineralization; (c) catagenetic stage, secondary mobilization of metals, formation of copper sulphides, calcite, barite and gypsum veinlets;

(7) Structure of the deposit: stratified ore within different lithologies (carbonates, shales, sandstones) with varied mineral composition (copper-polymetallic).

(8) Metal zonation: metal and mineral zonation dependent on distribution of oxidized (red-spotted) and reduced (grey-black) facies boundaries; distinct zonations—chalocite-bornite and chalcopyrite-pyrite plus indistinct zonation copper-lead-(zinc)-iron.

References

BISCHOFF, J.L. 1969. Red Sea geothermal brine deposits: their mineralogy, chemistry and genesis. In: DEGENS, E.T. & ROSS, D.A. (eds) *Hot Brines and Recent Heavy Metal Deposits in the Red Sea*, 363–7. Springer-Verlag, New York.

HARAŃCZYK, CZ. 1970. Zechstein lead-bearing shales in the Foresudetian Monocline in Poland. *Econ. Geol.* **65,** 481–95.

KONSTANTYNOWICZ, E., TOMASZEWSKI, J.B. & ZIMNY, W. 1963. The copper-ore deposit of the Foresudetic Zone (in Polish). Śląk Publ. Katowice.

OBERC, J. & SERKIES, J. 1968. Evolution of the Fore-Sudetian copper deposit. *Econ. Geol.* **63,** 372–9.

OBERC, J. & TOMASZEWSKI, J.B. 1963. Some problems of stratigraphy and subdivision of Zechstein in the Wroclaw Monocline (in Polish). *Przegl. geol.* No. 12.

RYDZEWSKI, A. 1976. Origin of the Lower Zechstein polymetallic mineralization (in Polish). *Przegl. geol.* No. 4.

TOMASZEWSKI, J.B. 1962a. Searching for the copper-ore deposits in the Zechstein (in Polish). *Rudy Metale.* No. 6.

—— 1962b. Foresudetic Monocline stratigraphy problems (in Polish). *Przegl. geol.* No. 12.

—— 1963a. The tectonics of the marginal part of the Foresudetic Monocline (in Polish) *Rudy Metale.* No. 6.

—— 1963b. Geological investigation on the Zechstein copper-ore deposits in Lower Silesia (in Polish). *SITG Katowice.*

—— 1963c. *Geological structure of the area of Lubin and Sieroszowice* (in Polish). Unpublished dissertation, University of Wroclaw.

—— 1966. Lithofacial characteristic of the Zechstein series in the marginal part of the Foresudetic Monocline (in Polish). *Z. Geol. Ziem Zachodnich,* PWN Wroclaw.

—— 1978. Nodules in the Weissliegendes (P1) of the Lubin area (in Polish). *Przegl. geol.* No. 7.

JAN B. TOMASZEWSKI, Instytut Gornictwa, Politechniki Wroclawskiej, Swierczewskiego 74, 50-020 Wroclaw, Poland.

Sedimentary environments of the lowest Zechstein sediments in the Lubin region, SW Poland

J.B. Tomaszewski

SUMMARY: The lowest Zechstein sediments in the Lubin area form the 'copper-bearing series' and contain the copper ore deposit of the Foresudetic Monocline. Observations, measurements and sampling of the lowest Zechstein sediments were taken over an area of *c*. 25 km^2 in the underground galleries of the Lubin copper mine. Chemical and mineralogical analysis of the samples was then carried out. Additional data on the ore body and the thickness of particular lithologies were supplied by the mine's geological service. These data enabled the construction of isopach maps for the copper bearing series, plus reconstruction of the sedimentary environments.

Lowest Zechstein sediments in the Lubin area

The study region lies within the Foresudetic Monocline, NW of Lubin (Fig. 1). The regional geological structure and Zechstein stratigraphy have been described by Tomaszewski (1962, 1963, 1978a). This paper deals with the development of clastic and carbonate sedimentation during the lowest Zechstein cycle (PZ1). The basal Zechstein sediments are shown in Fig. 2. The sediments discussed in this paper do not exceed 10 m in thickness and comprise about 5% of the total thickness of PZ1 cycle sediments. They were deposited by the initial transgression of the Zechstein Sea and directly overlie both red and bleached continental Rotliegendes sandstones. Their significance is increased by their genetic and spatial correlation with the copper mineralization (Tomaszewski, *this volume*). The sediments considered here are the white Zechstein sandstones, the border dolomite, the copper-bearing shales, the argillaceous dolomite and the streaked dolomite (Fig. 2). Analysis of the sedimentary conditions of the border dolomite and copper-bearing shale was facilitated by the construction of isopach maps (Figs 3 and 4).

One major difficulty in sedimentary and diagenetic construction is related to the white sandstones, which combine two different stratigraphic units (Oberc & Tomaszewski 1963). The lower unit of Rotliegendes age comprises white sandstones formed by diagenetic bleaching of red sandstones with no reworking and redeposition, whereas the upper, Zechstein unit, comprises white sandstones formed from reworking and redeposition of the top part of the Rotliegendes sandstones by the transgressing Zechstein Sea. The boundary between these units is indistinct and prevents stratigraphic separation in all localities. Only the redeposited sands are discussed below.

The white Zechstein sandstones

The white Zechstein sandstones (Fig. 2) comprise Rotliegendes and Weissliegendes sands reworked by the transgressing Zechstein Sea. These sands are of variable thickness from tens of centimetres to a few metres; in some places they are cross-laminated. The mineral composition is monotonous: quartz 75–85%, feldspars up to 15%, rock fragments average 8%, with few micas and heavy minerals. Sorting of grains is medial, with roundness index 0.70–0.86, and grain size 0.01–0.5 mm (with the dominant fraction 0.031–0.25 mm). Many quartz grains have glossy surfaces and these, plus other features of aeolian origin, indicate little transport and only brief reworking of the grains.

Characteristic sedimentary structures are flame structures, erosion channels and some clay laminae, testifying to periodic wave action on the basin floor. Measurable current directions indicate flow to the NE.

The sandstones are cemented by clays grading through mixed clay-carbonate cements to carbonate cements near the top. Characteristic components of the white Zechstein sandstones are grains of copper sulphides (chalcocite, bornite, chalcopyrite). Red and black ellipsoidal concretions a few centimetres long occur in the upper 2–4 m of the sandstones. The black concretions contain organic carbon plus pyrite, indicating a reducing environment during diagenesis, whereas the red concretions contain trivalent iron oxides, indicating oxidizing conditions (Tomaszewski 1978b and *this volume*). The reducing environments of the former may explain the presence of copper sulphides within the sandstones.

From Harwood, G.M. & Smith, D.B. (eds), 1986, *The English Zechstein and Related Topics*, Geological Society Special Publication No. 22, pp. 195–201.

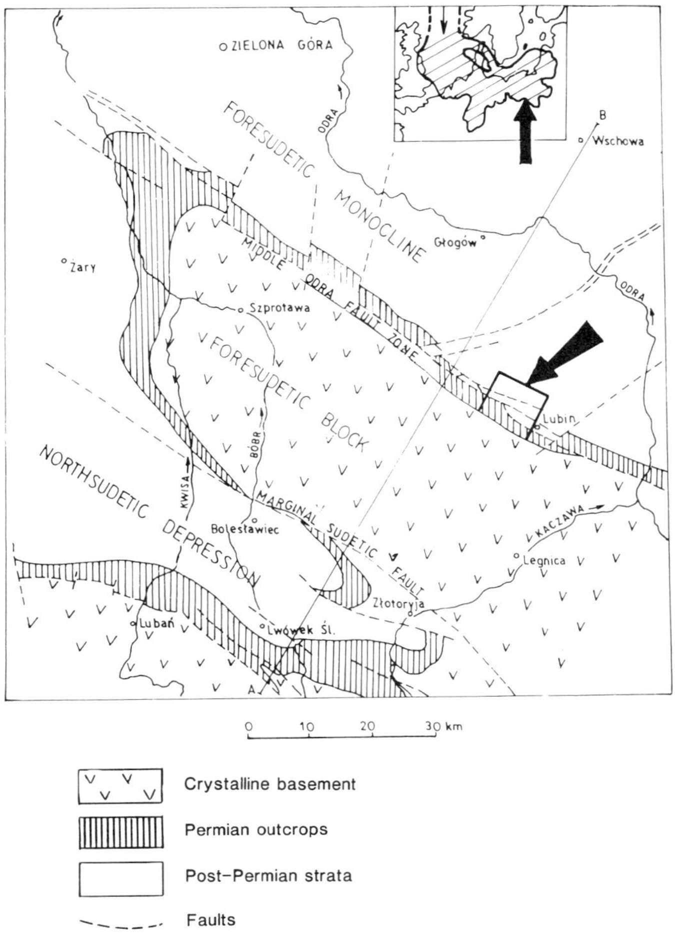

FIG. 1. Outline map of the Foresudetic region showing the location of the study area.

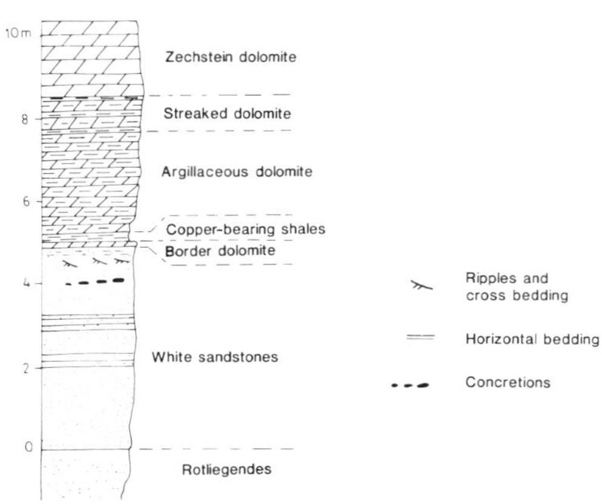

FIG. 2. Lithologies present in the Zechstein sediments of the Lubin area.

The border dolomite

Over most of the area the white sandstones are overlain by the border dolomite. This is a few to tens of centimetres thick (Figs 2 and 3) but wedges out completely in places (Fig. 3); the largest of these areas is between the western and central parts and is 4 km long and 0.1–0.7 km wide (Fig. 3) whereas a smaller (0.05–0.2 km in width and 1.5 km in length) zone trends north-westwards further east. Smaller isolated areas also lack border dolomite. The tapering of the border dolomite in these zones is a sedimentary feature. The lower surface of the border dolomite is, in places, uneven and undulating with amplitudes of up to 5 cm, whereas the upper surface is even and commonly forms a sharp contact with the overlying copper-bearing shales.

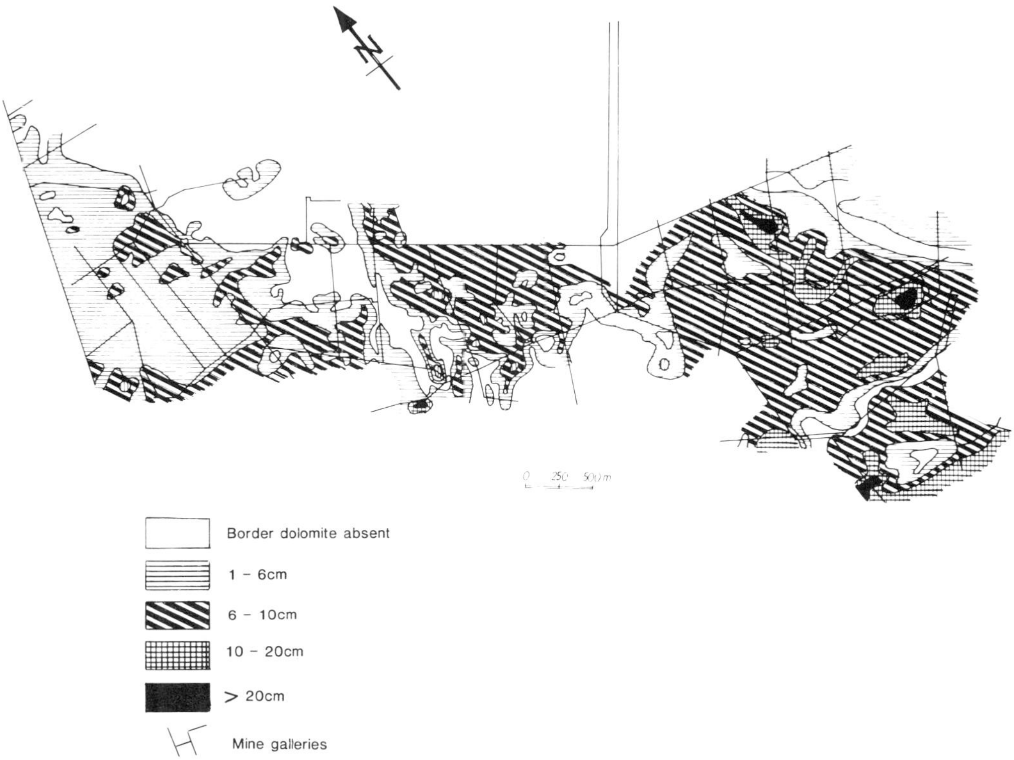

FIG. 3. Isopach map of the border dolomite in the Lubin area.

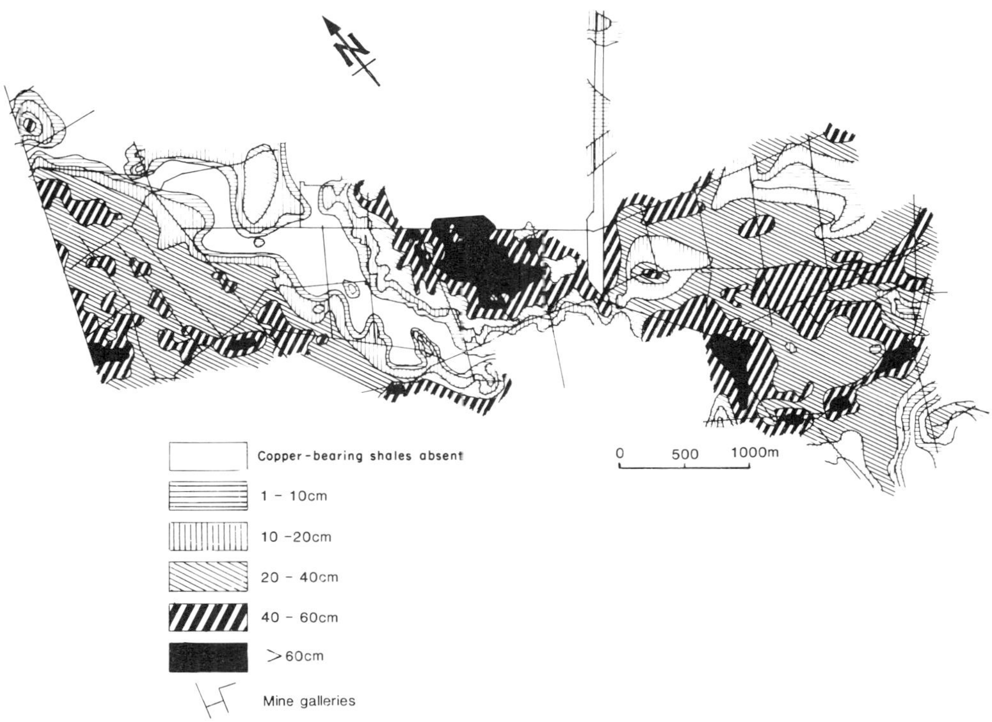

FIG. 4. Isopach map of the copper-bearing shales in the Lubin area.

The border dolomite is a compact, massive rock, commonly dark grey in colour and composed of dolomite or dolomitic limestone. It contains some argillaceous material, commonly in streaks. Fine (0.01–0.003 mm) poorly to subrounded quartz grains are common, although they decrease towards the top of the bed. Also present are fragments of bivalve shells, foraminifers and larger crystals of secondary dolomite or calcite. Copper sulphides are present although mineralization is generally lower than in the underlying sandstones or overlying shales.

The quartz grains, fossil debris plus absence of micrite indicates a moderate energy environment and diminishing delivery of clastic material from the shore.

The copper-bearing shales

The copper-bearing shales form a bed 0–30 cm thick with distinct shaly laminations (Fig. 2). The shales are absent in the same north–south trending zone where there is no border dolomite (Fig. 4). In the copper-bearing shales, this zone is about 4 km long and 0.2–0.6 km wide. The white sandstones are here interpreted to have formed a positive topographical feature on the seafloor, with both border dolomite and copper-bearing shales wedging against this high.

Within the copper-bearing shales it is possible to distinguish different lithological units both on the content of clay material, calcite and dolomite, and organic carbon matter and on their differing lamination. The basal layers are commonly black argillaceous shales with fine laminae above which appear argillo-dolomitic shales with less distinct laminae. There is some evidence of synsedimentary clastic veins.

The copper-bearing shales were formed in a pronounced reducing environment. This, with the small amount of terrigenous sediment, suggests deepening of the basin with low energy of sedimentation. Carbonate contents (both $CaCO_3$ and $CaMg(CO_3)_2$) increase upwards and in many places there is a gradual transition from organic carbon-rich black clay shales through clayey-dolomitic shales to a dark grey clay-rich dolomite, suggesting gradual decrease of reducing conditions. However, in some places a light grey clayey dolomite forms a distinct boundary over the clayey-dolomitic shale; this boundary may represent non-deposition or an abrupt change to more oxidizing conditions with consequent reduction of preserved organic carbon.

The copper-bearing shales are here intensely mineralized with copper sulphide with a distinct break below the overlying stratum, the argillaceous dolomite. In other areas, however, the transition to the argillaceous dolomite is more gradational.

The argillaceous dolomite

The argillaceous dolomite does not exceed 2–3 m in thickness (Fig. 2). It is dark grey, fine grained, compact and in places exhibits a fissility parallel to bedding. The content of argillaceous matter as well as the amount of organic carbon matter decreases upwards. Copper sulphide mineralization occurs in places.

Deposition of the argillaceous dolomite probably took place in a shallowing basin in a less reducing environment, enabling substantial deposition of carbonates and their dolomitization.

The streaked dolomite

The streaked dolomite is separated from the underlying argillaceous dolomite along an indistinct boundary. It has a variable thickness usually not exceeding 1 m (Fig. 2). The rock is fine grained, compact, and of a grey colour with alternating darker and lighter streaks. Argillaceous matter diminishes upwards and the streaked dolomite passes into the thick dolomite and limestone series (W1) which are 60–70 m thick in the Lubin area (Tomaszewski 1978a).

The streaked dolomite represents a transition from organic-rich sediments to the thick series of Zechstein limestones and dolomites, which were deposited on a shallow marine shelf in open marine water. There is a little copper sulphide mineralization in some parts of the streaked dolomite.

Diagenesis

The diagenesis of these lowest Zechstein deposits and formation of the resultant rich copper-polymetallic mineralization was a long and multistage process. The deeper beds of the white sandstones were compacted, with tight packing of quartz grains and small amounts of clay cement. The upper beds of the sandstones were cemented with calcite and dolomite early in diagenesis with the sand grains less tightly packed. Some sand grains have been corroded by the calcite cements.

Carbonate cements were also principal factors in the consolidation of the argillaceous layers of the copper-bearing shales and overlying beds. Diagenesis and transport of the sulphide mineralization was a separate stage. Rich primary synsedimentary sulphides accumulated within the reduced copper-bearing shales (Tomaszewski, *this volume*); this was the major source of the

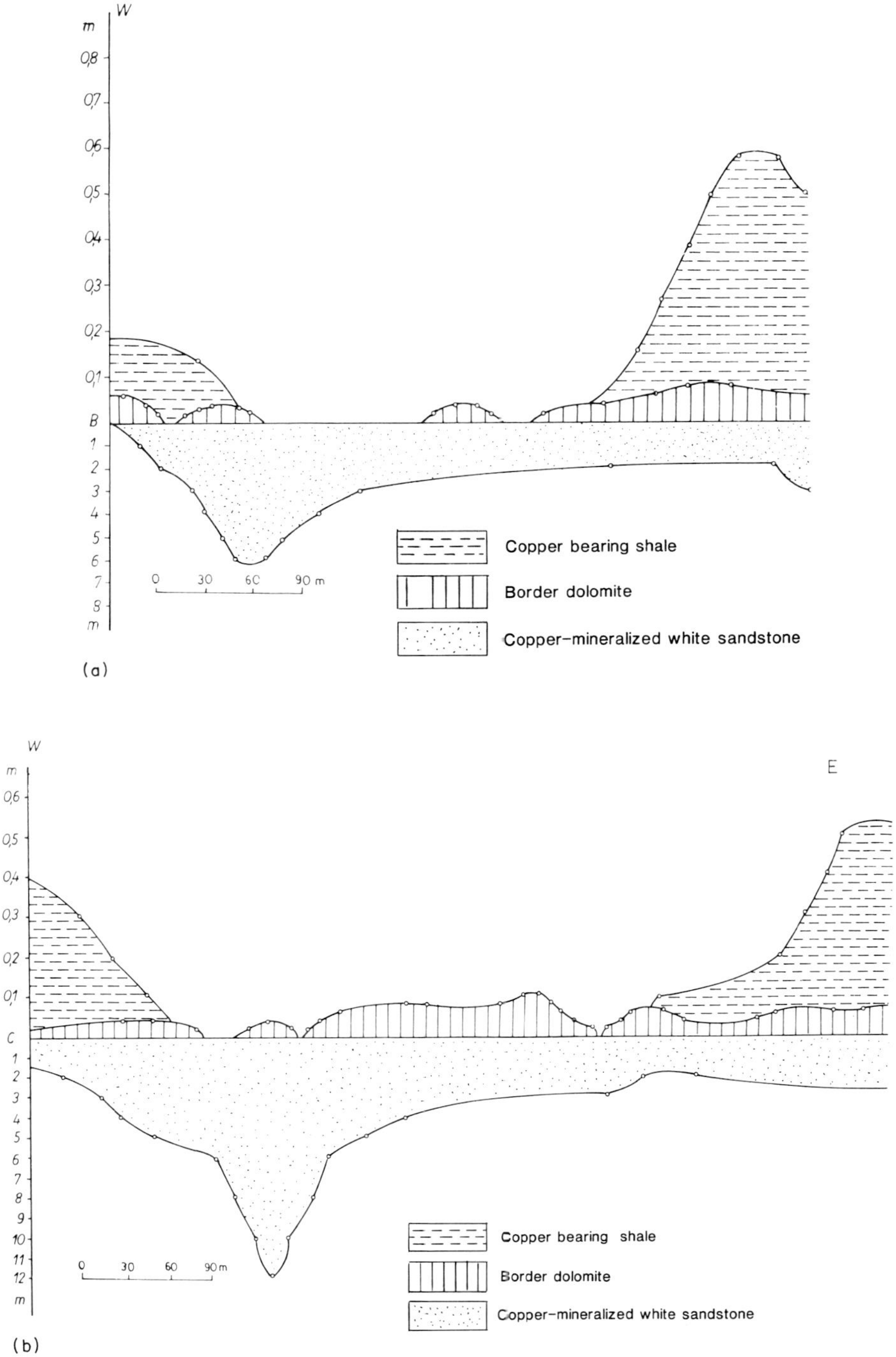

FIG. 5. (a,b) Examples of the increase of thickness of the copper-mineralized zone in the white sandstones (P1) with the thinning of the overlying border dolomite and copper-bearing shales.

mineralization. During consolidation of the underlying uncemented sandstones and overlying carbonates and continuing after their consolidation, the copper and other metals were remobilized and resulted in diagenetic mineralization within areas of the white sandstones, dolomites and limestones.

Selected problems of mineralization within these sediments

(i) Mineralization of the white sandstones

Where the border dolomite is absent, intense chalcocite mineralization, up to several metres thick, is found in the white sandstones (Fig. 5a,b). Individual sulphide aggregates reach 1.5 mm in diameter. Mineralization is also present, dispersed within the cement. A characteristic zone of bornite laminations, comprised of three to four laminae, each 0.5–1.0 cm, is present 0.5–1.5 m below the base of the copper-bearing shales. In the lower and middle laminae, chalcocite is also present. Lower in the profile, at a depth of 2–3 m below the base of the copper-bearing shales, there is a zone of dark laminae, up to 0.5 m wide, containing distinctly richer chalcocite-bornite-chalcopyrite mineralization.

(ii) Mineralization of the border dolomite

This mineralization is of lower grade than in the adjacent white sandstone and the copper-bearing shales (Fig. 6). The ore minerals, chalcocite, bornite, chalcopyrite and neodigenite occur in greatest concentrations near fossil debris; these areas are bounded above and below by black

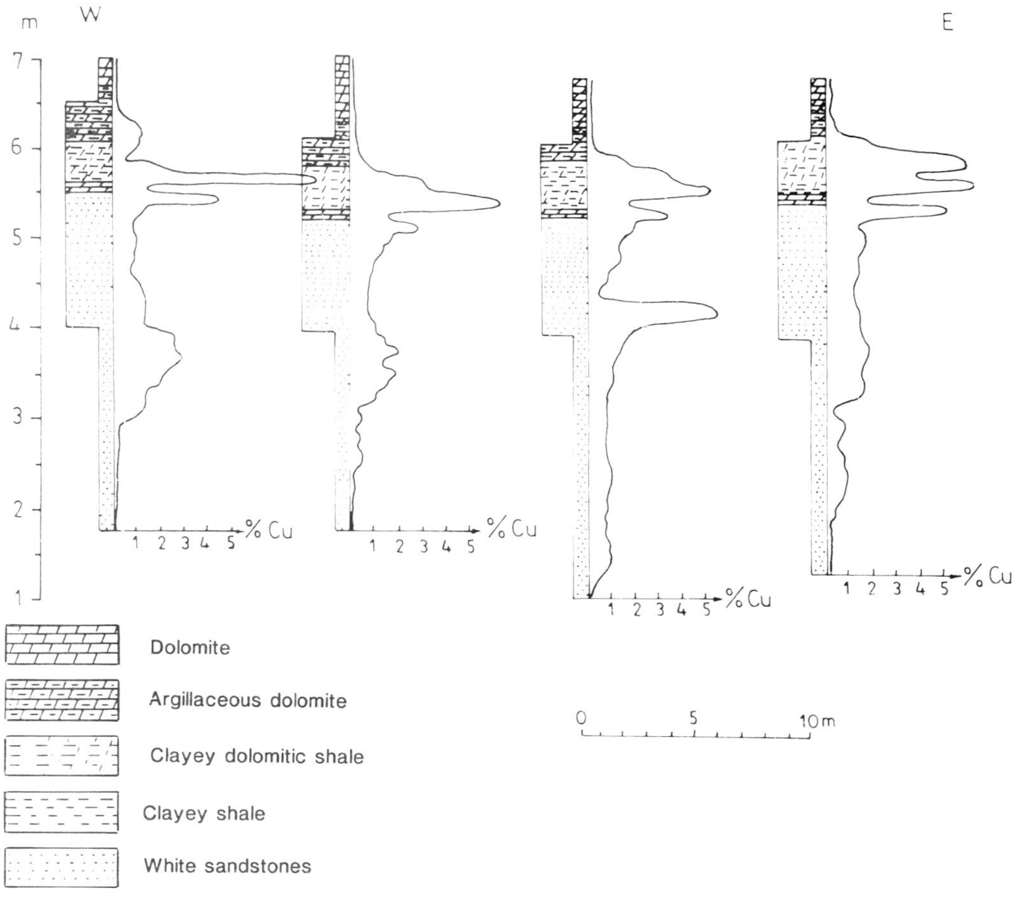

FIG. 6. Relative copper contents within the lowest Zechstein sediments in one gallery of the Lubin mine.

shales. Border dolomites with no fossil content contain only dispersed chalcocite mineralization.

(iii) The copper-bearing shales

These beds are intensely mineralized. The highest concentrations of sulphides occur in the more argillaceous beds where there is a high content of organic carbon (Fig. 6). Chalcocite and bornite are the principal minerals, while chalcopyrite, covellite, and enargite are subordinate. The primary mineralization appears dispersed in streaks conformable with bedding. Sulphide (chalcocite, less often bornite or chalcopyrite) veins and lenses of various shapes were formed during later diagenesis.

(iv) Mineralization of dolomite and limestones

This is mostly in the form of dispersed grains or aggregates, predominantly of chalcocite, but with subordinate bornite and chalcopyrite. Galena is present directly above the actual ore deposit.

Discussion

The distribution of mineralisation is thus highly variable. High copper contents are present in the copper-bearing shales (especially the argillaceous beds with high organic carbon contents), as well as in the white sandstones (about 1–3 m below their top) in certain areas. Where the border dolomite and the copper-bearing shales are both absent (Figs 3 and 4) and carbonate rocks directly overlie the topographic high of white sandstones, the thickness of the mineralized zone within the sandstones increases considerably (Figs 5a,b & 6).

Conclusions

The above facts show that both the sedimentology and the diagenesis of these lowest Zechstein strata are complex and, moreover, exhibit many variations within the area. There is thus no one synthesis which could combine deposition, sedimentary conditions and diagenetic environments throughout the region.

However, the following generalizations can be made: the primary (synsedimentary) copper mineralization in the copper-bearing shales and, subordinately, in the dolomites and white sandstones varied with:

(i) the physico-chemical conditions within the marine environments of the Lower Zechstein basin,
(ii) the direction of transport of metals in the basin, and
(iii) the sedimentation and diagenetic history, especially with respect to precipitation and remobilization of the ore minerals.

Within the area, variations in diagenetic processes were superimposed on the pattern of mineralization resulting from differential sedimentation. This caused significant displacement of primary, synsedimentary mineralization and changes of mineral composition within the sediments considered. The resultant complex picture encompasses a diverse lithological series as well as a range of mineral compositions.

It should also be emphasized that in view of the complexity of the mineralization reported by this reasearch on the lower Zechstein copper deposits of the Lubin mine, its size, and the scant human and technical means possessed by the author, the results obtained come nowhere near exhausting the issue, and many problems remain to be resolved.

References

Oberc, J. & Tomaszewski, J.B. 1963. Some problems of stratigraphy and subdivision of Zechstein in the Wroclaw Monocline (in Polish). *Przegl. geol.* **2,** No. 12.

Smith, D.B. 1981. The evolution of the English Zechstein Basin. *Proc. Symp. Central European Permian, Warszawa*, 1978, 9–47.

Tomaszewski, J.B. 1962. Foresudetic monocline stratigraphy problems (in Polish). *Przegl. geol*, No. 12.

—— 1963. The tectonics of the marginal part of the Foresudetic Monocline (in Polish). *Rudy Metale*, No. 6.

—— 1978a. The geological structure of the Lubin-Sieroszowice Region (Lower Silesia) (in Polish). *Geol. sudet.* **13,** No. 2.

—— 1978b. Nodules in the Weissliegendes P1 of the Lubin area (in Polish). *Przegl. geol.* **26,** No. 7.

—— (*this volume*). Comments on the genesis and structure of the copper-polymetallic ore deposit of the Foresudetic Monocline, SW Poland.

Jan B. Tomaszewski, Instytut Gornictwa, Politechnika Wroclawskiej, Swierczewskiego 74, 50–020 Wroclaw, Poland.

Chronostratigraphical and lithostratigraphical correlations of the Zechstein Limestone in Central Europe

T.M. Peryt

SUMMARY: A glacioeustatic mechanism governed the sedimentation of the PZ1 Zechstein Limestone and made it possible to correlate Zechstein Limestone deposits in different areas chronostratigraphically. The chronohorizons are surfaces (uncommon) or layers up to several centimetres thick that originated in a subaerial or shallow water environment and overlie deposits formed in a subtidal environment. These chronohorizons divide the Zechstein Limestone sequences into three parts or sub-cycles. The boundary between the previously established lower and upper lithostratigraphical units of the Zechstein Limestone corresponds with that between the first and the second sub-cycles or that between the second and the third sub-cycles, or both, depending on the local palaeogeography.

The most diagnostic indication of the subaerial exposure of carbonate deposits is the presence of structures and textures formed in near-surface environments. The recognition of fossil subaerial surfaces is important because their presence shows that fundamental changes in depositional and diagenetic environments have occurred; they have a considerable potential as aids to correlation. Recently such surfaces have been distinguished in many sequences of the Zechstein Limestone in western Poland (Table 1; Peryt 1981, 1984). The analysis of successions of depositional and diagenetic environments in the Zechstein Limestone has demonstrated that its sequence is distinctly cyclic; it is possible to distinguish up to three upward-shallowing sub-cycles, each several tens of centimetres to several tens of metres thick (Figs 1 and 2). The cyclic units are all asymmetric, each being formed mainly of open-sea sediments with a relatively thin regressive sequence at the top; the sharp upwards return to open sea sedimentation after emergence indicates rapid deepening or slow sedimentation during the transgression.

The cyclicity of the Zechstein Limestone probably resulted from glacioeustatic changes of sea-level, as earlier suggested by Smith (1970, 1974, 1980) and discussed in detail by Peryt (1984). The rapidity of glacioeustatic sea-level changes is of great value in facilitating chronostratigraphical correlation (Figs 1, 2 and 3). The chronohorizons in the Zechstein Limestone are up to several centimetres thick or, rarely, are surfaces between subaerial or extremely shallow water deposits and overlying subtidal rocks. In places where the progression from subtidal to subaerial sedimentation did not result purely from sediment build-up, sharp contacts between subtidal and very shallow water or subaerial deposits may be chronohorizons.

Firm chronostratigraphic correlation is possible only where the Zechstein floor emerged or nearly so during phases of low sea-level (Fig. 4) and the effects of sea-level changes are difficult to distinguish in the deepest parts of the Zechstein basin. In intermediate areas it is locally possible to link evidence of phases of shallowing to phases of emergence nearer the margins (Fig. 2) but such evidence is only reliable where borehole or surface evidence is good and closely spaced.

The scheme of chronostratigraphical correlation, summarizing the lithological data, is presented in Fig. 3. Because deposition of the Zechstein Limestone was generally regressive and because of the progradation of the carbonate platform, the second chronohorizon is more extensive than the first (Fig. 4). This, in turn, is of great importance in linking the chronostratigraphical and lithostratigraphical correlations, as shown below.

In west Poland, Alexandrowicz (1974) distinguished four lithostratigraphical units in the Zechstein Limestone. He referred to the subdivisions of the Zechstein Limestones in the Sangerhausen and Mansfeld troughs (Janowski & Jung 1962) and NE Mecklenburg (Helmuth 1968) and proved that it is possible to distinguish the same units in other peripheral and central parts of the Zechstein Limestone basin. Earlier, different subdivisions had been proposed for the central part of the Basin (Jankowski & Jung 1962) and for the peripheral part (Tzschorn 1963). The four units of the Zechstein Limestone recognized by Jankowski & Jung were distinguished by the structures and textures of the rocks and their chemical composition; where the uppermost part of the Zechstein Limestone is strongly diagenetically altered, the dependence on macroscopic observations alone meant that the boundary between units δ and γ as distinguished by Alexandrowicz

From HARWOOD, G.M. & SMITH, D.B. (eds), 1986, *The English Zechstein and Related Topics*, Geological Society Special Publication No. 22, pp. 203–209.

TABLE 1. The characteristics on which subaerial surfaces in the Zechstein Limestone deposits were distinguished: 3—major importance; 2—secondary importance; 1—minor importance.

	Platform	Slope of platform	Basin
Caliche profile and vadose layer	3	3	—
Vadose marine deposits	2	2	—
Vadose freshwater deposits	3	3	—
Sabkha sequence	2	—	—
Stromatolites	2	2	3
Fenestral fabric	2	2	3
Pattern of dolomitization	1	2	3
Evidence of dedolomitization	3	3	3
Presence of vadose cement	2	2	—
Presence of phreatic freshwater cement	3	—	—
Evidence of vadose compaction	2	2	—
Evidence of freshwater dissolution	2	2	—

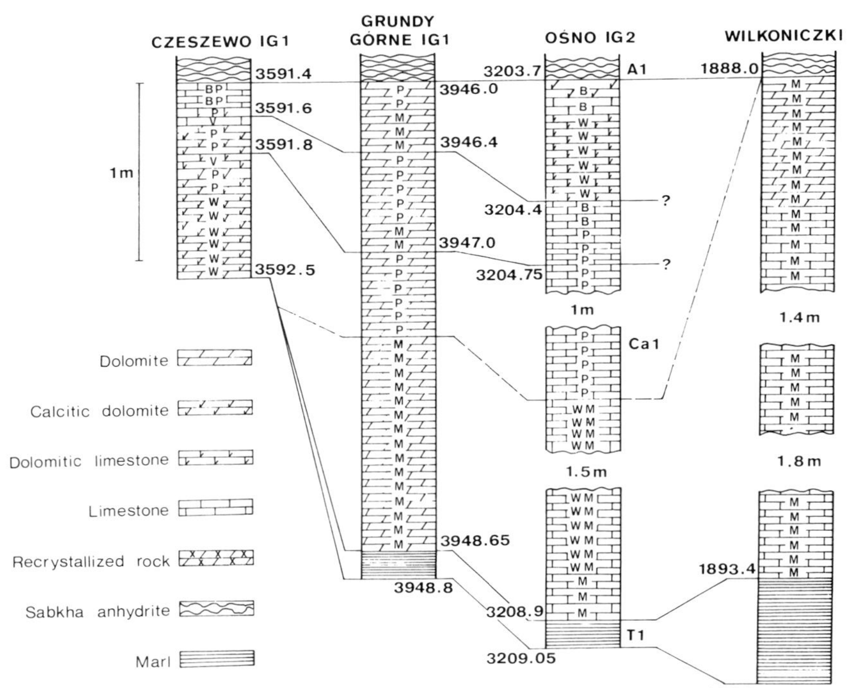

FIG. 1. Correlation of basinal sequences of the Zechstein Limestone in the Foresudetic area. Chronostratigraphical correlation lines are continuous, lithostratigraphical correlation lines are broken. M—mudstones; WM—wackestones/mudstones; W—wackestones; P—packstones; BP—boundstones/packstones; B—boundstones; V—vadose deposit.

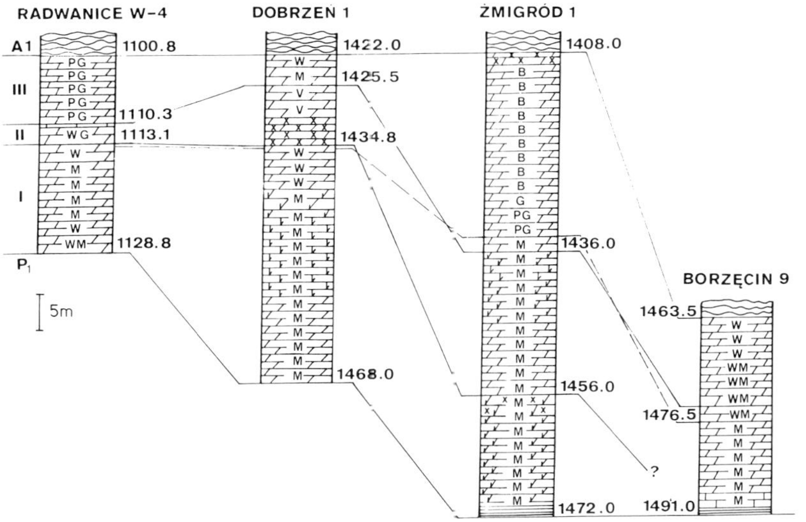

FIG. 2. Correlation of the Zechstein Limestone sequences from the peripheral part of the basin in the Foresudetic area; W-G—wackestones-grainstones; PG—packstones/grainstones; G—grainstones; other explanation as in Fig. 1.

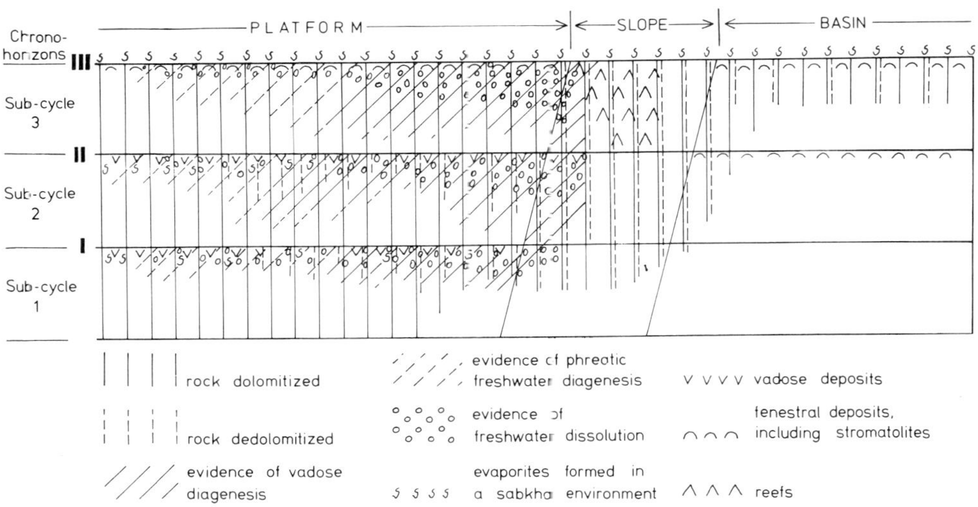

FIG. 3. Chronostratigraphical correlation of the Zechstein Limestone in western Poland.

(1974) was distinguished by other criteria by Jankowski & Jung (1962). It should be added, however, that the recognition of that boundary is difficult even in the diagenetically little-altered sequences of the Zechstein Limestone of the central part of the basin (cf. Langer 1977 and Gerlach & Knitzschke 1978).

Lorenc (1975) distinguished three petrographical units in his studies in the central part of the Foresudetic Monocline; his subdivision is a modification of that of Jankowski & Jung, taking the petrology into account. Motyl-Rakowska (1971) also referred to the subdivision by Jankowski & Jung (1962); she assigned the basinal Zechstein Limestone proved in boreholes Szklarka 2 in SW Poland and Wierzchowo 3 and Brda 1 in NW Poland to unit δ whereas Lorenc (1975) referred these sequences to the lowermost unit α as distinguished by him and equivalent to unit α and the lower part of unit β of Alexandrowicz (1974).

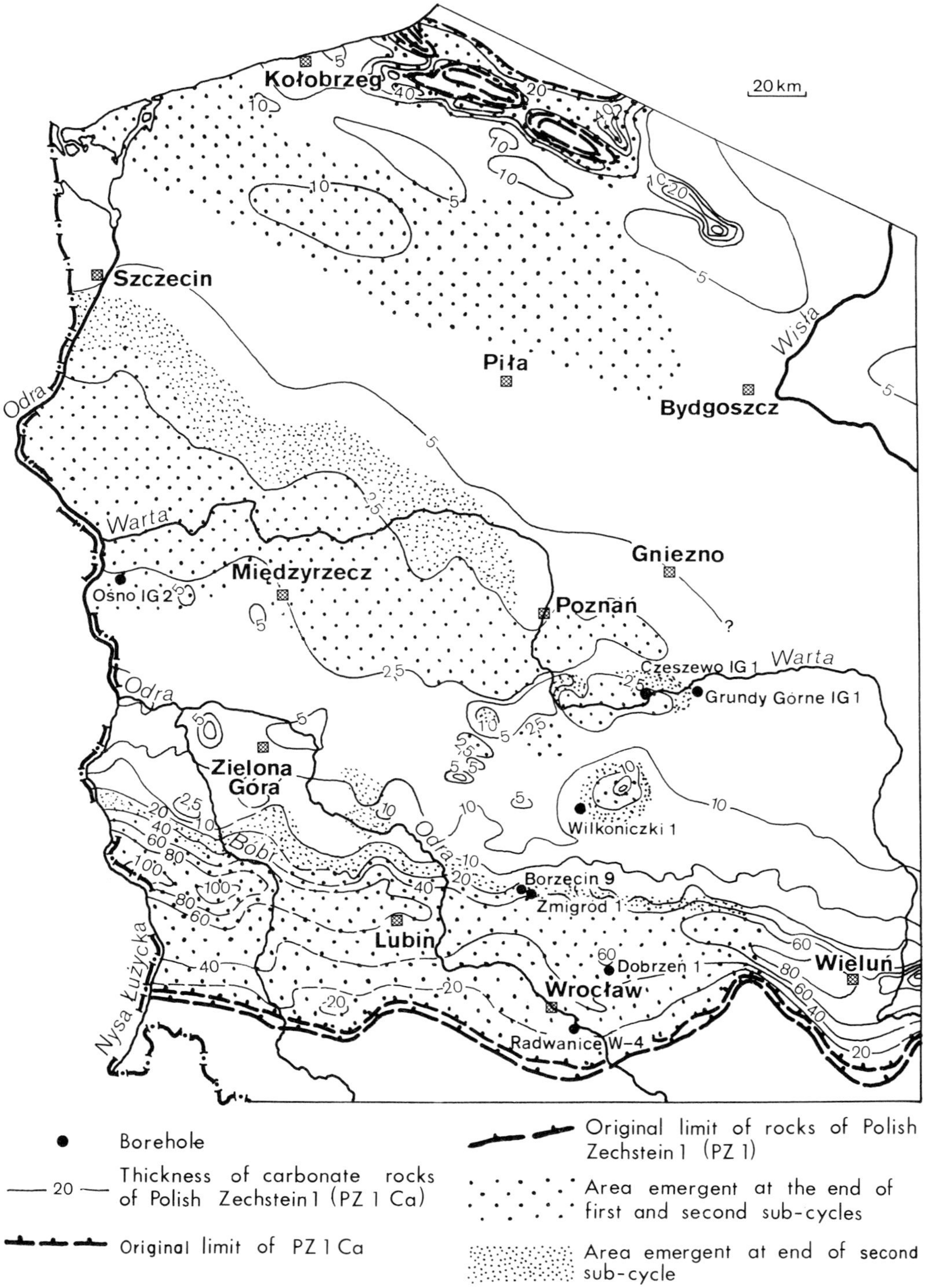

FIG. 4. Temporarily emergent areas during the deposition of the Zechstein Limestone. Note: at the end of the third sub-cycle almost the whole basin dried up.

TABLE 2. Lithostratigraphical correlation of the Zechstein Limestone sequences in Central and Western Europe

England (Smith 1980)	Denmark (Clark & Tallbacka 1980)	Westfalia (Füchtbauer 1980)	Harz, Thuringia			Mansfeld trough (Jankowski & Jung 1962)	Mecklenburg (Helmuth 1968)	Foresudetic Monocline (Peryt 1978)	Łeba elevation (Peryt & Piatkowski 1976)
			(Fulda 1935)	(Seifert 1972)	Langbein & Seidel 1980)				
Middle Magnesian Limestone	Vuggy packstone unit	Phase 2 deposits	Do1 Werradolomit	$\gamma_3, \gamma_2, \gamma_1$	δ, γ	δ, γ	δ, γ { $\gamma_3, \gamma_2, \gamma_1$ }	Supramicritic/ oncolitic/ complex	Pisolitic complex
Lower Magnesian Limestone	Mudstone unit	Phase 1 deposits	Ca1 Zechsteinkalk	β_2, β_1; α_2, α_1	β, α	β, α	β, α { α_2, α_1 }	Micritic complex	Micritic complex

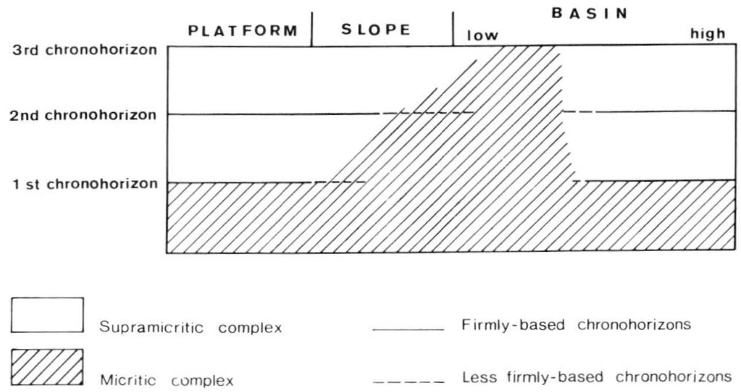

FIG. 5. Relationship between the chronostratigraphical and lithostratigraphical correlations in western Poland.

In the Foresudetic Monocline, near the margin of the basin, Peryt (1978) distinguished two microfacies complexes: a micritic and an overlying supramicritic complex in the barrier zone and an oncolitic complex in the lagoonal zone. The rocks of these complexes originated in similar environmental conditions although not necessarily at the same time. This modification of Jankowski & Jung's subdivision is also used by other authors working in the Foresudetic Zechstein Limestone (Oszczepalski 1980; Piatkowska 1978) and its usefulness has also been confirmed in the western part of the Peri-Baltic syneclise of northern Poland by Peryt & Piatkowski (1976).

The relationship between the chronostratigraphical and lithostratigraphical correlations in western Poland is presented in Fig. 5. Within the shallow zones the boundary between the micritic and supramicritic complexes and the first chronohorizon almost coincide, whereas towards the basin centre this boundary is clearly diachronous; in the outer zone of the peripheral part of the basin it may commonly be related to the second chronohorizon, although in places it cannot be recognized. In the deepest parts of the central basin the micritic complex constitutes the whole of the Zechstein Limestone.

I conclude that the lithostratigraphical correlation of the Zechstein Limestone of western Poland depends to a considerable extent on basinal changes of sedimentary environments, and the boundaries of the lithological units are diachronous in many places. This is also true of Zechstein Limestone sequences elsewhere, but it should be noted that there are only a few detailed sedimentological studies of this formation elsewhere.

Study of the literature shows that the Zechstein 1 carbonate of Poland comprises two units that may be distinguished throughout the basin (Table 2). At the contact of the units there is a fundamental change; the lower unit, commonly comprising mudstones and wackestones (locally with brachiopod banks) is separated from the upper unit of packstones, grainstones and boundstones, and this change is locally accentuated by the presence of sabkha deposits at the mutual contact (the Hampole Beds in England, Smith 1968) and in the Schale Z1 borehole in Westphalia (Füchtbauer 1980). Many authors regard this contact as evidence of a break in sedimentation; according to Smith (1968, 1970, p. 75) this break probably resulted from a substantial sea-level fall (leading to exposure of much of the shelf) followed by transgression.

Taking into account the observations made in western Poland, there is no doubt that the boundary between the lower and upper units corresponds with the boundary between the first and the second sub-cycles, or the second and the third cycles, or both, depending on the local palaeogeography. The relationships between the chronohorizons and the contact between the two units in western Poland are also typical of all other areas where more than one palaeogeographical zone is present.

References

ALEXANDROWICZ, S.W. 1974. Poziomy litostratygraficzne wapienia cechsztynskiego w zachodniej Polsce. *Annls. Soc. géol. Pologne,* **44,** 275–92.

BERES, B., KIENIG, E., KIJEWSKI, P. & TOMASZEWSKI, J.B. 1973. Rozpoziomowanie litostratygraficzne skal weglanowych cyklotemu P_2^1 w cechsztynie monokliny przedsudeckiej. *Przegl. geol.* **21,** 17–20.

CLARK, D.N. & TALLBACKA, L. 1980. The Zechstein deposits of southern Denmark. In: FÜCHTBAUER, H. & PERYT, T.M. (eds) (*q.v.*), 205–31.

FÜCHTBAUER, H. 1980. Composition and diagenesis of a stromatolitic bryozoan bioherm in the Zechstein 1 (northwestern Germany). In: FÜCHTBAUER, H. & PERYT, T.M. (eds) (*q.v.*), 233–51.

FÜCHTBAUER, H. & PERYT, T.M. (eds) 1981. *The Zechstein Basin with Emphasis on Carbonate Sequences. Contr. Sedim.* **9**.

FULDA, E. 1935. *Zechstein, Handbuch der vergleichenden Stratigraphie Deutschlands.* Gebruder Borntraeger, Berlin.

GERLACH, R. & KNITZSCHKE, G. 1978. Sedimentationszyklen an der Zechsteinbasis (Z1) im SE-Harzvorland und ihre Beziehungen zu einigen bergtechnischen Problemen. *Z. angew. Geol. DDR*, **24**, 214–21.

HELMUTH, H.-J. 1968. Zur Gliederung des Zechsteinkalks (Ca 1) im NE Mecklenburg. *Geologie*, **127**, 164–75.

JANKOWSKI, G. & JUNG, W. 1962. Zum Zechsteinkalk (Ca 1) im Bereich der Sangerhäuser und Mansfelder Mulde. *Geologie*, **11**, 943–53.

LANGBEIN, R. & SEIDEL, G. 1980. Zur Fazies des Zechsteinkalks im östlichen Thüringer Becken. *Z. geol. Wiss. DDR*, **8**, 835–51.

LANGER, M. 1977. Die Bedeutung der Algen für die Gliederung des Zechsteinkalks (Ca 1) in Beckenfazies und für die Korrelation mit den Zechsteinriffen. *Z. angew. Geol. DDR*, **23**, 83–7.

LORENC, S. 1975. Petrografia i zroznicowanie facjalne wapieni i anhydrytow Werra monokliny przedsudeckiej. *Geol. sudet.* **10**, 59–104.

MOTYL-RAKOWSKA, J. 1971. W sprawie podzialu wapienia cechsztynskiego w Polsce. *Geofiz. Geol. Naft. 174–75*, 1–4.

OSZCZEPALSKI, S. 1980. Paleography, sedimentation and mineralization of the Z1 carbonate series (Zechstein) in the western part of the Fore-Sudetic Monocline (western Poland). In: FÜCHTBAUER, H. & PERYT, T.M. (eds) (*q.v.*), 307–23.

PERYT, T.M. 1978. Sedimentology and paleoecology of the Zechstein Limestone (Upper Permian) in the Fore-Sudetic area (western Poland). *Sediment. Geol.* **20**, 217–43.

—— 1981. Former aragonitic submarine hemispheroids associated with vadose deposits, Zechstein Limestone (Upper Permian), Fore-Sudetic area, western Poland. *Neues Jb. Geol. Palaont. Mh*, 559–70.

—— 1984. Sedymentacja i wczesna diageneza utworow wapienią cechsztynskiego w Polsce zachodniej. *Prace Inst. geol. Warszawa.* **109**, 80 pp.

—— & PIATKOWSKI, T.S. 1976. Osady caliche w wapieniu cechsztynskim zachodniej czesci syneklizy perybaltyckiej. *Kwart. geol.* **20**, 525–38.

PIATKOWSKA, L. 1978. Wyksztalcenie mikrofacjalne wapienia cechsztynskiego w rejonie Trzebnicy na monoklinie przedsudeckiej. *Przegl. geol.* **26**, 186–8.

SEIFERT, J. 1972. Das Perm am Südostrand des Thüringer Beckens. *Jb. Geol.* **4**, 97–179.

SMITH, D.B. 1968. The Hampole Beds—a significant marker in the Lower Magnesian Limestone of Yorkshire, Derbyshire and Nottinghamshire. *Proc. Yorks. geol. Soc.* **36**, 463–77.

—— 1970. Permian and Trias. In: HICKLING, G. (ed.) *Geology of Durham County. Trans. nat. Hist. Soc. Northumb.* **41**, 66–91.

—— 1974. Sedimentation of Upper Artesia (Guadalupian) cyclic shelf deposits of the northern Guadalupe Mountains, New Mexico. *Bull. Am. Ass. Petrol. Geol.* **58**, 1699–1730.

—— 1980. The evolution of the English Zechstein basin. In: FÜCHTBAUER, H. & PERYT, T.M. (eds) (*q.v.*), 7–34.

TZSCHORN, G. 1963. Zur Geologie und Petrographie des Zechsteinkalks im Raum Spremberg-Weisswasser. *Z. angew. Geol. DDR*, **9**, 561–8.

T.M. PERYT, Instytut Geologiczny, 00–975, Warszawa, ul Rakowiecka 4, Poland.

Macrofauna of the Polish Zechstein: its occurrence and stratigraphy

L. Karwowski & J. Klapcinski

SUMMARY: This paper presents the characteristics of the Zechstein macrofauna of western and northern Poland. On the basis of the macrofauna and its variability, four biozones* are distinguished within the Zechstein succession. These are biostratigraphic units which correspond to both the basal clastics and the carbonate units of the succession. The faunal assemblages are characteristic of the individual carbonate horizons, and those assemblages typical of various palaeogeographical areas within the basin are reviewed. The macrofauna of the Polish Zechstein strata is also briefly compared with macrofauna recorded from the English Zechstein.

In Poland there has been much recent interest in the Zechstein macrofauna. Among the previous studies many were entirely devoted to this particular aspect of the Zechstein succession (e.g. Tokarski 1966; Alexandrowicz & Slupczynski 1970; Klapcinski 1971; Blaszczyk & Prymka 1973; Klapcinski & Karwowski 1978), while in several others the macrofauna was considered only in the background (e.g. Piekarskae & Kwiatkowski 1975; Peryt 1978). An evaluation of the biostratigraphic significance of this fauna has been attempted more recently by the present authors (Karwowski 1981; Klapcinski & Karwowski 1981). This paper is primarily intended to summarize the results obtained by the present authors in their last few years of research. In total, about 150 borehole cores of the Polish Zechstein (total thickness 3700 m) were studied (Fig. 1). Macrofauna was found to occur in 108 of the cores. About 2000 fauna specimens were collected, of which some 1100 were suitable for palaeontological identification. The results of the macrofauna analysis are reviewed below.

Review of the Polish Zechstein macrofauna

The macrofauna is present as internal moulds, or less commonly as external moulds. Its preservation state is variable, most often depending upon the type of host rock; in general, the fauna appears to be better preserved where present in fine-grained carbonates and in the Kupferschiefer while preservation is worse in clastic deposits and coarse-grained carbonates. The state of preservation also suggests that the macrofauna had been subjected to transport and redeposition, probably sometimes to the outside of the biotope area.

In the Polish Zechstein, the main groups of the macrofauna include bryozoans, brachiopods, bivalves, gastropods, and rare cephalopods (Table 1); in addition, some echinoderms were identified in thin sections although they could not be recognized macroscopically. It is worth noting that all of these macrofaunal groups are known to occur throughout the Central European Zechstein Basin.

In the study area (Fig. 1), as in many other parts of the Zechstein basin, the macrofauna occurs within the first three cycles of the Zechstein succession (i.e. in the cycles PZ1, PZ2 and PZ3). The complete Zechstein succession represents a major regressive episode (Pattison *et al.* 1973). In the lowest cycle (PZ1) the fauna is taxonomically richest and includes all the macrofaunal groups. In the overlying cycle (PZ2) only bivalves and gastropods are present, although the numbers of individuals are abundant. In the third cycle (PZ3) only bivalves are found with few species and scarce individuals.

Most of the bryozoa belong to the Cryptostomata The Trepostomata are represented by only one species (*Batostomella*). The bryozoans occur mostly near the top of the first-cycle carbonate (PZ1 Ca) but have also been identified microscopically in the second cycle carbonate. In many parts of the Zechstein basin, bryozoans are thought to have played an important role in the formation of reefs, including those described by Kerkmann (1964), Smith (1970) and Pattison *et al.* (1973). In the cores studied here, however, there is no evidence that the bryozoans ever formed true 'reefs'. The only recognized mass-occurrence of the bryozoans is in the first-cycle carbonate of the Peri-Beltic Syncline, northern Poland (Fig. 1); this is a stratified, biogenic carbonate-body which is a 'biostrome', rather than a true 'reef'. In the whole region studied, the

**Editor's note*: The authors originally used the term 'zonites' instead of 'biozones'. However, 'zonites' was not used as proposed by Henningsmoen (1961), *sensu stricto*; the editors and reviewers saw best to alter this term.

From HARWOOD, G.M. & SMITH, D.B. (eds), 1986, *The English Zechstein and Related Topics*, Geological Society Special Publication No. 22, pp. 211–216.

TABLE 1. Macrofauna in Polish Zechstein strata.

	1	*2*	*3*	*4*
BRYOZOA				
Acanthocladia anceps (Schlotheim)	?	+	?	
A. minor Korn		+		
A. laxa Korn		+		
Batostomella columnaris (Schlotheim)		+		
Thamniscus geometricus Korn	?	+		
Protoretepora ehrenbergi (Geinitz)		+		
Fenestella retiformis (Schlotheim)	?	+	?	
F. geinitzi (d'Orbigny)		+		
F. minuta (Korn)		+		
BRACHIOPODA				
Lingula credneri Geinitz	+	×		
Discina konincki Geinitz	×	×		
Streptorhynchus pelargonatus (Schlotheim)		×		
Strophalosia morrisiana King		×		
Dasyalosisia goldfussi (Münster)		?		
Orthothrix excavata (Geinitz)		×		
Craspedalosia lamellosa (Geinitz)		+		
Horridonia horrida (J. Sowerby)	×	+		
Stenoscisma humbletonensis (Howse)	?	+		
S. schlotheimi (von Buch)		+		
Rhynchospora geinitziana (de Verneuil)	+	+		
Pterospirifer alatus (Schlotheim)		+		
Trigonotreta permiana King		+		
T. undulata (j. de C. Sowerby)		×		
Dielasma elongatum (Schlotheim)	?	+		
NAUTLOIDEA				
Peripetoceras freieslebeni (Geinitz)	?	×		
GASTROPODA				
Straparollus permianus (King)		+	+	×
Cyclobathmus? obtusus (Brown)	?	+	+	
Loxonema? fasciata King		×	+	×
Naticopsis minima (Brown)	?	+	+	
Omphalotrochus? helicinus (Schlotheim)		×	+	×
Ianthinopsis? leighi (Brown)			+	
BIVALVIA				
Nuculites hollebeni Geinitz		?	+	
Pseudomonotis speluncaria Schlotheim	?	+	×	
P. laticostata Netschajew			+	
Nucula beyrichi Schauroth			×	
Liebea squamosa (J. de C. Sowerby)		×	+	+
L. septifer (King)			+	
Bakevellia ceratophaga (Schlotheim)		+	×	
B. antiqua (Münster)			×	
Aviculopecten cf. squamula Waagen		×	×	
Streblochondria pusilla (Schlotheim)		×	+	
S. sericea (de Verneuil)			×	
Elimata permiana (King)		×	×	
Netschajewia teplogfi (de Verneuil)			+	
N. alata (Netschajew)			+	
Modiomorpha cf. oblonga (Golowkinsky)			×	
Schizodus schlotheimi Geinitz	?	+	+	+

TABLE 1 (*cont.*)

	(1)	(2)	(3)	(4)
Palaeanodonta cf. castor (Eichwald)			+	×
P. verneuili (Amalitzky)			+	
Oligodon latus Netsachajew		×	×	
Astartella cf. vallisneriana (King)		×	×	
Permophorus costatus (Brown)		+	+	?
Panope mackrothi Geinitz		×	+	
Edmondia elongata Howse		+	+	
Janeia biarmica (de Verneuil)			+	
J. normalis (Howse)			+	
Wilkingia elegans (King)			×	
ECHINODERMATA				
Undet. crinoid remains		×		
PISCES				
Palaeoniscus sp.	+			

The vertical columns represent in order: (1) *Lingula credneri* biozone; (2) *Horridonia horrida* biozone; (3) *Liebea squamosa* biozone; and (4) *Schizodus schlotheimi* biozone. The presence of each species is indicated by + = common; × = recorded; ? = indicates a doubtful record.

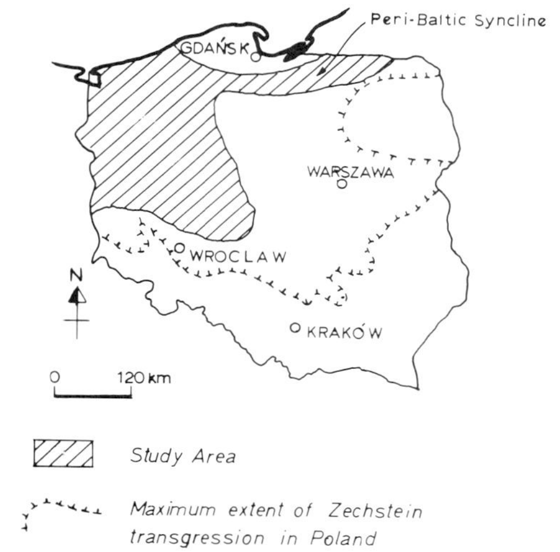

FIG. 1. Location map showing area of Zechstein strata sampled and discussed in this paper.

bryzoans tend to be concentrated in areas of pre-existing topographic highs within the basin.

Both inarticulate and articulate brachiopods are present; the former group includes the long-existent Lingulidae and Discinidae families. These are present mostly in the basal clastics of the first Zechstein cycle where *Lingula credneri* is the most abundant species. The genus *Lingula* is known to be characteristic of the early stages of marine transgressions (Ramsbottom *et al.* 1962) and is of primary importance as a facies indicator. Rare *Lingula* individuals and most of the Discinidae found occur at the base of the first-cycle carbonate.

The articulate brachipods, in turn, are represented by the typical Palaeozoic families of Strophalosidae, Dictyoclastidae, Strenoscismatidae and Syringothyridae. All these occur in the first-cycle carbonate where they display no significant vertical variation (apart from a slight vertical decrease in numbers). Among the predominant species are *Lingula credneri*, *Horridonia horrida*, *Stenoscima humbletonensis*, *S. schlotheimi*, *Pterospirifer alatus*, and *Dielasma elongatum* (Table 1).

Bivalves comprise about 80% of all the macrofaunal specimens collected. They are represented by 26 species which belong to 13 families. Most of them are epifaunal but some are infaunal (*Janeia*, *Edmondia*, *Schizodus*) and a few represent free swimming forms (*Streblochondria*, *Pseudomonotis*). Stenohaline bivalves predominate, but euryhaline forms are also present (*Liebea*, *Schizodus*, *Permophorus*).

Most of the bivalves occur within the Hauptdolomit, the Zechstein second-cycle carbonate (PZ2 Ca). All of the bivalve species known from the Polish Zechstein are present in the Hauptdolomit. The bivalve assemblage is dominated by *Liebea squamosa*, *Schizodus schlotheimi*, *Edmondia elongata*, *Janeia sp.*, *Nuculites hollebeni* and *Permophorus costatus*. All of the remaining species are much less common (Table 1).

The role of bivalves appears to be of minor importance in the Zechstein first cycle where brachiopods are dominant. From the first-cycle carbonate only 13 bivalve species have been described, most of which represent four genera; *Bakevellia*, *Pseudomonotis*, *Streblochondria* and *Edmondia*. Also the Zechstein third cycle carbonate (PZ3 Ca) is poor in both the number of bivalve species present and the number of individuals. The third-cycle carbonate contains bivalve genera such as *Liebea*, *Schizodus*, and *Palaeanodonta* (Table 1).

The remaining groups of macrofauna are poorly represented in the Polish Zechstein and include gastropods, cephalopods and echinoderms. The gastropods, largely euryhaline forms, occur in all three Zechstein cycles but are most frequent within the first- and second-cycle carbonates. The third cycle only contains two gastropod orders, archaeogastropods and caenogastropods. It is worth noting here that the taxonomy of the Zechstein gastropods has long been largely problematical, and still remains difficult to clarify because of the general scarcity and poor preservation of the fossil material. Because of their infrequent occurrence and problematical taxonomy the gastropods are of little importance in deducing the nature and stratigraphy of the Zechstein faunal communities.

The cephalopods are represented by only one nautiloid species, *Peripetoceras freieslebeni* which is occasionally present in the first-cycle carbonate. Echinoderms have also been identified to be present sporadically in the first-cycle carbonate where they are represented only by crinoid fragments. In the Kupferschiefer shale two fragments of presumed fish were identified, probably *Palaeoniscus*.

Stratigraphic significance of the Zechstein macrofauna

The position and stratigraphic significance of the Polish Zechstein macrofauna has previously been discussed by the present authors (Klapcinski & Karwowski 1981). In the present paper, therefore, these earlier inferences will merely be supplemented and an attempt made to state them more precisely.

The four faunal zones previously distinguished by the present authors (Klapcinski & Karwowski 1981) within the Zechstein succession of Poland, are redefined here as representing biostratigraphical units or biozones (Table 1). This is primarily because variability of the macrofauna as recorded throughout the Zechstein succession is due to progressive facies/salinity changes in the basin and hence the biozones are not due to faunal evolution.

Accordingly, the four formally-distinguished biozones are as follows:

(1) The *Lingula credneri* biozone, comprising the subzones of both white basal sandstones and overlying Kupferschiefer of the lowest Zechstein cycle, PZ1, is characterized by a single species.

(2) The *Horridonia horrida* biozone, comprising the first-cycle (PZ1) carbonate is a cenozonite, i.e. it has a characteristic faunal assemblage.

(3) The *Liebea squamosa* biozone, comprising the second-cycle (PZ2) carbonate, the Hauptdolomit, also has a characteristic assemblage.

(4) The *Schizodus schlotheimi* biozone comprising the Grauer Saltston and the Plattendolomit of the third cycle (PZ3) is characterized by a single species.

Palaeogeographical zoning of Polish Zechstein macrofaunas

As discussed above, there is a considerable variation in the macrofauna in the three lower Zechstein cycles (PZ1, PZ2 and P3). The faunal variability is in both taxonomic composition and frequency of occurrence, and is thought to have been controlled by primary palaeogeographical zoning within the basin. To evaluate this relationship in semi-quantitative terms, the abundance of macrofossils (calculated as the number of individuals per 1 m of core) was compared with the inferred palaeogeographical zoning of the basin (see map in Depowski 1978). The results should, however, be treated with caution because the original Zechstein faunal assemblages have been both largely modified and partially obscured by later redeposition. Reconstruction is therefore rather difficult. Consequently the present frequencies of the individual groups of fauna are not the same as their original growth environment frequencies, due to the different susceptibilities of each faunal group to transport and redeposition. Some of the frequencies appear therefore to be over-estimated (e.g. for brachiopods) while many others are probably under-estimated (e.g. bryozoa).

The basal Zechstein clastics of the first cyclothem are characterized by a poor faunal community which includes *Lingula credneri*, infrequent articulate brachiopods and some problematical bivalves. The calculated frequency of this assemblage is low, and averages 0.15. The community reveals no significant variability throughout the part of the basin studied.

The forereef carbonates of the first cycle are characterized by an assemblage which has the richest taxonomic composition and highest frequency. This community comprises all the bryozoans recorded and most of the articulate brachiopods, as well as free-swimming bivalve species and infrequent gastropods. Its overall frequency averages 0.60.

Perhaps the most typical occurrence of this particular community is in the forereef zone identified within the Peri-Baltic Syncline (referred to as 'facial zone B' in Piekarske & Kwiatkowski 1975). In many respects the community is similar to the 'Barrier Reef Top Community' described by Pattison (in McKerrow 1978) from the English Zechstein.

When present within broad elevated areas of the shallow intracontinental shelf, the PZ1 carbonates are characterized by a different assemblage: a community, low in diversity and numbers, consisting mostly of articulate brachiopods and the bryozoans *Acanthocladia anceps* and *A. minor* with infrequent bivalves (*Edmondia*, *Pseudomonotis*) and gastropods. The overall faunal frequency averages 0.40. This latter community is similar to the 'Back-reef Community' of Pattison (in McKerrow 1978).

In the remaining palaeogeographical zones of the PZ1 carbonate in the Peri-Baltic Syncline (i.e. the near-shore barrier, lagoon and shore zone clastic sedimentation areas), the first-cycle carbonate is characterized by a poor community composed of articulate brachiopods, bivalves, and rare bryozoans. The frequency is low, ranging between 0.05 and 0.30.

The macrofauna present in the second cycle carbonate, the Hauptdolomit (PZ2 Ca), appears to be of much lesser variability than that described above. In the Hauptdolomit the community is dominated by bivalves with infrequent gastropods. The community is essentially uniform throughout the part of the basin studied and has an almost constant faunal frequency which averages 0.40 throughout. In the inferred barrier and fore-barrier facies, this community is even more strongly dominated by the bivalves (mainly *Liebea squamosa*, *Schizodus schlotheimi*, *Edmondia elongata*, and *Permophorus costatus*) but has lower faunal frequencies (0.15–0.30 range; 0.25 average). The same community has a similar taxonomic composition, but yet lower faunal frequencies (0.0–0.20 range; 0.13 average) in both the lagoon facies and the coastal facies of mixed sulphate and clastic sedimentation; in this latter facies many cores are entirely devoid of fauna. In many respects (e.g. in the dominance of bivalves and in the lack of bryozoans and brachiopods), this community resembles the 'Hypersaline Landlocked Basin Community' described by Pattison (in McKerrow 1978).

The bivalve community present in the third-cycle carbonate, the Plattendolomit (PZ3 Ca), is entirely composed of euryhaline bivalves and has the lowest recorded faunal frequency (0.08 average). Unfortunately, there are not enough data on this community to allow further inferences.

Concluding remarks

(1) Four biostratigraphical units, or biozones, have been distinguished by the present authors in the Polish Zechstein.

(2) Because the differences between the respective faunal communities are of a strictly environmental nature and reflect changes in the sedimentary facies, it is concluded that the boundaries of these units may be heterochronous.

(3) The faunal communities of the first Zechstein cycle (PZ1) are largely dependent upon the inferred palaeogeographical zoning with the Zechstein basin.

(4) None of the communities in this first cycle have developed a sufficient ecological potential to produce true reefs within the area studied.

References

ALEXANDROWICZ, S.W. & SLUPCZYNSKI, K. 1970. O wystepowaniu *Lingula credneri Geinitz* w osadach cechsztynu okolic Ostrowa Wielkopolskiego. (On the presence of *Lingula credneri Geinitz* in the Lower Zechstein deposits in the vicinity of Ostrow Wielkopolski.) *Acta geol. Pol.* **20,** 693–707.

BLASZCZYK, J.K. & PRYMKA, W. 1973. Pierwsze stanowisko fauny w bialym spagowcu na monoklinie prazedsudeckiej. (In Polish only.) *Kwart. geol.* **17,** 276–84.

DEPOWSKI, S. 1978. *Lithofacies-Palaeogeographical Atlas of the Permian of Platform Area of Poland.* Geological Institute Publication, Warsaw.

KUTEK, J. 1972. Poziomy biostratygraficzne—zarys problematyki (in Polish). *Postepy N. Geol.* no. 4.

HENNINGSMOEN, G. 1961. Remarks on stratigraphical classification. *Norg. geol. Unders.* No. 213.

KARWOWSKI, L. 1981. The fauna and biofacies of the Zechstein deposits in Pomerania Trough, northern Poland (in press).

KERKMANN, K. 1969. Riffe und Algenbank im Zechstein von Thüringen. *Freiberger Forschhft.* **C252,** 1–85.

KLAPCINSKI, J. 1971. Litologia, fauna, stratygrafia permu monokliny Przedsudeckiej. (Lithology, fauna, stratigraphy and palaeogeography of the Permian in the Fore-Sudetic Monocline.) *Geol. sudet.* **5,** 77–126.

KLAPCINSKI, J. & KARWOWSKI, L. 1978. Fauna utworow cechsztynskich w polnocnej czesci monokliny przedsudeckiej. (The fauna of the Zechstein in the northern part of the Fore-Sudetic Monocline.) *Geol. sudet.* no. 2.

KLAPCINSKI, J. & KARWOWSKI, L. 1981. Stratigraphical analysis of the Zechstein deposits in western Poland based on macrofauna studies. In: *Proc. Int. Symp. Central European Permian*, 490–501.

MCKERROW, W.S. 1978. *The Ecology of Fossils.* Redwood Burn, Trowbridge.

PATTISON, J., SMITH, D.B. & WARRINGTON, G. 1973. A review of late Permian and early Triassic biostratigraphy in the British Isles. *Mem. Can. Soc. Petrol. Geol.* **2,** 200–60.

PERYT, P.M. 1978. Charakterystyka mikrofacjsaina cechsztynskich esadow weglanowych cyklotemu pierwszego i drugiego na obszarze monokliny przedsudeckiej. (Microfacies of the carbonate sediments of the Zechstein Werra and Stassfurt cyclothems in the Fore-Sudetic Monocline.) *Studia geol. Pol.* **54,** 1–38.

PIEKARSKAE, E. & KWIATKOWSKI, S. 1975. Microfacial analysis of the Zechstein Limestone in the eastern part of the Peri-Baltic syneclise. *Acta geol. pol.* **25,** 79–114.

RAMSBOTTOM, W.H.C., RHYS, E.W. & SMITH, E.G. 1962. Boreholes in the Carboniferous rocks of the Ashover district. Derbyshire. *Bull. geol. Surv. G.B.* **19,** 75–142.

SMITH, D.B. 1970. Permian and Trias. In: JOHNSON, G.A.L. & HICKLING, G. (eds) *The Geology of Durham County. Trans. nat. Hist. Soc., Northumb.* **41,** 66–91.

TOKARSKI, A. 1966. Lingula w bialym spagowcu monokliny przedsudeckiej. *Sprawozdania z posiedzen Komisji Oddz. PAN* Krakow.

LESZEK KARWOWSKI & JERZY KLAPCINSKI, Institute of Geological Sciences, University of Wroclaw, ul. Cybuskiego 30, 50-205 Wrocklaw, Poland.

The Leine Anhydrite of the Polish Zechstein: a significant lithostratigraphical marker-unit

J. Klapcinski

SUMMARY: Three lithostratigraphical members have been distinguished within the Leine Anhydrite of western Poland: (i) pseudospherulitic anhydrite, (ii) spherulitic anhydrite, and (iii) cryptospherulitic anhydrite. Because of its characteristic spherulitic texture and considerable lateral extent, the Leine anhydrite is considered to be a useful marker unit throughout a considerable part of the Central European Zechstein Basin.

There has been much recent interest in the Leine (Z3) Anhydrite of both the German and Polish parts of the Central European Zechstein basin. In previous studies much attention was paid to the lithostratigraphical division of this unit. On the basis of their textural characteristics, the anhydrite was subdivided into several members (e.g. alpha, beta, gamma, delta, epsilon, and zeta) by German authors (Behr 1960; Jung 1960; Heimlich 1964); in Poland, the lithostratigraphy of the Leine Anhydrite has been discussed by the author (Klapcinski 1966).

This paper is intended to present a recent version of the lithostratigraphy of the Leine Anhydrite ('Main' Anhydrite), and to discuss the significance of this unit as a possible marker in the Zechstein basin. The informal lithostratigraphy of the Polish Zechstein is given in Table 1 along with the relative position of the Leine, or 'Main', Anhydrite.

The present study of the Leine Anhydrite was based on borehole data derived from more than 400 exploratory boreholes located in western Poland, within the area from the Sudetic Mountains in the south to the Baltic Sea to the north (Fig. 1). Most of these boreholes were drilled during exploration for copper ore and natural fuels. The anhydrite was examined in detail in borehole samples from 17 selected representative cores, shown on Fig. 1.

Subdivision of the Leine Anhydrite

In the study area, one of the specific attributes of the Leine Anhydrite is the spherulitic texture. On the basis of this feature and its vertical variability, the anhydrite has been subdivided into the following lithostratigraphical members in ascending order: (i) pseudospherulitic anhydrite; (ii) spherulitic anhydrite; (iii) cryptospherulitic anhydrite; (iv) coarse-grained gypsum or rock salt.

Pseudospherulitic anhydrite

The pseudospherulitic anhydrite (Fig. 2a) occurs as a unit 2–11 m thick underlain by dolomite (the 'Platy' Dolomite), or locally by grey shales (Table 1). The rock consists of rod-like anhydrite crystals which are chaotically arranged, but in places reveal radial pattern with spherulites up to 1 cm in diameter. The spherulites are crystal aggregates with a grain size in the range of 0.01 to more than 2 mm. Among the anhydrite grains dolomite rhombohedra are commonly present; these may locally occur as irregular grain aggregates. The dolomite also occurs as streaks or bands, or filling pores and small cavities. The general abundance of dolomite in the anhydrite unit is explained by the close association of the two lithologies at the base of the Leine Anhydrite (Table 1).

The pseudospherulitic anhydrite grades upwards into the overlying spherulitic anhydrite; the contact is transitional with the spherulites becoming more abundant and better developed in the upper part of the pseudospherulitic member.

Spherulitic anhydrite

The spherulitic anhydrite (Figs 2b,c, 3) is 15–30 m thick and displays a typical spherulitic texture, with well-defined spherulites up to 2 cm or more in diameter. The spherulites often predominate in the rock, and consist of fine-grained (0.01–0.3 mm) aggregates displaying a radial arrangement; these are usually surrounded by coarser-grained (0.1–3 mm) aggregates. Dolomitic, or dolomitic-clay, streaks and bands are locally common. Many pores and small cavities in the rock are filled with halite.

Cryptospherulitic anhydrite

The cryptospherulitic anhydrite (Fig. 2d) is 6–15 m thick, with a transitional basal contact. In this

From HARWOOD, G.M. & SMITH, D.B. (eds), 1986, *The English Zechstein and Related Topics*, Geological Society Special Publication No. 22, pp. 217–222.

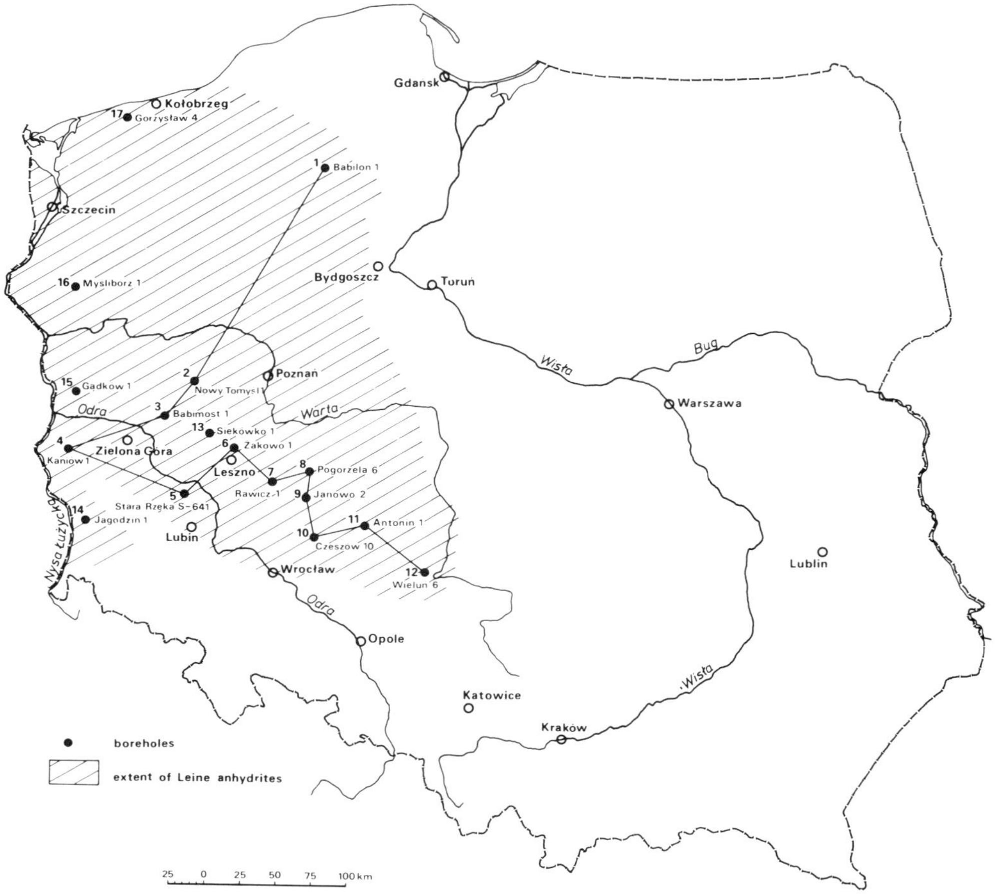

FIG. 1. Map showing location of the cores discussed in this paper from the Leine Anhydrite, western Poland.

member, the large spherulites are relatively rare. When viewed under a microscope, however, the rock consists of rod-shaped anhydrite crystals (2–10 mm long), many of which display a radial arrangement. The spherulite centre is constructed of a fine-grained (0.01–0.1 mm) aggregate. About 20–30% of the groundmass consists of coarser-grained (0.13 mm, or coarser) disorganized crystal aggregates, among which occur the spherulites, rarely more than 15 mm in diameter. Where overlain by rock salt, the anhydrite contains irregular dolomitic streaks or bands, and fine-grained, irregular aggregates of dolomite rhombohedra. Locally, aggregates of fibrous gypsum are present.

In most of the study area, the cryptospherulitic member of the Leine Anhydrite is overlain by rock salt (Table 1). In the southern part of the region, however, the anhydrite is covered by a unit of coarse-grained gypsum.

Coarse-grained gypsum

The coarse-grained gypsum occurs as a unit up to 3.5 m thick. Rock colour is from grey to brown-grey, with the brownish shade being due to admixtures of brown clay-sized matter. Gypsum crystals are large (up to 1.5 cm) and are usually plate- or rod-shaped. The coarse grains are scattered within a finer-grained (0.01–10 mm) mosaic of chaotically arranged crystals. Many grains display notched margins. Microstylolites occur locally among the gypsum grains. In places, the rock contains some light grey aggregates of fibrous gypsum; these latter consist of fine (0.1–1 mm), rod-shaped crystals whose longer axes are

FIG. 2. (a) Pseudospherulitic anhydrite, microphotograph, ×6, Czeszow 10 core, 1474.0 m.
(b) Spherulitic anhydrite, microphotograph, ×6, Czeszow 10 core, 1463.0 m.
(c) Spherulitic anhydrite, microphotograph, ×6, Czeszow 10 core, 1470.0 m.
(d) Cryptospherulitic anhydrite, microphotograph, ×6, Czeszow 10 core, 1456.0 m.

TABLE 1. Zechstein lithostratigraphy of western Poland

Cycles	Lithological units		German equivalents
Aller PZ 4	Upper brown-red shales Rock salt 'Pegmatitic' anhydrite Lower brown-red shales		Aller Salze Pegmatitanhydrit Roter Salzton
Leine PZ 3	Rock salt 'MAIN' ANHYDRITE 'Platy' dolomite Grey shales	(Leine Anhydrite)	Leine Salz Hauptanhydrit Plattendolomit Grauer Salzton
Stassfurzt PZ 2	Rock salt and/or potash salt 'Basal' anhydrite 'Main' dolomite		Stassfurt Salze Basalanhydrit Hauptdolomit
Werra PZ 1	Upper anhydrite Rock salt Lower anhydrite		Werraanhydrit
	Calcareous dolomite Dolomite limestone		Zechsteinkalk
	Marls and/or copper-shale		Kupferschiefer

oriented in certain preferred directions. Also occasionally present are fine grains of anhydrite and calcite.

Regional distribution of the Leine Anhydrite

The Leine Anhydrite has a great lateral extent (Fig. 1) and consists of three distinct members, as defined above. In Poland, this spherulitic anhydrite was first recognized in the copper mining district of Lubin, north-west of Wroclaw. In the numerous boreholes within this area the three anhydrite members described above were identified. These three members are also well developed both north and east of Wroclaw; their most easterly occurrence is in the Wielun-6 core. In south-western Poland, therefore, the east–west extent of the PZ3 spherulitic anhydrite, from the western state-border to the Warta River, is a distance of 250 km (Fig. 1).

The three members of the Leine Anhydrite are also present north of Wroclaw and Lubin, where they have been identified from cores located west of Poznan. The spherulitic anhydrite was also recognized in many boreholes located in northern Poland, as for example in the Babilon-1, Mysliborz-1 and Gorzysaw-4 cores. Thus, their south–north extent is over a distance of about 400 km between Wrocaw and the Baltic coast (Figs 1 and 4).

The Leine Anhydrite with spherulitic, or 'rosette' texture was also reported from the south-eastern Harz Foreland and from the Stassfurt region, Germany (Behr 1960; Jung 1960; Heimlich 1964). The spherulitic anhydrite may thus extend from the Warta River in Poland to at least the Saale River in Germany, a distance of 500 km. This wide extent, together with the characteristic textural appearance of these rocks, makes the spherulitic Leine Anhydrite a useful marker unit in this part of the Central European Zechstein basin.

FIG. 3. Spherulitic anhydrite, hand-specimen photograph, ×2, Kaniow 1 core, 994.5 m.

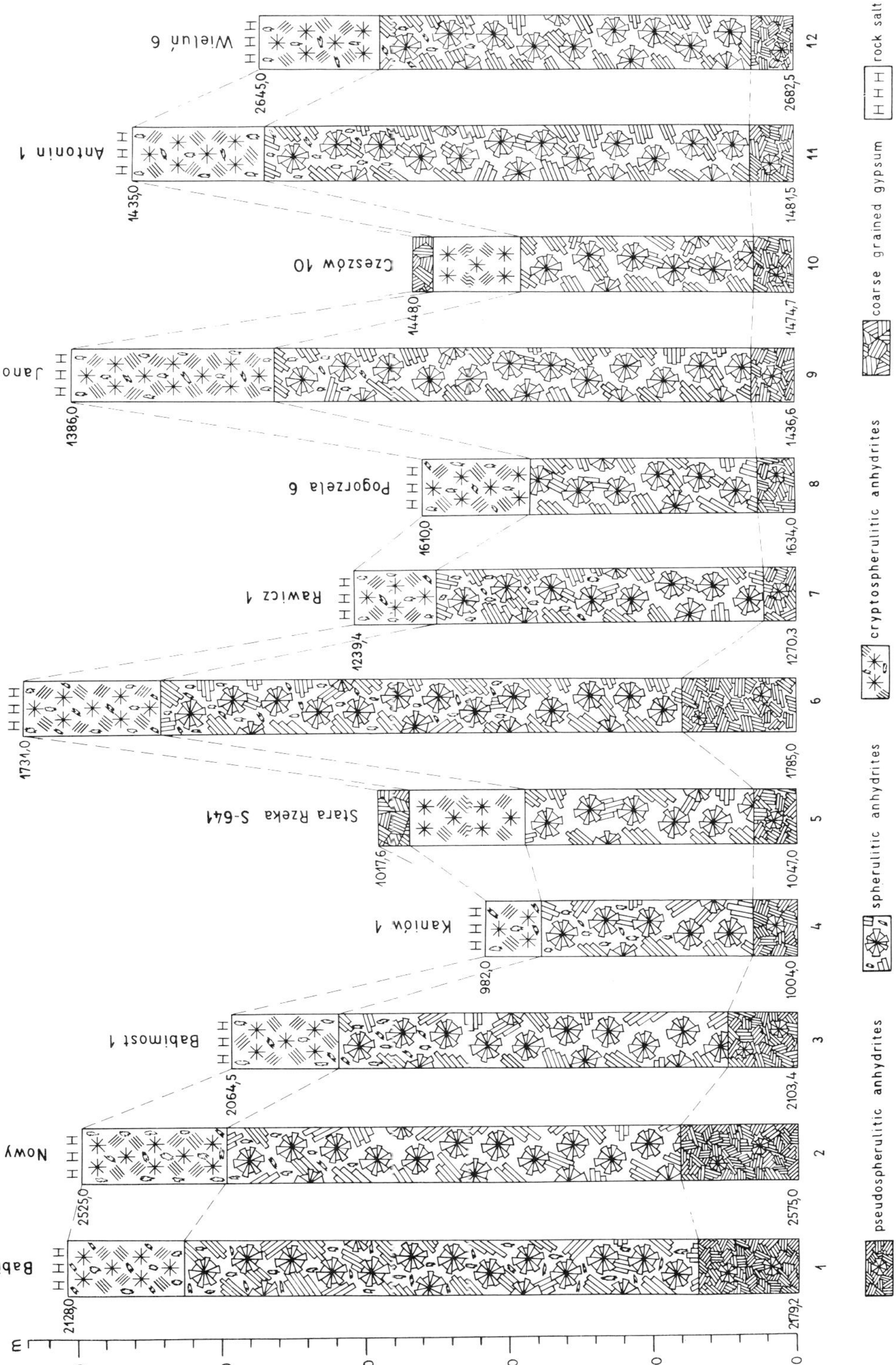

FIG. 4. Correlation of the three lithostratigraphical members of the Leine Anhydrite throughout western Poland. Base line is taken on the base of the Leine Anhydrite/top of the 'Platy' Dolomite (Plattendolomit).

This paper is concerned only with the lateral extent and vertical correlation of the Leine Anhydrite. The origin of the spherulites is problematical, although the author infers from the abundance of elongate, rod-shaped anhydrite crystals, that the textures are of secondary nature (cf. Carozzi 1960; Friedman 1965). This obviously presents an opening for future research on the Leine Anhydrite.

References

BEHR, E.M. 1960. Geologische Beobachtungen am Hauptanhydrit des Stassfurt—Egelner Sattels im Bereich des Kaliwerkes Stassfurt, Schacht VI/VII. *Freiberger Forschhft ser. C*, **90,** 88–105.

CAROZZI, A. 1960. *Microscopic Sedimentary Petrography*. Wiley, New York.

HEIMLICH, K. 1964. Zur Arbeitsmethodik bei Anhydrit-Untersuchungen und Gliederung des Hauptanhydrits (A3). *Geologie DDR*, **13,** 198–210.

FRIEDMAN, G. 1965. Terminology of crystallization textures and fabrics in sedimentary rocks. *J. sedim. Petrol.* **35,** 643–55.

JUNG, W. 1960. Zur Feingliederung des Basalanhydrits (Z2) und des Hauptanhydrits (Z3) im SE-Harzvorland. *Geologie, DDR*, **9,** 526–55.

KLAPCINSKI, J. 1960. Stratygrafia anhydrytow Leine w okolicach Lubina i Sieroszowic (monoklina przedsudecka). (Stratigraphy of the Leine anhydrite in the environs of Lubin and Sieroszowice.) *Annls. Soc. géol. Pol.* **36,** 315–21.

JERZY KLAPCINSKI, Institute of Geological Sciences, University of Wroclaw, ul. Cybulskiego 30, 50–205 Wroclaw, Poland.

USSR

The main types of carbonate rocks of the Kungurian evaporite basin of the Urals

B.I. Chuvashov

SUMMARY: The Kungurian basin of the Urals is a unique area for the study of evaporite sedimentation. All phases of evolution can be distinguished there, from a basin with a rich and diverse biota to a restricted one with high salt contents and scant biological communities. A similarity between English and German Zechstein sediments and the Kungurian strata of the Urals has long been noted. Some physical conditions of evaporite deposition may be inferred through a study of the associated carbonates.

Kungurian carbonates can be divided into three major groups: biogenic (organic), terrigenous and chemical. The organic carbonates are composed mostly of sediments trapped by blue-green algae and form small build-ups and algal mats, in both brackish-water (in the east) and high salinity (in the west) conditions. Coquinas are locally abundant. Clastic carbonates were transported as blocks, boulders, sands, muds, and in fine suspension. Great masses of micritic limestones accumulated by deposition from the latter. Numerous concretions, oolitic beds and packets of fine-grained dolomites without organic remains were laid down as chemical deposits and are related to the evaporites.

Geological setting

The Kungurian basin occupies a huge area within the European part of the USSR. Its sediments may be traced over 3000 km from the Caspian Sea in the south to the Arctic ocean in the north and from the Volga river in the west to the Urals in the east. The eastern part of the Kungurian basin, situated between the mountain chains of the Palaeourals and the border of the Russian platform, is the most remarkable. This belt, up to 200–250 km wide, is characterized by various types of sedimentation, related to the erosion of huge amounts of terrigenous material from the Palaeourals and to marked drops in sea-level. The latter formed an isolated basin with restricted circulation. Basinal areas were of higher salinity whereas, in shallow western parts of the basin, large amounts of fresh water came from the denuding Palaeourals. Brief ingressions of ocean waters from the north also took place. These factors combined with palaeogeographical factors to produce a complicated facies arrangement. Thick evaporite masses (gypsum, anhydrite, halite and potash salts) form complex interdigitations with sandstones, conglomerates, mudstones, limestones and dolomites (Fig. 1).

The stratigraphy and sedimentation of the Kungurian deposits have been studied by many geologists. The most important works were by Gerasimov & Tikhvinskaya (1934), Strakhov (1947) and Nalivkin (1949).

One hundred and fifty years ago the English geologist Murchison found much in common between the Kungurian sediments of the Urals and the Zechstein carbonates (Magnesian Limestones) of England (Murchison *et al.* 1845). The lower boundary of the Permian system, established by Murchison, was placed at the base of the Kungurian stage. Parallelism between the Zechstein and Kungurian deposits has long been defended by some Russian and Soviet geologists. Recent research has overtaken Murchison's stratigraphic ideas (Chuvashov 1981), but the similarity of the facies between the Zechstein and Kungurian basin remains.

I am very obliged to Professor P.A. Sofronitsky, who helped in field observations and the study of several of the Kungurian sections in the neighbourhood of Kungur city.

Stratigraphy and sedimentary environment of the strata

The Kungurian sediments are divided into two major lithostratigraphical groups—the Filippovskian and the Irenian (Fig. 1). Carbonate rock types in each facies zone are discussed below.

The Filippovskian Group

From east to west the lower unit of the Kungurian is composed of the Mysovsk, Leksk, Karnauchovsk and Filippovsk Divisions (Fig. 1). Each member comprises a facies belt several dozens of kilometres wide and readily traceable within the Middle Urals. In some other regions of Priuralye, however, they are less well defined.

From HARWOOD, G.M. & SMITH, D.B. (eds), 1986, *The English Zechstein and Related Topics*, Geological Society Special Publication No. 22, pp. 225–232.

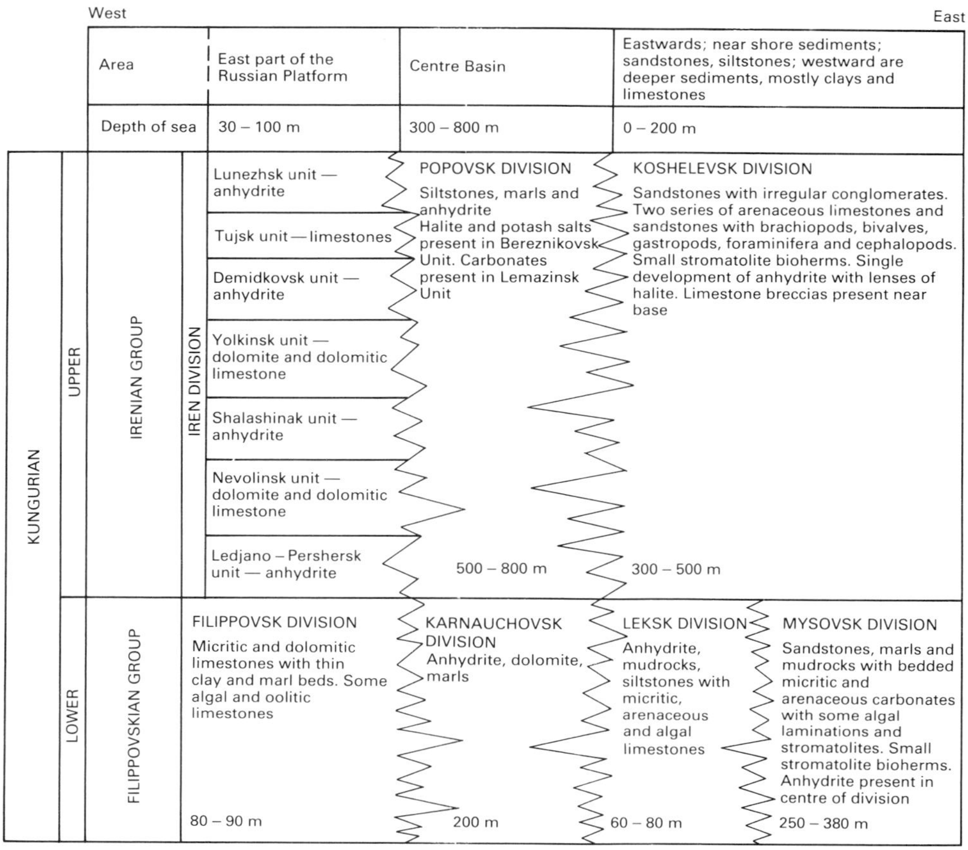

FIG. 1. Stratigraphical division of the facies zones within the Kungurian Stage in the Urals.

The Mysovsk Division

The Mysovsk Division, in the east, is represented by sandstones, siltstones, mudrocks, marls and limestones with gypsum and anhydrite, 25–40 m thick, present in the centre; rare lenses of gypsum are found elsewhere. Interbedded pebbly and boulder conglomerates occur in some sections. These sediments are interpreted to have been deposited in constantly-shifting shallow marine environments, with numerous islands, spits, bay-bars and vast mud lagoons with evaporites. Carbonates are common and are represented by different generic types; micritic limestones, algal mats, oncoid-rich carbonates, calcareous sandstones and bioherms.

Micritic limestones predominate; these are largely structureless but sometimes show microlaminations and graded bedding. Beds are usually up to 20 cm thick and are intercalated with mudrocks and sandstones of similar thickness. Similar limestones also occur as concretions which can be traced along the same horizon for up to dozens of metres until they pass laterally into a bed of limestone of similar thickness. Isolated, spherical, dense micritic concretions are also present with mudrocks and siltstones.

Some of the carbonate content of these sediments may result from the later stages of erosion of the Middle and Upper Palaeozoic limestones of the Palaeourals. The lithification of concretions and carbonate beds proceeded rapidly because derived angular fragments of concretions and underlying limestones are, in places, present in the upper beds of a section. Within the micritic limestones and concretions the shells of various organisms, particularly cephalopods, are preserved in their original form, although they occur deformed in argillites. The vacated and living camerae of nautiloids and ammonoids are commonly hollow or filled partly with radiaxial calcite. The cores of many concretions contain small wood fragments, plant tissues, brachiopods, gastropods or bivalve shells. Distinct chemical environments which promote the precipitation of calcium carbonate were therefore created around the organic remains.

Algal mats 10–15 cm thick are interpreted to have accumulated in quiet, shallow-water pools and near-shore bays. Their laminae were deformed in places by gas bubbles or jets which arose in the decaying algal mat (Fig. 2a).

At a number of levels beds composed of various oncoids occur (Fig. 2b). Their origin remains an enigma, as although the formation of oncoids appears to proceed in shallow water with a permanent turbulent flow, the oncoids here are usually confined to marl and limestone beds and were apparently deposited in quiet waters. The oncoids form complex grapestone-like aggregates

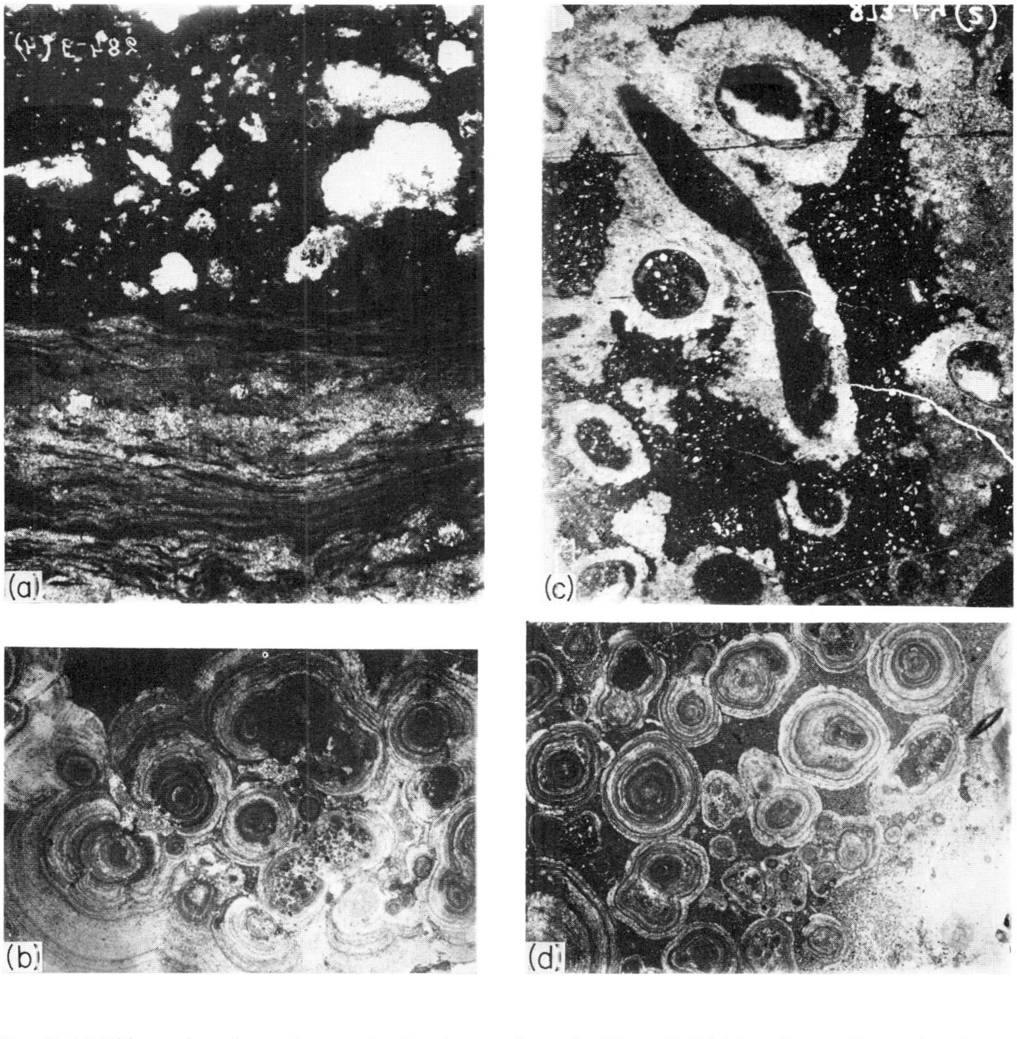

FIG. 2. (a) Thin-section photomicrograph of carbonate from the Mysovsk Division, Lower Kungurian. Lower part is stromatolitic limestone; upper portion is micritic limestone with possible 'birds eye' fenestrae. Lysva river basin, west slope of Middle Urals. Width of field of view 0.9 cm. (b,d) Thin-section photomicrographs of oncoid limestones with both simple and compound oncoids. Mysovsk Division. Sylva river, west slope of Middle Urals. Width of field of view 0.9 cm. (c) *Sylvaella tubularis* Tschuv., a problematic bioherm-forming organism. Mysovsk Division, Lysva river. Width of field of view 1.2 cm.

and have been reworked many times, finally accumulating in hollows on the seafloor. Oncoid formation took place in water of normal salinity as oncoid-rich layers contain a rich biota of foraminifers, small tetracorals, brachipods, bivalves, cephalopods, ostracods and algae. The shells of some organisms, such as foraminifers, may have an oncolitic coating.

Calcareous sandstones of the Mysovsk Division are minor; some contain fauna whereas others are barren. Interbedded with the former are rare, thin (up to 20 cm) beds, composed of shells of various organisms; foraminifers, bivalves, gastropods, ostracods and rarer cephalopods.

Three kinds of rare bioherm are present in this zone. The first type is composed of laminated or stacked stromatolites, whereas the second is composed of tubules of the problematic organism *Sylvella tubularis* Tchuv (Fig. 2c). Both types of bioherms average 3–5 m long and 1–1.5 m thick. The third type is represented by bioherms up to 9 m thick containing brachiopods, bryozoans and algae. Algae are represented by *Tubiphytes obscurus* Masl. and stromatolitic crusts.

The first, stromatolite, type of bioherm has much in common with the Zechstein build-ups of Western Germany (Kerkman 1967; Paul 1981) and Great Britain (Smith & Francis 1967; Smith 1981). Bioherms of the third type are comparable to the initial stages of the Zechstein build-ups of Thüringia (Kerkman 1967) and Poland (Peryt *et al.* 1978). *Sylvella* bioherms are typical only of the Urals.

The Leksk Division

The Leksk Division is situated further west and includes siltstones and mudrocks with lesser amounts of limestone, marl, anhydrite, gypsum and dolomite (Fig. 1). Evaporites occur in the middle of the member. Carbonates are mostly represented by micritic limestones deposited partly from a fine chemical suspension; they are commonly laminated. Other carbonates include thin-bedded calcarenites (10–15 cm thick) containing foraminifers, brachiopods, bivalves and ostracods. Stromatolites, oncoids and concretions are absent. Thin beds of micritic limestones and dolomites without organic remains occur among gypsum and anhydrite sediments.

The Karnauchovsk Division

The Karnauchovsk Division forms the next facies zone and contains mostly gypsum and anhydrite. Carbonates are rare but include fine-grained and micritic limestones and dolomites. The dolomites form both thin beds and nodules. Thin dolomite veins, which cross-cut bedding, are also present.

The Filippovsk Division

The Filippovsk Division marks the eastern border of the Russian platform. It comprises mostly carbonates; typical organisms include various small foraminifers, rare brachiopods and numerous bivalves. Fusulinids and tetracorals are absent. These faunal absences may be the result of abnormal salinities, unfavourable for the existence of some organisms.

In the Filippovsk Division oolitic limestone beds and lenses, 1–5 m thick, are present throughout the sequence (Fig. 3a). Oolites are indicative of shallow-water conditions which persisted over the Ufimian uplift. This uplift can be followed along the eastern border of the Russian platform for 300 km. The tops of the underlying Artinskian reefs are in many places situated immediately below areas of development of these oolites.

The major part of the Filippovsk Division is composed of white chalk-like marls and limestones, possible erosion products of carbonate islands which developed over the Ufimian uplift. Carbonate muds were deposited on the western slope of the basin.

These Filippovsk carbonates also contain bivalve coquinas, foraminiferal sands and platy stromatolites. Stromatolites, 5 m thick, consisting of large stacked hemispheroid stromatolites of the collenia type, are commonly found towards the top.

General interpretation

The interpretation of the various carbonates within the lower unit of the Kungurian stage makes it possible to conclude that in both western and eastern parts of the basin deposition of sediments took place under shallow-water conditions with a depth of perhaps 10–15 m, as indicated by the presence of blue-green algae. Water salinity varied from normal, or almost normal, in the west to brackish in the east, whereas in the deepest part of the basin higher salinities were found with deposition of gypsum, anhydrite and dolomite.

The Irenian Group

During Irenian time the Kungurian basin became more and more evaporitic. This process was complex and included ingression of normal salinity water from the north. The Irenian sediments belong to several facies zones (Fig. 1).

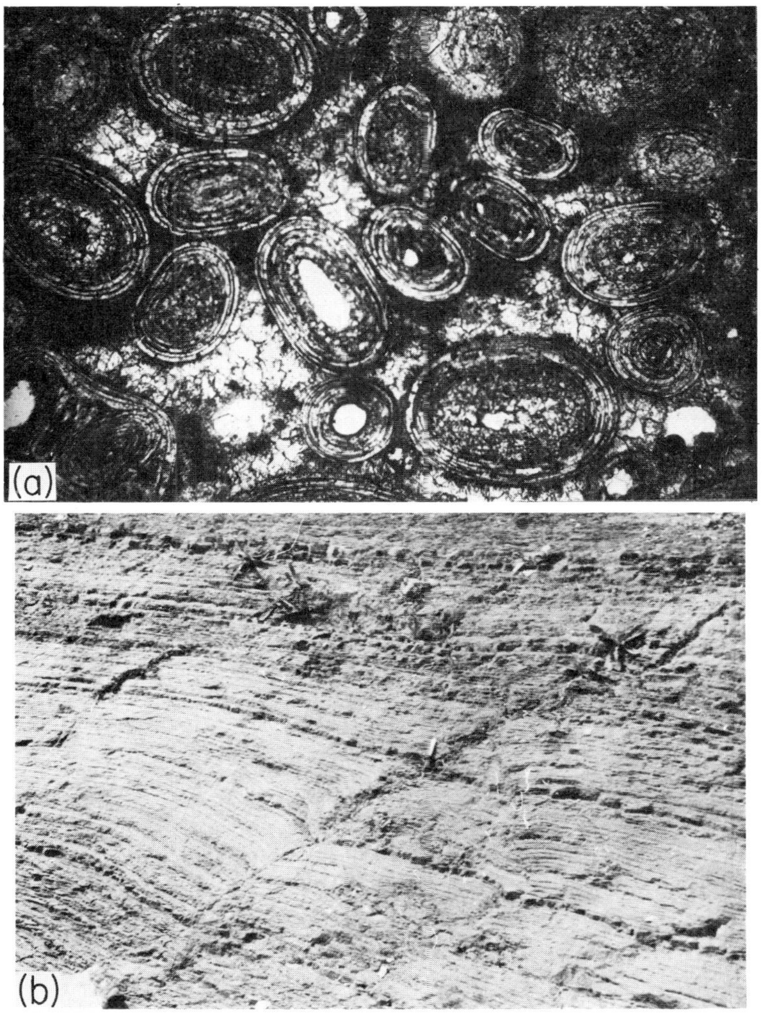

FIG. 3. (a) Oolitic limestone, base of Filippovsk Division, Sylva River, west slope of Middle Urals. Width of field of view 0.55 cm. (b) Varved marls from Popovsk Division, Irenian group. Chussovaja river, near Popor village. Pen in centre of photograph is 20 cm long.

The Koshelevsk Division

The Koshelevsk Division comprises an eastern belt of sandstones of changing width. The sandstones are interbedded with siltstones, mudrocks, marls, limestones, gypsum, anhydrite, boulder and pebble conglomerates and breccia. The carbonates within this facies zone are similar to those of the Mysovsk Division of the Lower Kungurian. However, the Irenian is characterized by an increase in tectonic activity and consequent erosion of coarser terrigenous material. This was accompanied by an increase of fresh water into the near-shore areas of the Kungurian basin and hence suppression of carbonate and evaporite sedimentation.

Sequences of marine clastic sediments several tens of metres thick are present at two levels and can be correlated with marine units further west. Thinner marine sediments include thin coquinas of brachiopods, bivalves, ostracods and cephalopods. Lenses and concretions of micritic limestones and marls, similar to those of the Mysovsk Division, are rare within the marine sediments.

The Koshelevsk Division also contains small stromatolitic bioherms 3–5 m long and 2–3 m

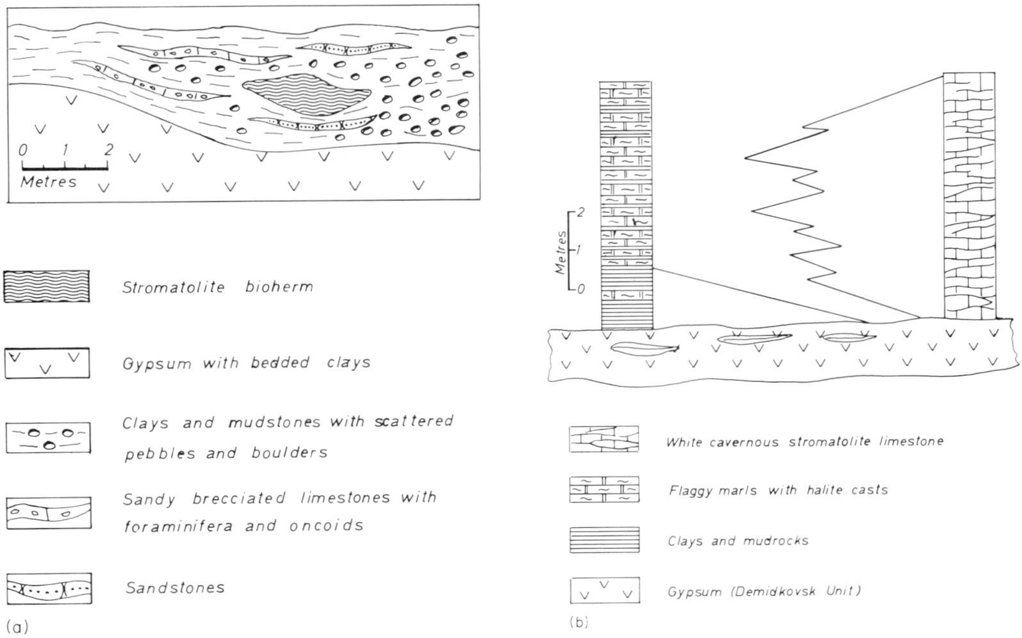

FIG. 4. (a) Stromatolite bioherm of the Koshelevsk Division, near Chussouoji. (b) Facies changes within the Tujsk Unit at quarry near Babinova gora.

thick (Fig. 4a). In some sections, in contrast, there are units of laminated flaggy marls with no marine organisms, but with numerous fossil plants and an abundant well-preserved insect fauna (for example, at the Sylva river). These marls are thus interpreted as freshwater lake sediments deposited on a large alluvial plain.

The Popovsk Division

Westwards the coarse-grained sediments of the Koshelevsk Division pass into a thick sequence of clayey rocks and marls with units of gypsum and anhydrite, halite and rare potash salt, the Popovsk Division. The sequence of halite and potash salts is known informally as the Bereznikovsk Unit. These sediments were deposited in the deep, central part of the basin away from the input of coarse sediment and from the influence of fresh water. Off-shore bars, which had been formed by coarse-grained sediments, divided the trough into a number of large depressions. Within individual depressions, salinity increased sufficiently for evaporite sedimentation to take place. Carbonate rocks in the deepest parts of the foredeep trough are mainly represented by monotonous varved marls (Fig. 3b).

The western slope of the trough has one additional interesting unit of carbonate rocks, the Lemazinsk Unit. At the beginning of the late Kungurian the eastern edge of the Russian platform was uplifted along faults which had an aggregate throw reaching 100–200 m. The consequent development of a large island or series of islands, 10–20 km wide and at least 300 km long, exposed the lower Kungurian and some underlying Artinskian carbonates.Their erosion products were deposited on the western slope of the trough and show a sharp decrease in thickness eastwards from a maximum of 100 m down to zero. Blocks up to 1 m in diameter are found in the lower part of the member; its middle is composed of carbonate fragments 1–2 cm in size, whereas the top consists only of the fine-grained material. Thin section petrography was used to prove the derived nature of these limestones.

The Iren Division

The western facies belt of the Irenian Group, composed of sulphate and carbonate rocks, may be divided into seven sequences, three of which are carbonates (Fig. 1). These sequences represent alternating marine–lagoonal sedimentation, similar to the Zechstein sequences in Europe.

The lower carbonate sequence within the Iren Division, the Nevolinsk Unit, comprises dolomites, dolomitic marls and thin argillite beds. In some places the Nevolinsk contains a central series of anhydrites up to 5 m thick. Carbonates of the Nevolinsk Unit include brachiopod coquinas, detrital dolomite, dolomitic sandstones, breccias, oolitic dolomites and flaggy micritic dolomites. The oolitic dolomites are confined to the basal 2–2.5 m of the Nevolinsk although they do overlie thin (10–15 cm) micritic dolomites.

The major part of the Nevolinsk Unit has a very rich fauna: fusulinids, plus other foraminifers, tetracorals, brachiopods, bivalves, bryozoans, cephalopods (rare ammonoids and, more often, nautiloids), ostracods, fish remains and scarce conodonts. This fauna indicates that the Nevolinsk sediments were deposited in water of normal salinity which persisted probably through only a short period of time.

Subsequent sedimentation reflects progressive salinity increase accompanied by impoverishment of biological communities with deposition of the Shalashninsk evaporites. The Yolkinsk Unit is composed of similar carbonates formed during a second, small-scale, ingression of normal salinity seawater. The base of the Yolkinsk Unit is also an oolitic dolomite, 1.5 m thick, with the overlying carbonates containing a similar fauna, but without fusulinids and cephalopods, to the Nevolinsk sediments.

The third carbonate packet of the Irenian Group, the Tujsk Unit, can be traced for over 400 km from south to north across an outcrop belt 170–175 km wide in the south to 90 km wide in the north. Throughout this area the Tujsk packet is divisible into three lithofacies (Oborin & Khursik 1970, 1973).

The Tujsk Unit is represented in the west by laminated and thin-bedded dolomites, rarely brecciated, with wax-like sepiolite inclusions. Dolomite is replaced by anhydrite in many places, becoming more distinct in the west. Exposures of the central lithofacies of the Tujsk Unit were studied in a number of quarries near Kungur town. There the main sedimentary facies can be traced within a short distance in one quarry (Fig. 4b). In the south-west of the quarry the Tujsk section consists of dark argillites with flaggy marls which pass in a north-easterly direction into flaggy marls with large (up to 2 cm) moulds of rock salt on bedding planes. However, over a distance of 150 m the whole section of the Tujsk Unit is replaced by white stromatolitic limestones (Fig. 4b) of possible algal origin.

There are no fossils in the Tujsk carbonates. The lack of fauna could be related to the water composition but although the presence of numerous imprints of rock salt crystals indicates periodic desiccation, the absence of primary gypsum or anhydrite shows that salinities were not very high. Hence it is possible that the lack of life within the basin could be due to the high concentration of certain chemical elements, particularly as there is a wide occurrence of ratofkite in the Tujsk rocks.

Conclusions

Carbonate formation took place in the eastern part of the Kungurian evaporite basin under diverse and complex conditions. Three main groups can be distinguished: biological, clastic and chemical.

Biological carbonates were mainly formed by blue-green algae, which gave rise to bioherms and algal mats, both in brackish waters in the east and under high salinity conditions in the west. Coquinas composed of foraminifera, brachiopods, bryozoa and bivalves are subordinate.

Clastic carbonates are the most abundant. Erosion products of Middle and Upper Palaeozoic carbonates were redeposited in the basin as large blocks, pebbles, boulders, sands and as a fine suspension. The eastern margin of the Kungurian basin is fringed with numerous beds, lenses and concretions of micritic limestones. Micritic limestones and marls were also deposited further west near the eastern boundary of the Russian platform. Basinal deposits were formed from precipitation and fine suspension. There is no definite evidence as to the chemical origin of some Kungurian carbonates. However, in the eastern part of the basin all concretions are of possible chemical origin.

References

CHUVASHOV, B.I. 1981. Sedimentation and development of the biological communities of the Kungurian basin of the Urals; its similarity and differences with Zechstein basin of the Central Europe. *Int. Symp. Central European Permian*, Warsaw, 324–40.

GERASIMOV, N.P. & TIKHVINSKAYA, E.U. 1934. The classical section of the Kungurian stage. *Zap. vseross. miner. Obshch. Moskva.* ser. II, part 63.

KERKMANN, K. 1967. Zur Kenntnis der Riffbildungen in der Werraserie des Thüringian Zechstein. *Freiberger Forschhft* **C-213**, *Paleont.* 123–42.

MURCHISON, R.I., DE VERNEUIL, E. & VON KEYSERLING, A. 1845. The Permian System. In: *The Geology of Europe and the Ural Mountains, 1 Geology*, 137–70.

NALIVKIN, D.V. 1949. Stratigraphy and tectonic of the Ufimian plateau and Uresan-Sylva depression. *Tr.*

Kamskogo filiala VNIGRI, **46,** Leningr, Moscow (in Russian).

OBORIN, A.A. & KHURSIK, V.Z. 1970. Some peculiarities of carbonate deposits of the Tujsk packet of the Permain Prikamje and their relation to tectonic uplift. *Tr. Kamskogo filiala VNIGRI*, **72,** 45–48 (in Russian).

OBORIN, A.A. & KHURSIK, V.Z. 1973. Lower Permian lithofacies of the Permian Preduralje. In: *Lower Permain Deposits of the Kamsk Preduralje. Tr. Kamskogo filiala VNIGNI*, **118,** 151–83 (in Russian).

PAUL, J. 1981. Upper Permian algal stromatolite reefs, Harz Mountains (F.R. Germany). In: FUCHTBAUER, H. & PERYT, T.M. (eds) *The Zechstein Basin with Emphasis on Carbonate Sequences, Contr. Sedim.* **9**.

PERYT, T.M., PIATKOWSKI, T.S. & WAGNER, R. 1978. Litologia i paleogeografia cechztynskich pozimow weglanowych. In: DEPOWSKI, S. (ed.) *Atlas litofacjalno-paleogeograficzny permu obsczarow platformo Atlas litofacjalno-paleogeograficzny permu obsczarow platformowych Polski. Inst. Geol. Warszawa.*

SMITH, D.B. 1981. Bryozoan-algal patch-reefs in the Upper Permian Lower Magnesian Limestone of Yorkshire, northeast England. In: TOOMEY, D.F. (ed.) *European Fossil Reef Models, Spec. Publ. Soc. econ. Paleont. Mineral., Tulsa*, **30,** 187–202.

SMITH, D.B. & FRANCIS, E.A. 1967. Geology of the country between Durham and West Hartlepool. *Mem. geol. Surv. G.B.* HMSO, London.

STRAKHOV, N.M. 1947. Geology of the Kungurian deposits of the Ishimbai oil-bearing region. *Byull. Mosk. Obshch. ispyt. Prir.* (in Russian).

BORIS J. CHUVASHOV, Pochtvyi per. 7, Institute of Geology and Geochemistry of the Urals, Scientific Centre of the Academy of Sciences of the USSR, Sverdlovsk, USSR.

Subject Index

(Figures in italics—Tables in bold)